Nanotechnology: A Comprehensive Introduction

Nanotechnology: A Comprehensive Introduction

Edited by
Cesar O'Neil

Larsen & Keller
www.larsen-keller.com

Nanotechnology: A Comprehensive Introduction
Edited by Cesar O'Neil
ISBN: 978-1-63549-194-4 (Hardback)

☰ Larsen & Keller

Published by Larsen and Keller Education,
5 Penn Plaza,
19th Floor,
New York, NY 10001, USA

Cataloging-in-Publication Data

Nanotechnology : a comprehensive introduction / edited by Cesar O'Neil.
    p. cm.
Includes bibliographical references and index.
ISBN 978-1-63549-194-4
1. Nanotechnology. 2. Nanostructured materials. 3. Technology. I. O'Neil, Cesar.
T174.7 .N36 2017
620.5--dc23

The publisher's policy is to use permanent paper from mills that operate a sustainable forestry policy. Furthermore, the publisher ensures that the text paper and cover boards used have met acceptable environmental accreditation standards.

Printed and bound in the United States of America.

For more information regarding Larsen and Keller Education and its products, please visit the publisher's website www.larsen-keller.com

Table of Contents

Preface

Nanotechnology is the science of studying and manipulating supra molecules, atoms and molecules. It is called 'nano' because it deals with the matter sized between 1 to 100 nanometers. This field is highly used in various areas like molecular engineering, organic chemistry, semiconductor physics, surface science, microfabrication, molecular biology, etc. This book presents this complex subject in the most comprehensible and easy to understand language. The topics included in it are of utmost significance and are bound to provide incredible insights to readers. This textbook is meant for students who are looking for an elaborate reference text on nanotechnology and its various uses.

Given below is the chapter wise description of the book:

Chapter 1- Nanotechnology is a very broad and diverse field. It is the study of matter on an atomic, molecular and supramolecular scale. This chapter is an overview of the subject matter incorporating all the major aspects of nanotechnology.

Chapter 2- Nanoelectronics is the practice of using nanotechnology in components that are electronic. The branches related to nanotechnology are molecular scale electronics, wet nanotechnology and nanolithography. This chapter elucidates the major branches of nanotechnology, providing a complete understanding.

Chapter 3- The growth of one-dimensional structures is known as vapor-liquid-solid method. The other fundamental concepts of nanotechnology that have been explained in this text are nanopore sequencing, molecular self-assembly, DNA nanotechnology and self- assembles monolayer. This section strategically encompasses and incorporates the major components and fundamental concepts of nanotechnology.

Chapter 4- Nanomaterials vary between 1-100 nanometres. Nanomaterials have become commercialized in the past few years. Several nanometres, very small in number form the quantum dots. Quantum dots is very central to nanotechnology. Hybrid material, metamaterial, metallogels, nanoparticles, silver nanoparticles and platinum nanoparticles are some of the topics that have been elucidated in this text.

Chapter 5- Carbon nanotubes are allotropes of carbon. They have unusual properties and they are very important for nanotechnology and other fields of material science. This chapter helps the reader in understanding the basic concepts of carbon nanotubes.

Chapter 6- Molecular nanotechnology is the technology that is based on mechanosynthesis. It involves biophysics, chemistry and other nanotechnologies. The aspects of molecular nanotechnology that have been explained in this text are mechanosynthesis, molecular assembler and molecular modelling. The chapter serves as a source to understand the major categories related to molecular nanotechnology.

Chapter 7- The devices related to nanotechnology are molecular machine, nanocar and nanomotor. Nanocar is not really a car, as it does not have a molecular motor whereas nanomotors are molecules that are capable of transferring energy into movement. The section serves as a source to understand the major devices related to nanotechnology.

Chapter 8- Nanotechnology has a number of impacts; it extends from medical to ethical to mental to environmental applications. The main benefits of nanotechnology are water purification systems, energy systems, physical enhancement and nanomedicine. In order to clearly understand nanotechnology it is necessary to understand the impacts of nanotechnology.

Chapter 9- Over the past several decades, technology has developed immensely. The applications of nanotechnology have improved life all over the world. In order to move forward from the present level of technology, scientists have been developing energy applications of nanotechnology. Some of the applications are industrial applications of nanotechnology, nanomedicine, nanobiotechnology and green nanotechnology. The diverse applications of nanotechnology in the current scenario have been thoroughly discussed in this chapter.

Chapter 10- History of nanotechnology concerns itself with the development that has taken place over a period of years in the field of nanotechnology. To develop a better understanding of nanotechnology one must understand the evolution of nanotechnology as well. The following section helps the readers in understanding the evolution of nanotechnology.

Indeed, my job was extremely crucial and challenging as I had to ensure that every chapter is informative and structured in a student-friendly manner. I am thankful for the support provided by my family and colleagues during the completion of this book.

Editor

Introduction to Nanotechnology

Nanotechnology is a very broad and diverse field. It is the study of matter on an atomic, molecular and supramolecular scale. This chapter is an overview of the subject matter incorporating all the major aspects of nanotechnology.

Nanotechnology

Nanotechnology ("nanotech") is manipulation of matter on an atomic, molecular, and supramolecular scale. The earliest, widespread description of nanotechnology referred to the particular technological goal of precisely manipulating atoms and molecules for fabrication of macroscale products, also now referred to as molecular nanotechnology. A more generalized description of nanotechnology was subsequently established by the National Nanotechnology Initiative, which defines nanotechnology as the manipulation of matter with at least one dimension sized from 1 to 100 nanometers. This definition reflects the fact that quantum mechanical effects are important at this quantum-realm scale, and so the definition shifted from a particular technological goal to a research category inclusive of all types of research and technologies that deal with the special properties of matter which occur below the given size threshold. It is therefore common to see the plural form "nanotechnologies" as well as "nanoscale technologies" to refer to the broad range of research and applications whose common trait is size. Because of the variety of potential applications (including industrial and military), governments have invested billions of dollars in nanotechnology research. Until 2012, through its National Nanotechnology Initiative, the USA has invested 3.7 billion dollars, the European Union has invested 1.2 billion and Japan 750 million dollars.

Nanotechnology as defined by size is naturally very broad, including fields of science as diverse as surface science, organic chemistry, molecular biology, semiconductor physics, microfabrication, molecular engineering, etc. The associated research and applications are equally diverse, ranging from extensions of conventional device physics to completely new approaches based upon molecular self-assembly, from developing new materials with dimensions on the nanoscale to direct control of matter on the atomic scale.

Scientists currently debate the future implications of nanotechnology. Nanotechnology may be able to create many new materials and devices with a vast range of applications, such as in nanomedicine, nanoelectronics, biomaterials energy production, and consumer products. On the other hand, nanotechnology raises many of the same issues as any new technology, including concerns about the toxicity and environmental impact of nanomaterials, and their potential effects on global economics, as well as speculation about various doomsday scenarios. These concerns have led to a debate among advocacy groups and governments on whether special regulation of nanotechnology is warranted.

Origins

The concepts that seeded nanotechnology were first discussed in 1959 by renowned physicist Richard Feynman in his talk *There's Plenty of Room at the Bottom*, in which he described the possibility of synthesis via direct manipulation of atoms. The term "nano-technology" was first used by Norio Taniguchi in 1974, though it was not widely known.

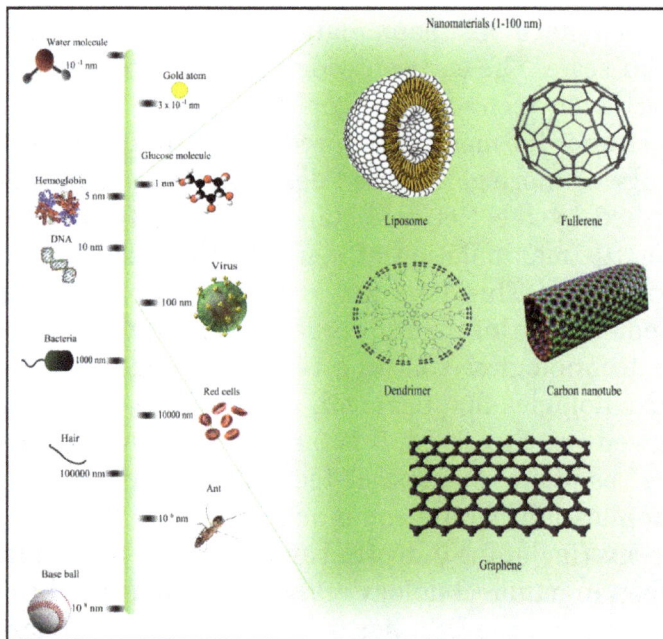

Comparison of Nanomaterials Sizes

Inspired by Feynman's concepts, K. Eric Drexler used the term "nanotechnology" in his 1986 book *Engines of Creation: The Coming Era of Nanotechnology*, which proposed the idea of a nanoscale "assembler" which would be able to build a copy of itself and of other items of arbitrary complexity with atomic control. Also in 1986, Drexler co-founded The Foresight Institute (with which he is no longer affiliated) to help increase public awareness and understanding of nanotechnology concepts and implications.

Thus, emergence of nanotechnology as a field in the 1980s occurred through convergence of Drexler's theoretical and public work, which developed and popularized a conceptual framework for nanotechnology, and high-visibility experimental advances that drew additional wide-scale attention to the prospects of atomic control of matter. In the 1980s, two major breakthroughs sparked the growth of nanotechnology in modern era.

First, the invention of the scanning tunneling microscope in 1981 which provided unprecedented visualization of individual atoms and bonds, and was successfully used to manipulate individual atoms in 1989. The microscope's developers Gerd Binnig and Heinrich Rohrer at IBM Zurich Research Laboratory received a Nobel Prize in Physics in 1986. Binnig, Quate and Gerber also invented the analogous atomic force microscope that year.

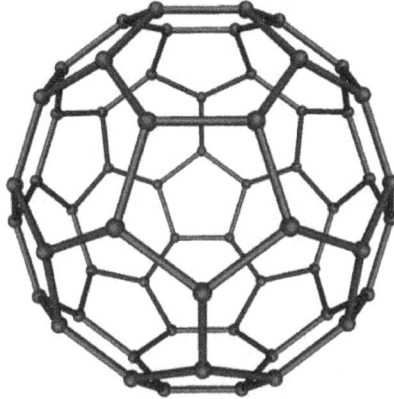

Buckminsterfullerene C_{60}, also known as the buckyball, is a representative member of the carbon structures known as fullerenes. Members of the fullerene family are a major subject of research falling under the nanotechnology umbrella.

Second, Fullerenes were discovered in 1985 by Harry Kroto, Richard Smalley, and Robert Curl, who together won the 1996 Nobel Prize in Chemistry. C_{60} was not initially described as nanotechnology; the term was used regarding subsequent work with related graphene tubes (called carbon nanotubes and sometimes called Bucky tubes) which suggested potential applications for nanoscale electronics and devices.

In the early 2000s, the field garnered increased scientific, political, and commercial attention that led to both controversy and progress. Controversies emerged regarding the definitions and potential implications of nanotechnologies, exemplified by the Royal Society's report on nanotechnology. Challenges were raised regarding the feasibility of applications envisioned by advocates of molecular nanotechnology, which culminated in a public debate between Drexler and Smalley in 2001 and 2003.

Meanwhile, commercialization of products based on advancements in nanoscale technologies began emerging. These products are limited to bulk applications of nanomaterials and do not involve atomic control of matter. Some examples include the Silver Nano platform for using silver nanoparticles as an antibacterial agent, nanoparticle-based transparent sunscreens, carbon fiber strengthening using silica nanoparticles, and carbon nanotubes for stain-resistant textiles.

Governments moved to promote and fund research into nanotechnology, such as in the U.S. with the National Nanotechnology Initiative, which formalized a size-based definition of nanotechnology and established funding for research on the nanoscale, and in Europe via the European Framework Programmes for Research and Technological Development.

By the mid-2000s new and serious scientific attention began to flourish. Projects emerged to produce nanotechnology roadmaps which center on atomically precise manipulation of matter and discuss existing and projected capabilities, goals, and applications.

Fundamental Concepts

Nanotechnology is the engineering of functional systems at the molecular scale. This covers both current work and concepts that are more advanced. In its original sense, nanotechnology refers to the projected ability to construct items from the bottom up, using techniques and tools being developed today to make complete, high performance products.

One nanometer (nm) is one billionth, or 10^{-9}, of a meter. By comparison, typical carbon-carbon bond lengths, or the spacing between these atoms in a molecule, are in the range 0.12–0.15 nm, and a DNA double-helix has a diameter around 2 nm. On the other hand, the smallest cellular life-forms, the bacteria of the genus Mycoplasma, are around 200 nm in length. By convention, nanotechnology is taken as the scale range 1 to 100 nm following the definition used by the National Nanotechnology Initiative in the US. The lower limit is set by the size of atoms (hydrogen has the smallest atoms, which are approximately a quarter of a nm diameter) since nanotechnology must build its devices from atoms and molecules. The upper limit is more or less arbitrary but is around the size below which phenomena not observed in larger structures start to become apparent and can be made use of in the nano device. These new phenomena make nanotechnology distinct from devices which are merely miniaturised versions of an equivalent macroscopic device; such devices are on a larger scale and come under the description of microtechnology.

To put that scale in another context, the comparative size of a nanometer to a meter is the same as that of a marble to the size of the earth. Or another way of putting it: a nanometer is the amount an average man's beard grows in the time it takes him to raise the razor to his face.

Two main approaches are used in nanotechnology. In the "bottom-up" approach, materials and devices are built from molecular components which assemble themselves chemically by principles of molecular recognition. In the "top-down" approach, nano-objects are constructed from larger entities without atomic-level control.

Areas of physics such as nanoelectronics, nanomechanics, nanophotonics and nano-ionics have evolved during the last few decades to provide a basic scientific foundation of nanotechnology.

Larger to Smaller: A Materials Perspective

Several phenomena become pronounced as the size of the system decreases. These include statistical mechanical effects, as well as quantum mechanical effects, for example the "quantum size effect" where the electronic properties of solids are altered with great reductions in particle size. This effect does not come into play by going from macro to micro dimensions. However, quantum effects can become significant when the nanometer size range is reached, typically at distances of 100 nanometers or less, the so-called quantum realm. Additionally, a number of physical (mechanical, electrical, optical, etc.) properties change when compared to macroscopic systems. One example is the increase in surface area to volume ratio altering mechanical, thermal and catalytic properties of materials. Diffusion and reactions at nanoscale, nanostructures materials and nanodevices with fast ion transport are generally referred to nanoionics. *Mechanical* properties of nanosystems are of interest in the nanomechanics research. The catalytic activity of nanomaterials also opens potential risks in their interaction with biomaterials.

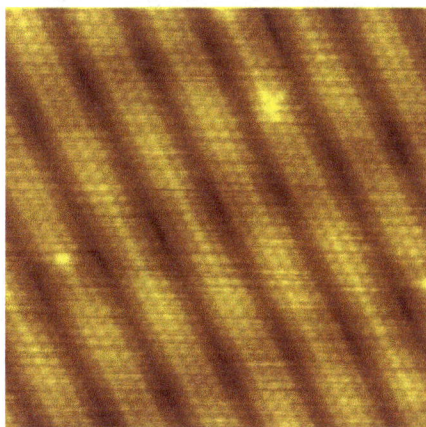

Image of reconstruction on a clean Gold(100) surface, as visualized using scanning tunneling microscopy. The positions of the individual atoms composing the surface are visible.

Materials reduced to the nanoscale can show different properties compared to what they exhibit on a macroscale, enabling unique applications. For instance, opaque substances can become transparent (copper); stable materials can turn combustible (aluminium); insoluble materials may become soluble (gold). A material such as gold, which is chemically inert at normal scales, can serve as a potent chemical catalyst at nanoscales. Much of the fascination with nanotechnology stems from these quantum and surface phenomena that matter exhibits at the nanoscale.

Simple to Complex: A Molecular Perspective

Modern synthetic chemistry has reached the point where it is possible to prepare small molecules to almost any structure. These methods are used today to manufacture a wide variety of useful chemicals such as pharmaceuticals or commercial polymers. This ability raises the question of extending this kind of control to the next-larger level, seeking methods to assemble these single molecules into supramolecular assemblies consisting of many molecules arranged in a well defined manner.

These approaches utilize the concepts of molecular self-assembly and/or supramolecular chemistry to automatically arrange themselves into some useful conformation through a bottom-up approach. The concept of molecular recognition is especially important: molecules can be designed so that a specific configuration or arrangement is favored due to non-covalent intermolecular forces. The Watson–Crick basepairing rules are a direct result of this, as is the specificity of an enzyme being targeted to a single substrate, or the specific folding of the protein itself. Thus, two or more components can be designed to be complementary and mutually attractive so that they make a more complex and useful whole.

Such bottom-up approaches should be capable of producing devices in parallel and be much cheaper than top-down methods, but could potentially be overwhelmed as the size and complexity of the desired assembly increases. Most useful structures require complex and thermodynamically unlikely arrangements of atoms. Nevertheless, there are many examples of self-assembly based on molecular recognition in biology, most notably Watson–Crick basepairing and enzyme-substrate interactions. The challenge for nanotechnology is whether these principles can be used to engineer new constructs in addition to natural ones.

Molecular Nanotechnology: A Long-term View

Molecular nanotechnology, sometimes called molecular manufacturing, describes engineered nanosystems (nanoscale machines) operating on the molecular scale. Molecular nanotechnology is especially associated with the molecular assembler, a machine that can produce a desired structure or device atom-by-atom using the principles of mechanosynthesis. Manufacturing in the context of productive nanosystems is not related to, and should be clearly distinguished from, the conventional technologies used to manufacture nanomaterials such as carbon nanotubes and nanoparticles.

When the term "nanotechnology" was independently coined and popularized by Eric Drexler (who at the time was unaware of an earlier usage by Norio Taniguchi) it referred to a future manufacturing technology based on molecular machine systems. The premise was that molecular scale biological analogies of traditional machine components demonstrated molecular machines were possible: by the countless examples found in biology, it is known that sophisticated, stochastically optimised biological machines can be produced.

It is hoped that developments in nanotechnology will make possible their construction by some other means, perhaps using biomimetic principles. However, Drexler and other researchers have proposed that advanced nanotechnology, although perhaps initially implemented by biomimetic means, ultimately could be based on mechanical engineering principles, namely, a manufacturing technology based on the mechanical functionality of these components (such as gears, bearings, motors, and structural members) that would enable programmable, positional assembly to atomic specification. The physics and engineering performance of exemplar designs were analyzed in Drexler's book *Nanosystems*.

In general it is very difficult to assemble devices on the atomic scale, as one has to position atoms on other atoms of comparable size and stickiness. Another view, put forth by Carlo Montemagno, is that future nanosystems will be hybrids of silicon technology and biological molecular machines. Richard Smalley argued that mechanosynthesis are impossible due to the difficulties in mechanically manipulating individual molecules.

This led to an exchange of letters in the ACS publication Chemical & Engineering News in 2003. Though biology clearly demonstrates that molecular machine systems are possible, non-biological molecular machines are today only in their infancy. Leaders in research on non-biological molecular machines are Dr. Alex Zettl and his colleagues at Lawrence Berkeley Laboratories and UC Berkeley. They have constructed at least three distinct molecular devices whose motion is controlled from the desktop with changing voltage: a nanotube nanomotor, a molecular actuator, and a nanoelectromechanical relaxation oscillator.

An experiment indicating that positional molecular assembly is possible was performed by Ho and Lee at Cornell University in 1999. They used a scanning tunneling microscope to move an individual carbon monoxide molecule (CO) to an individual iron atom (Fe) sitting on a flat silver crystal, and chemically bound the CO to the Fe by applying a voltage.

Current Research

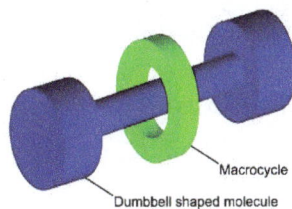

Macrocycle
Dumbbell shaped molecule

Graphical representation of a rotaxane, useful as a molecular switch.

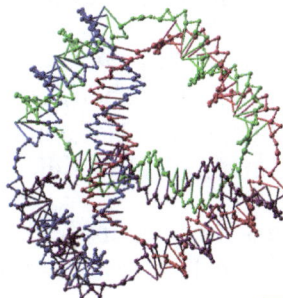

This DNA tetrahedron is an artificially designed nanostructure of the type made in the field of DNA nanotechnology. Each edge of the tetrahedron is a 20 base pair DNA double helix, and each vertex is a three-arm junction.

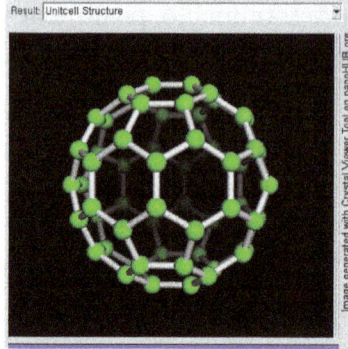

Rotating view of C_{60}, one kind of fullerene.

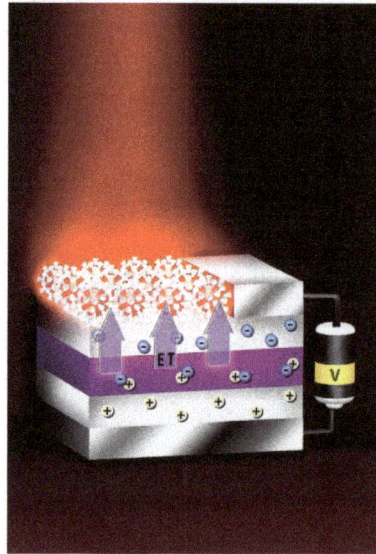

This device transfers energy from nano-thin layers of quantum wells to nanocrystals above them, causing the nanocrystals to emit visible light.

Nanomaterials

The nanomaterials field includes subfields which develop or study materials having unique properties arising from their nanoscale dimensions.

- Interface and colloid science has given rise to many materials which may be useful in nanotechnology, such as carbon nanotubes and other fullerenes, and various nanoparticles and nanorods. Nanomaterials with fast ion transport are related also to nanoionics and nanoelectronics.

- Nanoscale materials can also be used for bulk applications; most present commercial applications of nanotechnology are of this flavor.

- Progress has been made in using these materials for medical applications.

- Nanoscale materials such as nanopillars are sometimes used in solar cells which combats the cost of traditional Silicon solar cells.

- Development of applications incorporating semiconductor nanoparticles to be used in the next generation of products, such as display technology, lighting, solar cells and biological imaging.

- Recent application of nanomaterials include a range of biomedical applications, such as tissue engineering, drug delivery, and biosensors.

Bottom-up Approaches

These seek to arrange smaller components into more complex assemblies.

- DNA nanotechnology utilizes the specificity of Watson–Crick basepairing to construct well-defined structures out of DNA and other nucleic acids.

- Approaches from the field of "classical" chemical synthesis (Inorganic and organic synthesis) also aim at designing molecules with well-defined shape (e.g. bis-peptides).

- More generally, molecular self-assembly seeks to use concepts of supramolecular chemistry, and molecular recognition in particular, to cause single-molecule components to automatically arrange themselves into some useful conformation.

- Atomic force microscope tips can be used as a nanoscale "write head" to deposit a chemical upon a surface in a desired pattern in a process called dip pen nanolithography. This technique fits into the larger subfield of nanolithography.

Top-down Approaches

These seek to create smaller devices by using larger ones to direct their assembly.

- Many technologies that descended from conventional solid-state silicon methods for fabricating microprocessors are now capable of creating features smaller than 100 nm, falling under the definition of nanotechnology. Giant magnetoresistance-based hard drives already on the market fit this description, as do atomic layer deposition (ALD) techniques. Peter Grünberg and Albert Fert received the Nobel Prize in Physics in 2007 for their discovery of Giant magnetoresistance and contributions to the field of spintronics.

- Solid-state techniques can also be used to create devices known as nanoelectromechanical systems or NEMS, which are related to microelectromechanical systems or MEMS.

- Focused ion beams can directly remove material, or even deposit material when suitable precursor gasses are applied at the same time. For example, this technique is used routinely to create sub-100 nm sections of material for analysis in Transmission electron microscopy.

- Atomic force microscope tips can be used as a nanoscale "write head" to deposit a resist, which is then followed by an etching process to remove material in a top-down method.

Functional Approaches

These seek to develop components of a desired functionality without regard to how they might be assembled.

- Magnetic assembly for the synthesis of anisotropic superparamagnetic materials such as recently presented magnetic nanochains.

- Molecular scale electronics seeks to develop molecules with useful electronic properties. These could then be used as single-molecule components in a nano-electronic device.

- Synthetic chemical methods can also be used to create synthetic molecular motors, such as in a so-called nanocar.

Biomimetic Approaches

- Bionics or biomimicry seeks to apply biological methods and systems found in nature, to the study and design of engineering systems and modern technology. Biomineralization is one example of the systems studied.

- Bionanotechnology is the use of biomolecules for applications in nanotechnology, including use of viruses and lipid assemblies. Nanocellulose is a potential bulk-scale application.

Speculative

These subfields seek to anticipate what inventions nanotechnology might yield, or attempt to propose an agenda along which inquiry might progress. These often take a big-picture view of nanotechnology, with more emphasis on its societal implications than the details of how such inventions could actually be created.

- Molecular nanotechnology is a proposed approach which involves manipulating single molecules in finely controlled, deterministic ways. This is more theoretical than the other subfields, and many of its proposed techniques are beyond current capabilities.

- Nanorobotics centers on self-sufficient machines of some functionality operating at the nanoscale. There are hopes for applying nanorobots in medicine, but it may not be easy to do such a thing because of several drawbacks of such devices. Nevertheless, progress on innovative materials and methodologies has been demonstrated with some patents granted about new nanomanufacturing devices for future commercial applications, which also progressively helps in the development towards nanorobots with the use of embedded nanobioelectronics concepts.

- Productive nanosystems are "systems of nanosystems" which will be complex nanosystems that produce atomically precise parts for other nanosystems, not necessarily using novel nanoscale-emergent properties, but well-understood fundamentals of manufacturing. Because of the discrete (i.e. atomic) nature of matter and the possibility of exponential growth, this stage is seen as the basis of another industrial revolution. Mihail Roco, one of the architects of the USA's National Nanotechnology Initiative, has proposed four states of nanotechnology that seem to parallel the technical progress of the Industrial Revolution, progressing from passive nanostructures to active nanodevices to complex nanomachines and ultimately to productive nanosystems.

- Programmable matter seeks to design materials whose properties can be easily, reversibly and externally controlled though a fusion of information science and materials science.

- Due to the popularity and media exposure of the term nanotechnology, the words picotechnology and femtotechnology have been coined in analogy to it, although these are only used rarely and informally.

Dimensionality in Nanomaterials

Nanomaterials can be classified in 0D, 1D, 2D and 3D nanomaterials. The dimensionality play a major role in determining the characteristic of nanomaterials including physical, chemical and biological characteristics. With the decrease in dimensionality, an increase in surface-to-volume ratio is observed. This indicate that smaller dimensional nanomaterials have higher surface area compared to 3D nanomaterials. Recently, two dimensional (2D) nanomaterials are extensively investigated for electronic, biomedical, drug delivery and biosensor applications.

Tools and Techniques

There are several important modern developments. The atomic force microscope (AFM) and the Scanning Tunneling Microscope (STM) are two early versions of scanning probes that launched nanotechnology. There are other types of scanning probe microscopy. Although conceptually similar to the scanning confocal microscope developed by Marvin Minsky in 1961 and the scanning acoustic microscope (SAM) developed by Calvin Quate and coworkers in the 1970s, newer scanning probe microscopes have much higher resolution, since they are not limited by the wavelength of sound or light.

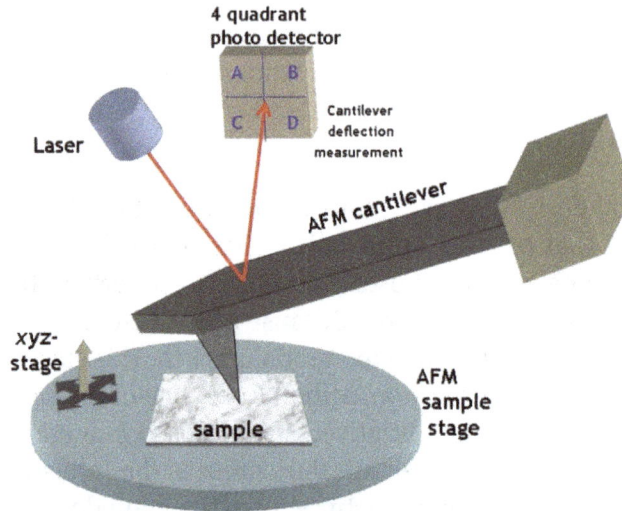

Typical AFM setup. A microfabricated cantilever with a sharp tip is deflected by features on a sample surface, much like in a phonograph but on a much smaller scale. A laser beam reflects off the backside of the cantilever into a set of photodetectors, allowing the deflection to be measured and assembled into an image of the surface.

The tip of a scanning probe can also be used to manipulate nanostructures (a process called positional assembly). Feature-oriented scanning methodology may be a promising way to implement these nanomanipulations in automatic mode. However, this is still a slow process because of low scanning velocity of the microscope.

Various techniques of nanolithography such as optical lithography, X-ray lithography dip pen nanolithography, electron beam lithography or nanoimprint lithography were also developed. Lithography is a top-down fabrication technique where a bulk material is reduced in size to nanoscale pattern.

Another group of nanotechnological techniques include those used for fabrication of nanotubes and nanowires, those used in semiconductor fabrication such as deep ultraviolet lithography, electron beam lithography, focused ion beam machining, nanoimprint lithography, atomic layer deposition, and molecular vapor deposition, and further including molecular self-assembly techniques such as those employing di-block copolymers. The precursors of these techniques preceded the nanotech era, and are extensions in the development of scientific advancements rather than techniques which were devised with the sole purpose of creating nanotechnology and which were results of nanotechnology research.

The top-down approach anticipates nanodevices that must be built piece by piece in stages, much as manufactured items are made. Scanning probe microscopy is an important technique both for characterization and synthesis of nanomaterials. Atomic force microscopes and scanning tunneling microscopes can be used to look at surfaces and to move atoms around. By designing different tips for these microscopes, they can be used for carving out structures on surfaces and to help guide self-assembling structures. By using, for example, feature-oriented scanning approach, atoms or molecules can be moved around on a surface with scanning probe microscopy techniques. At present, it is expensive and time-consuming for mass production but very suitable for laboratory experimentation.

In contrast, bottom-up techniques build or grow larger structures atom by atom or molecule by molecule. These techniques include chemical synthesis, self-assembly and positional assembly. Dual polarisation interferometry is one tool suitable for characterisation of self assembled thin films. Another variation of the bottom-up approach is molecular beam epitaxy or MBE. Researchers at Bell Telephone Laboratories like John R. Arthur. Alfred Y. Cho, and Art C. Gossard developed and implemented MBE as a research tool in the late 1960s and 1970s. Samples made by MBE were key to the discovery of the fractional quantum Hall effect for which the 1998 Nobel Prize in Physics was awarded. MBE allows scientists to lay down atomically precise layers of atoms and, in the process, build up complex structures. Important for research on semiconductors, MBE is also widely used to make samples and devices for the newly emerging field of spintronics.

However, new therapeutic products, based on responsive nanomaterials, such as the ultradeformable, stress-sensitive Transfersome vesicles, are under development and already approved for human use in some countries.

Applications

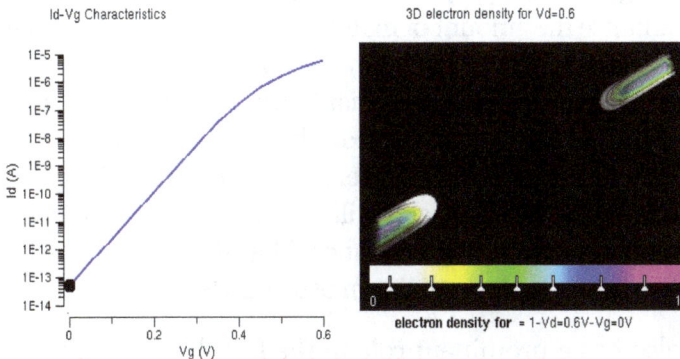

One of the major applications of nanotechnology is in the area of nanoelectronics with MOSFET's being made of small nanowires ~10 nm in length. Here is a simulation of such a nanowire.

Nanostructures provide this surface with superhydrophobicity, which lets water droplets roll down the inclined plane.

As of August 21, 2008, the Project on Emerging Nanotechnologies estimates that over 800 manufacturer-identified nanotech products are publicly available, with new ones hitting the market at a pace of 3–4 per week. The project lists all of the products in a publicly accessible online database. Most applications are limited to the use of "first generation" passive nanomaterials which includes titanium dioxide in sunscreen, cosmetics, surface coatings, and some food products; Carbon allotropes used to produce gecko tape; silver in food packaging, clothing, disinfectants and household appliances; zinc oxide in sunscreens and cosmetics, surface coatings, paints and outdoor furniture varnishes; and cerium oxide as a fuel catalyst.

Further applications allow tennis balls to last longer, golf balls to fly straighter, and even bowling balls to become more durable and have a harder surface. Trousers and socks have been infused with nanotechnology so that they will last longer and keep people cool in the summer. Bandages are being infused with silver nanoparticles to heal cuts faster. Video game consoles and personal computers may become cheaper, faster, and contain more memory thanks to nanotechnology. Nanotechnology may have the ability to make existing medical applications cheaper and easier to use in places like the general practitioner's office and at home. Cars are being manufactured with nanomaterials so they may need fewer metals and less fuel to operate in the future.

Scientists are now turning to nanotechnology in an attempt to develop diesel engines with cleaner exhaust fumes. Platinum is currently used as the diesel engine catalyst in these engines. The catalyst is what cleans the exhaust fume particles. First a reduction catalyst is employed to take nitrogen atoms from NOx molecules in order to free oxygen. Next the oxidation catalyst oxidizes the hydrocarbons and carbon monoxide to form carbon dioxide and water. Platinum is used in both the reduction and the oxidation catalysts. Using platinum though, is inefficient in that it is expensive and unsustainable. Danish company InnovationsFonden invested DKK 15 million in a search for new catalyst substitutes using nanotechnology. The goal of the project, launched in the autumn of 2014, is to maximize surface area and minimize the amount of material required. Objects tend to minimize their surface energy; two drops of water, for example, will join to form one drop and decrease surface area. If the catalyst's surface area that is exposed to the exhaust fumes is maximized, efficiency of the catalyst is maximized. The team working on this project aims to create nanoparticles that will not merge. Every time the surface is optimized, material is saved. Thus, creating these nanoparticles will increase the effectiveness of the resulting diesel engine catalyst—in turn leading to cleaner exhaust fumes—and will decrease cost. If successful, the team hopes to reduce platinum use by 25%.

Nanotechnology also has a prominent role in the fast developing field of Tissue Engineering. When designing scaffolds, researchers attempt to the mimic the nanoscale features of a Cell's microenvironment to direct its differentiation down a suitable lineage. For example, when creating scaffolds to support the growth of bone, researchers may mimic osteoclast resorption pits.

Researchers have successfully used DNA origami-based nanobots capable of carrying out logic functions to achieve targeted drug delivery in cockroaches. It is said that the computational power of these nanobots can be scaled up to that of a Commodore 64.

Implications

An area of concern is the effect that industrial-scale manufacturing and use of nanomaterials would have on human health and the environment, as suggested by nanotoxicology research. For these reasons, some groups advocate that nanotechnology be regulated by governments. Others counter that overregulation would stifle scientific research and the development of beneficial innovations. Public health research agencies, such as the National Institute for Occupational Safety and Health are actively conducting research on potential health effects stemming from exposures to nanoparticles.

Some nanoparticle products may have unintended consequences. Researchers have discovered that bacteriostatic silver nanoparticles used in socks to reduce foot odor are being released in the wash. These particles are then flushed into the waste water stream and may destroy bacteria which are critical components of natural ecosystems, farms, and waste treatment processes.

Public deliberations on risk perception in the US and UK carried out by the Center for Nanotechnology in Society found that participants were more positive about nanotechnologies for energy applications than for health applications, with health applications raising moral and ethical dilemmas such as cost and availability.

Experts, including director of the Woodrow Wilson Center's Project on Emerging Nanotechnologies David Rejeski, have testified that successful commercialization depends on adequate oversight, risk research strategy, and public engagement. Berkeley, California is currently the only city in the United States to regulate nanotechnology; Cambridge, Massachusetts in 2008 considered enacting a similar law, but ultimately rejected it. Relevant for both research on and application of nanotechnologies, the insurability of nanotechnology is contested. Without state regulation of nanotechnology, the availability of private insurance for potential damages is seen as necessary to ensure that burdens are not socialised implicitly.

Health and Environmental Concerns

Nanofibers are used in several areas and in different products, in everything from aircraft wings to tennis rackets. Inhaling airborne nanoparticles and nanofibers may lead to a number of pulmonary diseases, e.g. fibrosis. Researchers have found that when rats breathed in nanoparticles, the particles settled in the brain and lungs, which led to significant increases in biomarkers for inflammation and stress response and that nanoparticles induce skin aging through oxidative stress in hairless mice.

A two-year study at UCLA's School of Public Health found lab mice consuming nano-titanium dioxide showed DNA and chromosome damage to a degree "linked to all the big killers of man, namely cancer, heart disease, neurological disease and aging".

A major study published more recently in Nature Nanotechnology suggests some forms of carbon nanotubes – a poster child for the "nanotechnology revolution" – could be as harmful as asbestos if inhaled in sufficient quantities. Anthony Seaton of the Institute of Occupational Medicine in Edinburgh, Scotland, who contributed to the article on carbon nanotubes said "We know that some of them probably have the potential to cause mesothelioma. So those sorts of materials need to be handled very carefully." In the absence of specific regulation forthcoming from governments, Paull and Lyons (2008) have called for an exclusion of engineered nanoparticles in food. A newspaper article reports that workers in a paint factory developed serious lung disease and nanoparticles were found in their lungs.

Regulation

Calls for tighter regulation of nanotechnology have occurred alongside a growing debate related to the human health and safety risks of nanotechnology. There is significant debate about who is responsible for the regulation of nanotechnology. Some regulatory agencies currently cover some nanotechnology products and processes (to varying degrees) – by "bolting on" nanotechnology to existing regulations – there are clear gaps in these regimes. Davies (2008) has proposed a regulatory road map describing steps to deal with these shortcomings.

Stakeholders concerned by the lack of a regulatory framework to assess and control risks associated with the release of nanoparticles and nanotubes have drawn parallels with bovine spongiform encephalopathy ("mad cow" disease), thalidomide, genetically modified food, nuclear energy, reproductive technologies, biotechnology, and asbestosis. Dr. Andrew Maynard, chief science advisor to the Woodrow Wilson Center's Project on Emerging Nanotechnologies, concludes that there is insufficient funding for human health and safety research, and as a result there is currently limited understanding of the human health and safety risks associated with nanotechnology. As a result, some academics have called for stricter application of the precautionary principle, with delayed marketing approval, enhanced labelling and additional safety data development requirements in relation to certain forms of nanotechnology.

The Royal Society report identified a risk of nanoparticles or nanotubes being released during disposal, destruction and recycling, and recommended that "manufacturers of products that fall under extended producer responsibility regimes such as end-of-life regulations publish procedures outlining how these materials will be managed to minimize possible human and environmental exposure" (p. xiii).

The Center for Nanotechnology in Society has found that people respond to nanotechnologies differently, depending on application – with participants in public deliberations more positive about nanotechnologies for energy than health applications – suggesting that any public calls for nano regulations may differ by technology sector.

References

- Drexler, K. Eric (1992). Nanosystems: Molecular Machinery, Manufacturing, and Computation. New York: John Wiley & Sons. ISBN 0-471-57547-X.

- Allhoff, Fritz; Lin, Patrick; Moore, Daniel (2010). What is nanotechnology and why does it matter?: from science to ethics. John Wiley and Sons. pp. 3–5. ISBN 1-4051-7545-1.

- Gaharwar, A.K. [et al., edited by] (2013). Nanomaterials in tissue engineering : fabrication and applications. Oxford: Woodhead Publishing. ISBN 978-0-85709-596-1.

- Lapshin, R. V. (2011). "Feature-oriented scanning probe microscopy". In H. S. Nalwa. Encyclopedia of Nanoscience and Nanotechnology (PDF). 14. USA: American Scientific Publishers. pp. 105–115. ISBN 1-58883-163-9.

- Murray R.G.E. (1993) Advances in Bacterial Paracrystalline Surface Layers. T. J. Beveridge, S. F. Koval (Eds.). Plenum Press. ISBN 978-0-306-44582-8.

- "CDC – Nanotechnology – NIOSH Workplace Safety and Health Topic". National Institute for Occupational Safety and Health. June 15, 2012. Retrieved 2012-08-24.

- "CDC – NIOSH Publications and Products – Filling the Knowledge Gaps for Safe Nanotechnology in the Workplace". National Institute for Occupational Safety and Health. November 7, 2012. Retrieved 2012-11-08.

Branches of Nanotechnology

Nanoelectronics is the practice of using nanotechnology in components that are electronic. The branches related to nanotechnology are molecular scale electronics, wet nanotechnology and nanolithography. This chapter elucidates the major branches of nanotechnology, providing a complete understanding.

Nanoelectronics

Nanoelectronics refer to the use of nanotechnology in electronic components. The term covers a diverse set of devices and materials, with the common characteristic that they are so small that inter-atomic interactions and quantum mechanical properties need to be studied extensively. Some of these candidates include: hybrid molecular/semiconductor electronics, one-dimensional nanotubes/nanowires, or advanced molecular electronics. Recent silicon CMOS technology generations, such as the 22 nanometer node, are already within this regime. Nanoelectronics are sometimes considered as disruptive technology because present candidates are significantly different from traditional transistors.

Fundamental Concepts

In 1965 Gordon Moore observed that silicon transistors were undergoing a continual process of scaling downward, an observation which was later codified as Moore's law. Since his observation transistor minimum feature sizes have decreased from 10 micrometers to the 28-22 nm range in 2011. The field of nanoelectronics aims to enable the continued realization of this law by using new methods and materials to build electronic devices with feature sizes on the nanoscale.

The volume of an object decreases as the third power of its linear dimensions, but the surface area only decreases as its second power. This somewhat subtle and unavoidable principle has huge ramifications. For example, the power of a drill (or any other machine) is proportional to the volume, while the friction of the drill's bearings and gears is proportional to their surface area. For a normal-sized drill, the power of the device is enough to handily overcome any friction. However, scaling its length down by a factor of 1000, for example, decreases its power by 1000^3 (a factor of a billion) while reducing the friction by only 1000^2 (a factor of only a million). Proportionally it has 1000 times less power per unit friction than the original drill. If the original friction-to-power ratio was, say, 1%, that implies the smaller drill will have 10 times as much friction as power; the drill is useless.

For this reason, while super-miniature electronic integrated circuits are fully functional, the same technology cannot be used to make working mechanical devices beyond the scales where frictional forces start to exceed the available power. So even though you may see microphotographs of delicately etched silicon gears, such devices are currently little more than curiosities with limited real world applications, for example, in moving mirrors and shutters. Surface tension increases in much the same way, thus magnifying the tendency for very small objects to stick together. This could possibly make any kind of "micro factory" impractical: even if robotic arms and hands could be scaled down, anything they pick up will tend to be impossible to put down. The above being said, molecular evolution has resulted in working cilia, flagella, muscle fibers and rotary motors in aqueous environments, all on the nanoscale. These machines exploit the increased frictional forces found at the micro or nanoscale. Unlike a paddle or a propeller which depends on normal frictional forces (the frictional forces perpendicular to the surface) to achieve propulsion, cilia develop motion from the exaggerated drag or laminar forces (frictional forces parallel to the surface) present at micro and nano dimensions. To build meaningful "machines" at the nanoscale, the relevant forces need to be considered. We are faced with the development and design of intrinsically pertinent machines rather than the simple reproductions of macroscopic ones.

All scaling issues therefore need to be assessed thoroughly when evaluating nanotechnology for practical applications.

Approaches to Nanoelectronics

Nanofabrication

For example, single electron transistors, which involve transistor operation based on a single electron. Nanoelectromechanical systems also fall under this category. Nanofabrication can be used to construct ultradense parallel arrays of nanowires, as an alternative to synthesizing nanowires individually.

Nanomaterials Electronics

Besides being small and allowing more transistors to be packed into a single chip, the uniform and symmetrical structure of nanotubes allows a higher electron mobility (faster electron movement in the material), a higher dielectric constant (faster frequency), and a symmetrical electron/hole characteristic.

Also, nanoparticles can be used as quantum dots.

Molecular Electronics

Single molecule devices are another possibility. These schemes would make heavy use of molecular self-assembly, designing the device components to construct a larger structure or even a complete system on their own. This can be very useful for reconfigurable computing, and may even completely replace present FPGA technology.

Molecular electronics is a new technology which is still in its infancy, but also brings hope for truly atomic scale electronic systems in the future. One of the more promising applications of molecular electronics was proposed by the IBM researcher Ari Aviram and the theoretical chemist Mark Ratner in their 1974 and 1988 papers *Molecules for Memory, Logic and Amplification.*

This is one of many possible ways in which a molecular level diode / transistor might be synthesized by organic chemistry. A model system was proposed with a spiro carbon structure giving a molecular diode about half a nanometre across which could be connected by polythiophene molecular wires. Theoretical calculations showed the design to be sound in principle and there is still hope that such a system can be made to work.

Other Approaches

Nanoionics studies the transport of ions rather than electrons in nanoscale systems.

Nanophotonics studies the behavior of light on the nanoscale, and has the goal of developing devices that take advantage of this behavior.

Nanoelectronic Devices

Current high-technology production processes are based on traditional top down strategies, where nanotechnology has already been introduced silently. The critical length scale of integrated circuits is already at the nanoscale (50 nm and below) regarding the gate length of transistors in CPUs or DRAM devices.

Computers

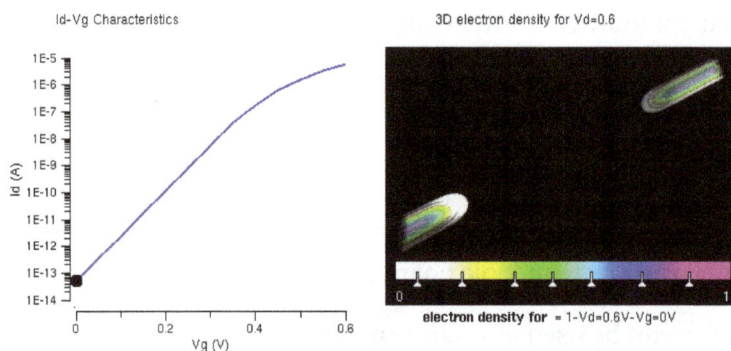

Simulation result for formation of inversion channel (electron density) and attainment of threshold voltage (IV) in a nanowire MOSFET. Note that the threshold voltage for this device lies around 0.45V.

Nanoelectronics holds the promise of making computer processors more powerful than are possible with conventional semiconductor fabrication techniques. A number of approaches are currently being researched, including new forms of nanolithography, as well as the use of nanomaterials such as nanowires or small molecules in place of traditional CMOS components. Field effect transistors have been made using both semiconducting carbon nanotubes and with heterostructured semiconductor nanowires.

In 1999, the CMOS transistor developed at the Laboratory for Electronics and Information Technology in Grenoble, France, tested the limits of the principles of the MOSFET transistor with a diameter of 18 nm (approximately 70 atoms placed side by side). This was almost one tenth the size of the smallest industrial transistor in 2003 (130 nm in 2003, 90 nm in 2004, 65 nm in 2005 and 45 nm in 2007). It enabled the theoretical integration of seven billion junctions on a €1 coin. However, the CMOS transistor, which was created in 1999, was not a simple research experiment to study how CMOS technology functions, but rather a demonstration of how this technology functions now that we ourselves are getting ever closer to working on a molecular scale. Today it would be impossible to master the coordinated assembly of a large number of these transistors on a circuit and it would also be impossible to create this on an industrial level.

Memory Storage

Electronic memory designs in the past have largely relied on the formation of transistors. However, research into crossbar switch based electronics have offered an alternative using reconfigurable interconnections between vertical and horizontal wiring arrays to create ultra high density memories. Two leaders in this area are Nantero which has developed a carbon nanotube based crossbar memory called Nano-RAM and Hewlett-Packard which has proposed the use of memristor material as a future replacement of Flash memory.

An example of such novel devices is based on spintronics. The dependence of the resistance of a material (due to the spin of the electrons) on an external field is called magnetoresistance. This effect can be significantly amplified (GMR - Giant Magneto-Resistance) for nanosized objects, for example when two ferromagnetic layers are separated by a nonmagnetic layer, which is several nanometers thick (e.g. Co-Cu-Co). The GMR effect has led to a strong increase in the data storage density of hard disks and made the gigabyte range possible. The so-called tunneling magnetoresistance (TMR) is very similar to GMR and based on the spin dependent tunneling of electrons through adjacent ferromagnetic layers. Both GMR and TMR effects can be used to create a non-volatile main memory for computers, such as the so-called magnetic random access memory or MRAM.

Novel Optoelectronic Devices

In the modern communication technology traditional analog electrical devices are increasingly replaced by optical or optoelectronic devices due to their enormous bandwidth and capacity, respectively. Two promising examples are photonic crystals and quantum dots. Photonic crystals are materials with a periodic variation in the refractive index with a lattice constant that is half the wavelength of the light used. They offer a selectable band gap for the propagation of a certain wavelength, thus they resemble a semiconductor, but for light or photons instead of electrons. Quantum dots are nanoscaled objects, which can be used, among many other things, for the construction of lasers. The advantage of a quantum dot laser over the traditional semiconductor laser is that their emitted wavelength depends on the diameter of the dot. Quantum dot lasers are cheaper and offer a higher beam quality than conventional laser diodes.

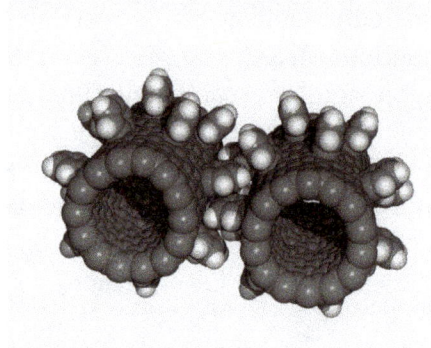

Displays

The production of displays with low energy consumption might be accomplished using carbon nanotubes (CNT). Carbon nanotubes are electrically conductive and due to their small diameter of several nanometers, they can be used as field emitters with extremely high efficiency for field emission displays (FED). The principle of operation resembles that of the cathode ray tube, but on a much smaller length scale.

Quantum Computers

Entirely new approaches for computing exploit the laws of quantum mechanics for novel quantum computers, which enable the use of fast quantum algorithms. The Quantum computer has quantum bit memory space termed "Qubit" for several computations at the same time. This facility may improve the performance of the older systems.

Radios

Nanoradios have been developed structured around carbon nanotubes.

Energy Production

Research is ongoing to use nanowires and other nanostructured materials with the hope to create cheaper and more efficient solar cells than are possible with conventional planar silicon solar cells. It is believed that the invention of more efficient solar energy would have a great effect on satisfying global energy needs.

There is also research into energy production for devices that would operate *in vivo*, called bio-nano generators. A bio-nano generator is a nanoscale electrochemical device, like a fuel cell or galvanic cell, but drawing power from blood glucose in a living body, much the same as how the body generates energy from food. To achieve the effect, an enzyme is used that is capable of stripping glucose of its electrons, freeing them for use in electrical devices. The average person's body could, theoretically, generate 100 watts of electricity (about 2000 food calories per day) using a bio-nano generator. However, this estimate is only true if all food was converted to electricity, and the human body needs some energy consistently, so possible power generated is likely much lower. The electricity generated by such a device could power devices embedded in the body (such as pacemakers), or sugar-fed nanorobots. Much of the research done on bio-nano generators is still experimental, with Panasonic's Nanotechnology Research Laboratory among those at the forefront.

Medical Diagnostics

There is great interest in constructing nanoelectronic devices that could detect the concentrations of biomolecules in real time for use as medical diagnostics, thus falling into the category of nanomedicine. A parallel line of research seeks to create nanoelectronic devices which could interact with single cells for use in basic biological research. These devices are called nanosensors. Such miniaturization on nanoelectronics towards in vivo proteomic sensing should enable new approaches for health monitoring, surveillance, and defense technology.

Molecular Scale Electronics

Molecular scale electronics, also called single-molecule electronics, is a branch of nanotechnology that uses single molecules, or nanoscale collections of single molecules, as electronic components. Because single molecules constitute the smallest stable structures imaginable, this miniaturization is the ultimate goal for shrinking electrical circuits.

The field is often termed simply as "molecular electronics", but this term is also used to refer to the distantly related field of conductive polymers and organic electronics, which uses the properties of molecules to affect the bulk properties of a material. A nomenclature distinction has been suggested so that *molecular materials for electronics* refers to this latter field of bulk applications, while *molecular scale electronics* refers to the nanoscale single-molecule applications treated here.

Fundamental Concepts

Conventional electronics have traditionally been made from bulk materials. Ever since their invention in 1958, the performance and complexity of integrated circuits has undergone exponential growth, a trend named Moore's law, as feature sizes of the embedded components have shrunk accordingly. As the structures shrink, the sensitivity to deviations increases. In a few technology generations, when the minimum feature sizes reaches 13 nm, the composition of the devices must be controlled to a precision of a few atoms for the devices to work. With bulk methods growing increasingly demanding and costly as they near inherent limits, the idea was born that the components could instead be built up atom by atom in a chemistry lab (bottom up) versus carving them out of bulk material (top down). This is the idea behind molecular electronics, with the ultimate miniaturization being components contained in single molecules.

In single-molecule electronics, the bulk material is replaced by single molecules. Instead of forming structures by removing or applying material after a pattern scaffold, the atoms are put together in a chemistry lab. In this way, billions of billions of copies are made simultaneously (typically more than 10^{20} molecules are made at once) while the composition of molecules are controlled down to the last atom. The molecules used have properties that resemble traditional electronic components such as a wire, transistor or rectifier.

Single-molecule electronics is an emerging field, and entire electronic circuits consisting exclusively of molecular sized compounds are still very far from being realized. However, the unceasing demand for more computing power, along with the inherent limits of lithographic methods as of 2016, make the transition seem unavoidable. Currently, the focus is on discovering molecules with interesting properties and on finding ways to obtain reliable and reproducible contacts between the molecular components and the bulk material of the electrodes.

Theoretical Basis

Molecular electronics operates in the quantum realm of distances less than 100 nanometers. The miniaturization down to single molecules brings the scale down to a regime where quantum mechanics effects are important. In conventional electronic components, electrons can be filled in or drawn out more or less like a continuous flow of electric charge. In contrast, in molecular electronics the transfer of one electron alters the system significantly. For example, when an electron has been transferred from a source electrode to a molecule, the molecule gets charged up, which makes it far harder for the next electron to transfer . The significant amount of energy due to charging must be accounted for when making calculations about the electronic properties of the setup, and is highly sensitive to distances to conducting surfaces nearby.

The theory of single-molecule devices is especially interesting since the system under consideration is an open quantum system in nonequilibrium (driven by voltage). In the low bias voltage regime, the nonequilibrium nature of the molecular junction can be ignored, and the current-voltage traits of the device can be calculated using the equilibrium electronic structure of the system. However, in stronger bias regimes a more sophisticated treatment is required, as there is no longer a variational principle. In the elastic tunneling case (where the passing electron does not exchange energy with the system), the formalism of Rolf Landauer can be used to calculate the transmission through the system as a function of bias voltage, and hence the current. In inelastic tunneling, an elegant formalism based on the non-equilibrium Green's functions of Leo Kadanoff and Gordon Baym, and independently by Leonid Keldysh was advanced by Ned Wingreen and Yigal Meir. This Meir-Wingreen formulation has been used to great success in the molecular electronics community to examine the more difficult and interesting cases where the transient electron exchanges energy with the molecular system (for example through electron-phonon coupling or electronic excitations).

Further, connecting single molecules reliably to a larger scale circuit has proven a great challenge, and constitutes a significant hindrance to commercialization.

Examples

Common for molecules used in molecular electronics is that the structures contain many alternating double and single bonds. This is done because such patterns delocalize the molecular orbitals, making it possible for electrons to move freely over the conjugated area.

Wires

The sole purpose of molecular wires is to electrically connect different parts of a molecular electrical circuit. As the assembly of these and their connection to a macroscopic circuit is still not mastered, the focus of research in single-molecule electronics is primarily on the functionalized molecules: molecular wires are characterized by containing no functional groups and are hence composed of plain repetitions of a conjugated building block. Among these are the carbon nanotubes that are quite large compared to the other suggestions but have shown very promising electrical properties.

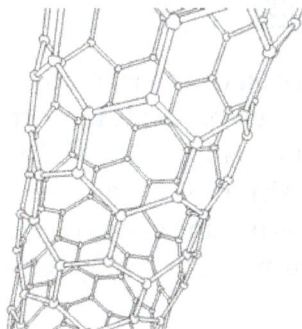

This animation of a rotating carbon nanotube shows its 3D structure.

The main problem with the molecular wires is to obtain good electrical contact with the electrodes so that electrons can move freely in and out of the wire.

Transistors

Single-molecule transistors are fundamentally different from the ones known from bulk electronics. The gate in a conventional (field-effect) transistor determines the conductance between the source and drain electrode by controlling the density of charge carriers between them, whereas the gate in a single-molecule transistor controls the possibility of a single electron to jump on and off the molecule by modifying the energy of the molecular orbitals. One of the effects of this difference is that the single-molecule transistor is almost binary: it is either *on* or *off*. This opposes its bulk counterparts, which have quadratic responses to gate voltage.

It is the quantization of charge into electrons that is responsible for the markedly different behavior compared to bulk electronics. Because of the size of a single molecule, the charging due to a single electron is significant and provides means to turn a transistor *on* or *off*. For this to work, the electronic orbitals on the transistor molecule cannot be too well integrated with the orbitals on the electrodes. If they are, an electron cannot be said to be located on the molecule or the electrodes and the molecule will function as a wire.

A popular group of molecules, that can work as the semiconducting channel material in a molecular transistor, is the oligopolyphenylenevinylenes (OPVs) that works by the Coulomb blockade mechanism when placed between the source and drain electrode in an appropriate way. Fullerenes work by the same mechanism and have also been commonly used.

Semiconducting carbon nanotubes have also been demonstrated to work as channel material but although molecular, these molecules are sufficiently large to behave almost as bulk semiconductors.

The size of the molecules, and the low temperature of the measurements being conducted, makes the quantum mechanical states well defined. Thus, it is being researched if the quantum mechanical properties can be used for more advanced purposes than simple transistors (e.g. spintronics).

Physicists at the University of Arizona, in collaboration with chemists from the University of Madrid, have designed a single-molecule transistor using a ring-shaped molecule similar to benzene. Physicists at Canada's National Institute for Nanotechnology have designed a single-molecule transistor using styrene. Both groups expect (the designs were experimentally unverified as of June 2005) their respective devices to function at room temperature, and to be controlled by a single electron.

Rectifiers (Diodes)

Hydrogen can be removed from individual tetraphenylporphyrin (H_2TPP) molecules by applying excess voltage to the tip of a scanning tunneling microscope (STAM, a);

this removal alters the current-voltage (I-V) curves of TPP molecules, measured using the same STM tip, from diode-like (red curve in b) to resistor-like (green curve). Image (c) shows a row of TPP, H_2TPP and TPP molecules. While scanning image (d), excess voltage was applied to H_2TPP at the black dot, which instantly removed hydrogen, as shown in the bottom part of (d) and in the re-scan image (e). Such manipulations can be used in single-molecule electronics.

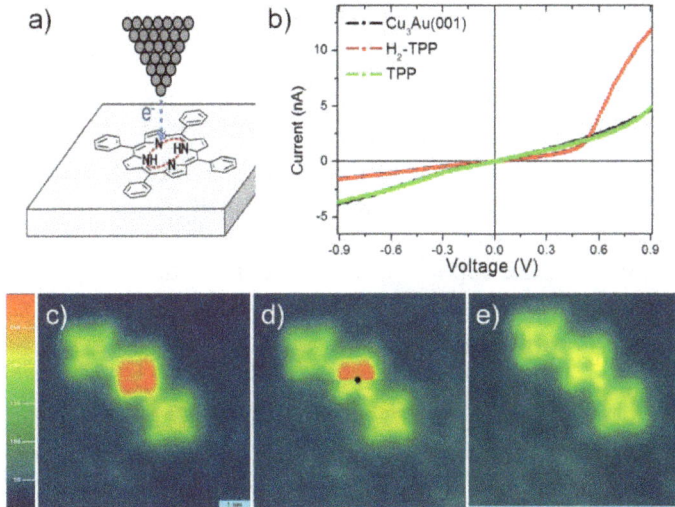

Molecular rectifiers are mimics of their bulk counterparts and have an asymmetric construction so that the molecule can accept electrons in one end but not the other. The molecules have an electron donor (D) in one end and an electron acceptor (A) in the other. This way, the unstable state $D^+ - A^-$ will be more readily made than $D^- - A^+$. The result is that an electric current can be drawn through the molecule if the electrons are added through the acceptor end, but less easily if the reverse is attempted.

Methods

One of the biggest problems with measuring on single molecules is to establish reproducible electrical contact with only one molecule and doing so without shortcutting the electrodes. Because the current photolithographic technology is unable to produce electrode gaps small enough to contact both ends of the molecules tested (on the order of nanometers), alternative strategies are applied.

Molecular Gaps

One way to produce electrodes with a molecular sized gap between them is break junctions, in which a thin electrode is stretched until it breaks. Another is electromigration. Here a current is led through a thin wire until it melts and the atoms migrate to produce the gap. Further, the reach of conventional photolithography can be enhanced by chemically etching or depositing metal on the electrodes.

Probably the easiest way to conduct measurements on several molecules is to use the tip of a scanning tunneling microscope (STM) to contact molecules adhered at the other end to a metal substrate.

Anchoring

A popular way to anchor molecules to the electrodes is to make use of sulfur's high chemical affinity to gold. In these setups, the molecules are synthesized so that sulfur atoms are placed strategically to function as crocodile clips connecting the molecules to the gold electrodes. Though useful, the anchoring is non-specific and thus anchors the molecules randomly to all gold surfaces. Further, the contact resistance is highly dependent on the precise atomic geometry around the site of anchoring and thereby inherently compromises the reproducibility of the connection.

To circumvent the latter issue, experiments has shown that fullerenes could be a good candidate for use instead of sulfur because of the large conjugated π-system that can electrically contact many more atoms at once than one atom of sulfur.

Fullerene Nanoelectronics

In polymers, classical organic molecules are composed of both carbon and hydrogen (and sometimes additional compounds such as nitrogen, chlorine or sulphur). They are obtained from petrol and can often be synthesized in large amounts. Most of these molecules are insulating when their length exceeds a few nanometers. However, naturally occurring carbon is conducting, especially graphite recovered from coal or encountered otherwise. From a theoretical viewpoint, graphite is a semi-metal, a category in between metals and semi-conductors. It has a layered structure, each sheet being one atom thick. Between each sheet, the interactions are weak enough to allow an easy manual cleavage.

Tailoring the graphite sheet to obtain well defined nanometer-sized objects remains a challenge. However, by the close of the twentieth century, chemists were exploring methods to fabricate extremely small graphitic objects that could be considered single molecules. After studying the interstellar conditions under which carbon is known to form clusters, Richard Smalley's group (Rice University, Texas) set up an experiment in which graphite was vaporized via laser irradiation. Mass spectrometry revealed that clusters containing specific *magic numbers* of atoms were stable, especially those clusters of 60 atoms. Harry Kroto, an English chemist who assisted in the experiment, suggested a possible geometry for these clusters – atoms covalently bound with the exact symmetry of a soccer ball. Coined buckminsterfullerenes, buckyballs, or C_{60}, the clusters retained some properties of graphite, such as conductivity. These objects were rapidly envisioned as possible building blocks for molecular electronics.

Problems

Artifacts

When trying to measure electronic traits of molecules, artificial phenomena can occur that can be hard to distinguish from truly molecular behavior. Before they were discovered, these artifacts have mistakenly been published as being features pertaining to the molecules in question.

Applying a voltage drop on the order of volts across a nanometer sized junction results in a very strong electrical field. The field can cause metal atoms to migrate and eventually close the gap by a thin filament, which can be broken again when carrying a current. The two levels of conductance imitate molecular switching between a conductive and an isolating state of a molecule.

Another encountered artifact is when the electrodes undergo chemical reactions due to the high field strength in the gap. When the voltage bias is reversed, the reaction will cause hysteresis in the measurements that can be interpreted as being of molecular origin.

A metallic grain between the electrodes can act as a single electron transistor by the mechanism described above, thus resembling the traits of a molecular transistor. This artifact is especially common with nanogaps produced by the electromigration method.

Commercialization

One of the biggest hindrances for single-molecule electronics to be commercially exploited is the lack of methods to connect a molecular sized circuit to bulk electrodes in a way that gives reproducible results. At the current state, the difficulty of connecting single molecules vastly outweighs any possible performance increase that could be gained from such shrinkage. The difficulties grow worse if the molecules are to have a certain spatial orientation and/or have multiple poles to connect.

Also problematic is that some measurements on single molecules are carried out in cryogenic temperatures (near absolute zero), which is very energy consuming. This is done to reduce signal noise enough to measure the faint currents of single molecules.

History and Recent Progress

In their treatment of so-called *donor-acceptor* complexes in the 1940s, Robert Mulliken and Albert Szent-Györgyi advanced the concept of charge transfer in molecules. They subsequently further refined the study of both charge transfer and energy transfer in molecules. Likewise, a 1974 paper from Mark Ratner and Ari Aviram illustrated a theoretical molecular rectifier. In 1988, Aviram described in detail a theoretical single-molecule field-effect transistor. Further concepts were proposed by Forrest Carter of the Naval Research Laboratory, including single-molecule logic gates. A wide range of ideas were presented, under his aegis, at a conference entitled *Molecular Electronic Devices* in 1988. These were all theoretical constructs and not concrete devices. The *direct* measurement of the electronic traits of individual molecules awaited the development of methods for making molecular-scale electrical contacts. This was no easy task. Thus, the first experiment directly-measuring the conductance of a single molecule was only reported in 1995 on a single C_{60} molecule by C. Joachim and J. K. Gimzewsky in their seminal Physical Revie Letter paper and later in 1997 by Mark Reed and co-workers on a few hundred molecules. Since then, this branch of the field has advanced rapidly. Likewise, as it has grown possible to measure such properties directly, the theoretical predictions of the early workers have been confirmed substantially.

Recent progress in nanotechnology and nanoscience has facilitated both experimental and theoretical study of molecular electronics. Development of the scanning tunneling microscope (STM) and later the atomic force microscope (AFM) have greatly facilitated manipulating single-molecule electronics. Also, theoretical advances in molecular electronics have facilitated further understanding of non-adiabatic charge transfer events at electrode-electrolyte interfaces.

The concept of molecular electronics was first published in 1974 when Aviram and Ratner suggested an organic molecule that could work as a rectifier. Having both huge commercial and fundamental interest, much effort was put into proving its feasibility, and 16 years later in 1990, the first demonstration of an intrinsic molecular rectifier was realized by Ashwell and coworkers for a thin film of molecules.

The first measurement of the conductance of a single molecule was realised in 1994 by C. Joachim and J. K. Gimzewski and published in 1995. This was the conclusion of 10 years of research started at IBM TJ Watson, using the scanning tunnelling microscope tip apex to switch a single molecule as already explored by A. Aviram, C. Joachim and M. Pomerantz at the end of the 80's. The trick was to use an UHV Scanning Tunneling microscope to allow the tip apex to gently touch the top of a single C 60 molecule adsorbed on an Au(110) surface. A resistance of 55 MOhms was recorded along with a low voltage linear I-V. The contact was certified by recording the I-z current distance property, which allows measurement of the deformation of the C 60 cage under contact. This first experiment was followed by the reported result using a mechanical break junction method to connect two gold electrodes to a sulfur-terminated molecular wire by Mark Reed and James Tour in 1997.

A single-molecule amplifier was implemented by C. Joachim and J.K. Gimzewski in IBM Zurich. This experiment, involving one C60 molecule, demonstrated that one such molecule can provide gain in a circuit via intramolecular quantum interference effects alone.

A collaboration of researchers at Hewlett-Packard (HP) and University of California, Los Angeles (UCLA), led by James Heath, Fraser Stoddart, R. Stanley Williams, and Philip Kuekes, has developed molecular electronics based on rotaxanes and catenanes.

Work is also occurring on the use of single-wall carbon nanotubes as field-effect transistors. Most of this work is being done by International Business Machines (IBM).

Some specific reports of a field-effect transistor based on molecular self-assembled monolayers were shown to be fraudulent in 2002 as part of the Schön scandal.

Until recently entirely theoretical, the Aviram-Ratner model for a unimolecular rectifier has been confirmed unambiguously in experiments by a group led by Geoffrey J. Ashwell at Bangor University, UK. Many rectifying molecules have so far been identified, and the number and efficiency of these systems is growing rapidly.

Supramolecular electronics is a new field involving electronics at a supramolecular level.

An important issue in molecular electronics is the determination of the resistance of a single molecule (both theoretical and experimental). For example, Bumm, et al. used STM to analyze a single molecular switch in a self-assembled monolayer to determine how conductive such a molecule can be. Another problem faced by this field is the difficulty of performing direct characterization since imaging at the molecular scale is often difficult in many experimental devices.

Wet Nanotechnology

Wet nanotechnology (also known as wet nanotech) involves working up to large masses from small ones.

Wet nanotechnology requires water in which the process occurs. The process also involves chemists and biologists trying to reach larger scales by putting together individual molecules. While Eric Drexler put forth the idea of nano-assemblers working dry, wet nanotech appears to be the likely first area in which something like a nano-assembler may achieve economic results. Pharmaceuticals and bioscience are central features of most nanotech start-ups. Richard A.L. Jones calls nanotechnology that steals bits of natural nanotechnology and puts them in a synthetic structure *biokleptic nanotechnology*. He calls building with synthetic materials according to nature's design principles *biomimetic nanotechnology*.

Using these guiding principles could lead to trillions of nanotech robots, that resemble bacteria in structural properties, entering a person's blood stream to do medical treatments.

Background

Wet nanotechnology is a fairly new area of nanotech that is going to mostly be dominated by the different forms of wet engineering. The processes that will be used are going to take place in aqueous solutions and are very close to that of bio-molecular manufacturing; this deals greatly with the creation of things like proteins found in our body such as DNA. In order for mankind to mimic these nanoscale machines in a way that they could be produced with some efficiency, they must look into bottom-up manufacturing. Bottom-up manufacturing deals with manipulating individual atoms during the manufacturing process, so that there is absolute control of their placement and interactions.

Then from the atomic scale, nanomachines could be made and even be designed to self-replicate themselves as long as they are designed in an environment with copious amount of the needed materials. Because individual atoms are being manipulated in the process, bottom-up manufacturing is often referred to as "atom by atom" manufacturing. If the manufacturing of nanomachines can be made more readily available through improved techniques, there could be a large economic and social impact. This would start with improvements in making microelectromechanical systems and then would allow for the creation of nanoscale biological sensors along with things that have not been thought of yet. This is because "wet" nanotech is only in the beginning of its life. Scientists and engineers alike feel that biomimetics is a great way to start looking at creating nanoscale machines. Humans have only had a few thousand years to try to learn about the mechanics of things at really small scales. However, nature has been working on perfecting the design and functionality of nanomachines for millions of years. This is why there are already nanomachines, such as ATP synthase, working in our bodies that have on unheard of 95% efficiency.

"Wet" vs. "Dry"

Wet nanotechnology is a form of wet engineering as opposed to dry engineering. There are different fields that deal with those two types of engineering. Molecular scientists, genetic engineers, and biologists all deal with wet engineering. They study processes that happen in life, and for the most part those processes take place in aqueous environments. Our bodies are made up mostly of water. Electrical and mechanical engineers are on the other side of the line in dry engineering.

They are involved with processes and manufacturing that does not occur in aqueous environments. For the most part, wet engineering deals with "soft" materials that allow for flexibility which is vital at the nanoscale in biological manufacturing. Dry engineers mostly handle things with rigid structures and parts. These differences stem for the fact that the forces that the two types of engineers must deal with are very different. At a larger scale, most things are dominated by Newtonian physics. However, when one looks at the nanoscale, especially in biological matters, the dominating force is Brownian motion.

Because nanotechnology in the new age is going to most likely deal with both dry and wet in conjunction with each other, there is going to have to be a change in the way society looks at engineering and manufacturing. People will have to be not only well educated in engineering but also in biology because the integration of the two is how there will be the largest improvements in nanotechnology.

Brownian Motion as It Relates to Wet Nanotech

With the existence of natural nanomachines, "a complex precision microscopic-sized machine that fits the standard definition of a machine", such as ATP synthase and T4 bacteriophage, scientists and biologists know that they are capable of making similar types of machines at the same scale. However, nature has had a long time to perfect the building and creation of these nanomachines and humankind has only just begun to look into them with greater interest.

This interest may have been sparked because of the existence of nanomachines such as ATP synthase (adenosine triphosphate), which is the "second in importance only to DNA". ATP is the main energy converter that our bodies contain and without it, life as we know it would not be able to flourish or even survive.

What Does Brownian Motion have to do With Complex Nanomachines?

Brownian motion is a random, constantly fluctuating force that acts on a body in environments that are at a microscale. This force is one that mechanical engineers and physicists are not used to dealing with because, at the larger scale that humankind tends to think of things, this force is not one that needs to be taken into account. People think of gravity, inertia, and other physics based forces that act on us all the time, however at the nanoscale those forces are mostly "negligible".

In order for nanomachines to be recreated by man, either there will need to be discoveries that allow us to understand how to "exploit" Brownian motion as nature does or find a way to work around it by using materials that are rigid enough to stand up to these forces. The way that nature has been able to exploit Brownian motion is through self-assembly. This force pushes and pulls all of the proteins and amino acids around in our bodies and sticks them together in all sorts of combinations. The combinations that do not work separate and continue with their random attachment however, the combinations that do work produce things like ATP synthase. Through this process nature has been able to make a nanomachine that is 95% efficient, which is a feat that man has not been able to accomplish yet. This is all because nature does not try to work around the forces it uses them at its advantage.

Growing cells in culture to take advantage of their internal chemical synthesis machinery can be considered a form of nanotechnology but this machinery has also been manipulated outside of living cells.

Nanolithography

Nanolithography is the branch of nanotechnology concerned with the study and application of fabricating nanometer-scale structures, meaning patterns with at least one lateral dimension between 1 and 100 nm. Different approaches can be categorized in serial or parallel, mask or maskless/direct-write, top-down or bottom-up, beam or tip-based, resist-based or resist-less methods. As of 2015, nanolithography is a very active area of research in academia and in industry. Applications of nanolithography include among others: Multigate devices such as Field effect transistors (FET), Quantum dots, Nanowires, Gratings, Zone plates and Photomasks, nanoelectromechanical systems (NEMS), or semiconductor integrated circuits (nanocircuitry).

Optical Lithography

Optical lithography, which has been the predominant patterning technique since the advent of the semiconductor age, is capable of producing sub-100-nm patterns with the use of very short optical wavelengths. Several optical lithography techniques require the use of liquid immersion and a host of resolution enhancement technologies like phase-shift masks (PSM) and optical proximity correction (OPC). Multiple patterning is a method of increasing the resolution by printing features in between pre-printed features on the same layer by etching or creating sidewall spacers, and has been used in commercial production of microprocessors since the 32 nm process node e.g. by directed self-assembly (DSA). Extreme ultraviolet lithography (EUVL) uses ultrashort wavelengths (13.5 nm) and as of 2015, is the most popularly considered Next-generation lithography (NGL) technique for mass-fabrication.

Electron-beam Lithography

Electron beam lithography or *Electron-Beam Direct-Write Lithography* (EBDW) scans a focused beam of electrons on a surface covered with an electron-sensitive film or resist (e.g. PMMA or HSQ) to draw custom shapes. By changing the solubility of the resist and subsequent selective removal of material by immersion in a solvent, sub-10 nm resolutions have been achieved. This form of direct-write, maskless lithography has high resolution and low throughput, limiting single-column e-beams to photomask fabrication, low-volume production of semiconductor devices, and research&development. Multiple-electron beam approaches have as a goal an increase of throughput for semiconductor mass-production.

Nanoimprint Lithography

Nanoimprint lithography (NIL), and its variants, such as Step-and-Flash Imprint Lithography, LISA and LADI are promising nanopattern replication technologies where patterns are created by mechanical deformation of imprint resist, typically a monomer or polymer formulation that is cured by heat or UV light during imprinting. This technique can be combined with contact printing and cold welding.

Multiphoton Lithography

Multiphoton lithography (also known as *direct laser lithography* or *direct laser writing*) patterns surfaces without the use of a photomask, whereby two-photon absorption is utilized to induce a change in the solubility of the resist.

Scanning Probe Lithography

Scanning probe lithography (SPL) is a tool for patterning at the nanometer-scale down to individual atoms using scanning probes. Dip-pen nanolithography is an additive, diffusive method, thermochemical nanolithography triggers chemical reactions, thermal scanning probe lithography creates 3D surfaces from polymers, and local oxidation nanolithography employs a local oxidation reaction for patterning purposes.

References

- Petty, M.C.; Bryce, M.R.; Bloor, D. (1995). An Introduction to Molecular Electronics. London: Edward Arnold. ISBN 0-19-521156-1.

- Petty, M.C.; Bryce, M.R. & Bloor, D. (1995). Introduction to Molecular Electronics. New York: Oxford University Press. pp. 1–25. ISBN 0-19-521156-1.

- Waser, Rainer; Lüssem, B. & Bjørnholm, T. (2008). "Chapter 8: Concepts in Single-Molecule Electronics". Nanotechnology. Volume 4: Information Technology II. Weinheim: Wiley-VCH Verlag GmbH & Co. KGaA. pp. 175–212. ISBN 978-3-527-31737-0.

- "ASML: Press - Press Releases - ASML reaches agreement for delivery of minimum of 15 EUV lithography systems". www.asml.com. Retrieved 2015-05-11.

- Jacoby, Mitch (27 January 2003). "Molecule-based circuitry revisited". Chemical and Engineering News. Retrieved 24 February 2011.

Fundamental Concepts of Nanotechnology

The growth of one-dimensional structures is known as vapor-liquid-solid method. The other fundamental concepts of nanotechnology that have been explained in this text are nanopore sequencing, molecular self-assembly, DNA nanotechnology and self- assembles monolayer. This section strategically encompasses and incorporates the major components and fundamental concepts of nanotechnology.

Vapor–Liquid–Solid Method

The vapor–liquid–solid method (VLS) is a mechanism for the growth of one-dimensional structures, such as nanowires, from chemical vapor deposition. The growth of a crystal through direct adsorption of a gas phase on to a solid surface is generally very slow. The VLS mechanism circumvents this by introducing a catalytic liquid alloy phase which can rapidly adsorb a vapor to supersaturation levels, and from which crystal growth can subsequently occur from nucleated seeds at the liquid–solid interface. The physical characteristics of nanowires grown in this manner depend, in a controllable way, upon the size and physical properties of the liquid alloy.

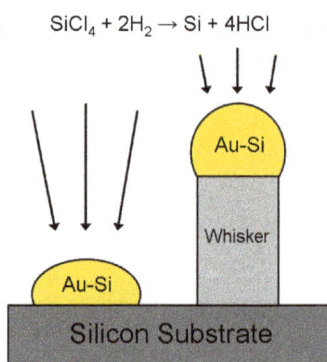

$$SiCl_4 + 2H_2 \rightarrow Si + 4HCl$$

Figure 1: Schematic illustration of Si whisker growth from the reaction of $SiCl_4$ and H_2 vapor phases. This reaction is catalyzed by gold-silicon droplet deposited on the wafer surface prior to whisker growth.

Historical Background

The VLS mechanism was proposed in 1964 as an explanation for silicon whisker growth from the gas phase in the presence of a liquid gold droplet placed upon a silicon substrate. The explanation was motivated by the absence of axial screw dislocations in the whiskers (which in themselves are a growth mechanism), the requirement of the gold droplet for growth, and the presence of the droplet at the tip of the whisker during the entire growth process.

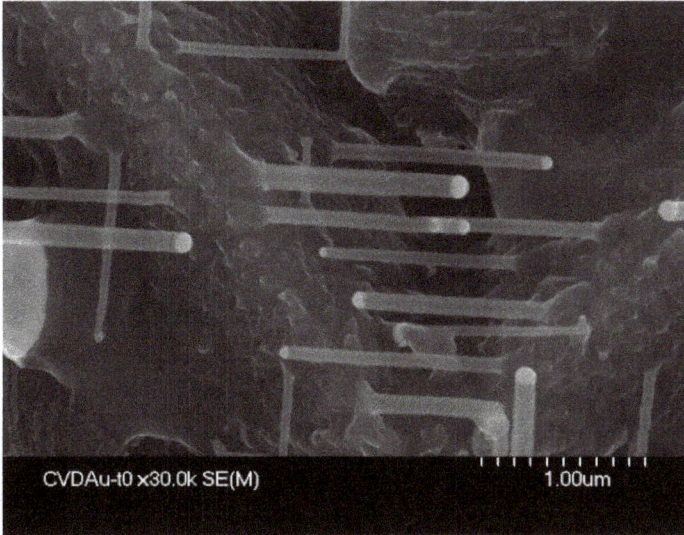

Figure 2: CVD Growth of Si nanowires using Au particle catalysts

Introduction

The VLS mechanism is typically described in three stages:

- Preparation of a liquid alloy droplet upon the substrate from which a wire is to be grown

- Introduction of the substance to be grown as a vapor, which adsorbs on to the liquid surface, and diffuses into the droplet

- Supersaturation and nucleation at the liquid/solid interface leading to axial crystal growth

Experimental Technique

The VLS process takes place as follows:

1. A thin (~1–10 nm) Au film is deposited onto a silicon (Si) wafer substrate by sputter deposition or thermal evaporation.

2. The wafer is annealed at temperatures higher than the Au-Si eutectic point, creating Au-Si alloy droplets on the wafer surface (the thicker the Au film, the larger the droplets). Mixing Au with Si greatly reduces the melting temperature of the alloy as compared to the alloy constituents. The melting temperature of

the Au:Si alloy reaches a minimum (~363 °C) when the ratio of its constituents is 4:1 Au:Si, also known as the Au:Si eutectic point.

3. Lithography techniques can also be used to controllably manipulate the diameter and position of the droplets.

4. One-dimensional crystalline nanowires are then grown by a liquid metal-alloy droplet-catalyzed chemical or physical vapor deposition process, which takes place in a vacuum deposition system. Au-Si droplets on the surface of the substrate act to lower the activation energy of normal vapor-solid growth. For example, Si can be deposited by means of a $SiCl_4$:H_2 gaseous mixture reaction (chemical vapor deposition), only at temperatures above 800 °C, in normal vapor-solid growth. Moreover, below this temperature almost no Si is deposited on the growth surface. However, Au particles can form Au-Si eutectic droplets at temperatures above 363 °C and adsorb Si from the vapor state (because Au can form a solid-solution with all Si concentrations up to 100%) until reaching a supersaturated state of Si in Au. Furthermore, nanosized Au-Si droplets have much lower melting points (ref) because the surface area-to-volume ratio is increasing, becoming energetically unfavorable, and nanometer-sized particles act to minimize their surface energy by forming droplets (spheres or half-spheres).

5. Si has a much higher melting point (~1414 °C) than that of the eutectic alloy, therefore Si atoms precipitate out of the supersaturated liquid-alloy droplet at the liquid-alloy/solid-Si interface, and the droplet rises from the surface. This process is illustrated in figure 1.

Typical Features of The VLS Method

- Greatly lowered reaction energy compared to normal vapor-solid growth.

- Wires grow only in the areas activated by the metal catalysts and the size and position of the wires are determined by that of the metal catalysts.

- This growth mechanism can also produce highly anisotropic nanowire arrays from a variety of material.

Requirements for Catalyst Particles

The requirements for catalysts are:

- It must form a liquid solution with the crystalline material to be grown at the nanowire growth temperature.

- The solid solubility of the catalyzing agent is low in the solid and liquid phases of the substrate material.

- The equilibrium vapor pressure of the catalyst over the liquid alloy must be small so that the droplet does not vaporize, shrink in volume (and therefore radius), and decrease the radius of the growing wire until, ultimately, growth is terminated.

- The catalyst must be inert (non-reacting) to the reaction products (during CVD nanowire growth).

- The vapor–solid, vapor–liquid, and liquid–solid interfacial energies play a key role in the shape of the droplets and therefore must be examined before choosing a suitable catalyst; small contact angles between the droplet and solid are more suitable for large area growth, while large contact angles result in the formation of smaller (decreased radius) whiskers.

- The solid-liquid interface must be well-defined crystallographically in order to produce highly directional growth of nanowires. It is also important to point out that the solid-liquid interface cannot, however, be completely smooth. Furthermore, if the solid liquid interface was atomically smooth, atoms near the interface trying to attach to the solid would have no place to attach to until a new island nucleates (atoms attach at step ledges), leading to an extremely slow growth process. Therefore, "rough" solid surfaces, or surfaces containing a large number of surface atomic steps (ideally 1 atom wide, for large growth rates) are needed for deposited atoms to attach and nanowire growth to proceed.

Growth Mechanism

Catalyst Droplet Formation

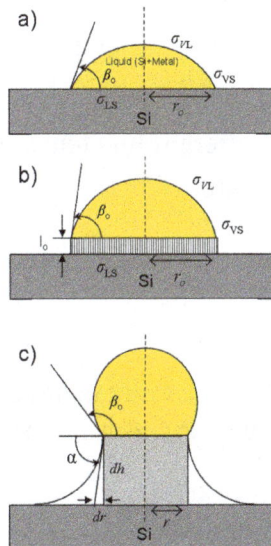

Figure 4: Schematic illustration of metal-alloy catalyzed whisker growth depicting the catalyst droplet formation during the early stages of whisker growth.

The materials system used, as well as the cleanliness of the vacuum system and therefore the amount of contamination and/or the presence of oxide layers at the droplet and wafer surface during the experiment, both greatly influence the absolute magnitude of the forces present at the droplet/surface interface and, in turn, determine the shape of the droplets. The shape of the droplet, i.e. the contact angle (β_o, see Figure 4) can, be modeled mathematically, however, the actual forces present during growth are extremely difficult to measure experimentally. Nevertheless, the shape of a catalyst particle at the surface of a crystalline substrate is determined by a balance of the forces of surface tension and the liquid–solid interface tension. The radius of the droplet varies with the contact angle as:

$$R = \frac{r_o}{\sin(\beta_o)},$$

where r_o is the radius of the contact area and β_o is defined by a modified Young's equation:

$$\sigma_1 \cos(\beta_o) = \sigma_s - \sigma_{ls} - \frac{\tau}{r_o},$$

It is dependent on the surface (σ_s) and liquid–solid interface (σ_{ls}) tensions, as well as an additional line tension (τ) which comes into effect when the initial radius of the droplet is small (nanosized). As a nanowire begins to grow, its height increases by an amount dh and the radius of the contact area decreases by an amount dr (see Figure 4). As the growth continues, the inclination angle at the base of the nanowires (α, set as zero before whisker growth) increases, as does β_o:

$$\sigma_1 \cos(\beta_o) = \sigma_s \cos(\alpha) - \sigma_{ls} - \frac{\tau}{r_o}.$$

The line tension therefore greatly influences the catalyst contact area. The most import result from this conclusion is that different line tensions will result in different growth modes. If the line tensions are too large, nanohillock growth will result and thus stop the growth.

Nanowhisker Diameter

The diameter of the nanowire which is grown depends upon the properties of the alloy droplet. The growth of nano-sized wires requires nano-size droplets to be prepared on the substrate. In an equilibrium situation this is not possible as the minimum radius of a metal droplet is given by

$$R_{min} = \frac{2V_l}{RTln(s)} \sigma_{lv}$$

where V_l is the molar volume of the droplet, σ_{lv} the liquid-vapor surface energy, and s is the degree of supersaturation of the vapor. This equations restricts the minimum diameter of the droplet, and of any crystals which can be grown from it, under typically conditions to well above the nanometer level. Several techniques to generate smaller droplets have been developed, including the use of monodispersed nanoparticles spread in low dilution on the substrate, and the laser ablation of a substrate-catalyst mixture so to form a plasma which allows well-separated nanoclusters of the catalyst to form as the systems cools.

Whisker Growth Kinetics

During VLS whisker growth, the rate at which whiskers grow is dependent on the whisker diameter: the larger the whisker diameter, the faster the nanowire grows axially. This is because the supersaturation of the metal-alloy catalyst ($Ä\mu$) is the main driving force for nanowhisker growth and decreases with decreasing whisker diameter (also known as the Gibbs-Thomson effect):

$$\Delta\mu = \Delta\mu_o - \frac{4\alpha\Omega}{d}.$$

Again, $\Delta\mu$ is the main driving force for nanowhisker growth (the supersaturation of the metal droplet). More specifically, $\Delta\mu_o$ is the difference between the chemical potential of the depositing species (Si in the above example) in the vapor and solid whisker phase. $\Delta\mu_o$ is the initial difference proceeding whisker growth (when $d \to \infty$), while $\dot{U}$ is the atomic volume of Si and α the specific free energy of the wire surface. Examination of the above equation, indeed reveals that small diameters (<100 nm) exhibit small driving forces for whisker growth while large wire diameters exhibit large driving forces.

Related Growth Techniques

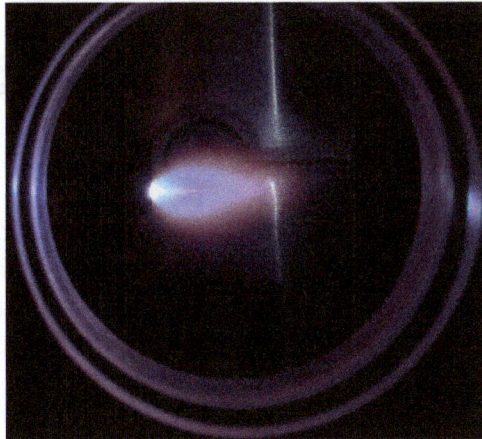

A plasma plume ejected from a target during pulsed laser deposition.

Laser-assisted Growth

Involves the removal of material from metal-containing solid targets by irradiating the surface with high-powered (~100 mJ/pulse) short (10 Hz) laser pulses, usually with wavelengths in the ultraviolet (UV) region of the light spectrum. When such a laser pulse is adsorbed by a solid target, material from the surface region of the target absorbs the laser energy and either (a) evaporates or sublimates from the surface or is (b) converted into a plasma. These particles are easily transferred to the substrate where they can nucleate and grow into nanowires. The laser-assisted growth technique is particularly useful for growing nanowires with high melting temperatures, multicomponent or doped nanowires, as well as nanowires with extremely high crystalline quality. The high intensity of the laser pulse incident at the target allows the deposition of high melting point materials, without having to try to evaporate the material using extremely high temperature resistive or electron bombardment heating. Furthermore, targets can simply be made from a mixture of materials or even a liquid. Finally, the plasma formed during the laser absorption process allows for the deposition of charged particles as well as a catalytic means to lower the activation barrier of reactions between target constituents.

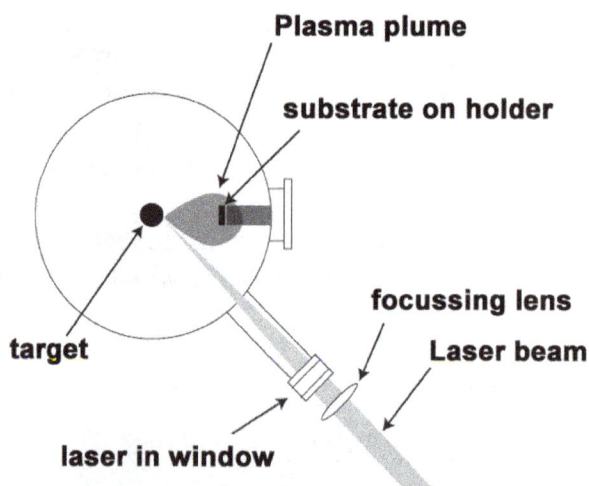

One possible configuration of a PLD deposition chamber.

Thermal Evaporation

Some very interesting nanowires microstructures can be obtained by simply thermally evaporating solid materials. This technique can be carried out in a relatively simple set-up composed of a dual-zone vacuum furnace. The hot end of the furnace contains the evaporating source material, while the evaporated particles are carrier downstream, (by way of a carrier gas) to the colder end of the furnace where they can absorb, nucleate, and grow on a desired substrate.

Metal-catalyzed Molecular Beam Epitaxy

Molecular beam epitaxy (MBE) has been used since 2000 to create high-quality semiconductor wires based on the VLS growth mechanism. However, in metal-catalyzed MBE the metal particles do not catalyze a reaction between precursors but rather adsorb vapor phase particles. This is because the chemical potential of the vapor can be drastically lowered by entering the liquid phase.

MBE is carried out under ultra-high vacuum (UHV) conditions where the mean-free-path (distance between collisions) of source atoms or molecules is on the order of meters. Therefore, evaporated source atoms (from, say, an effusion cell) act as a beam of particles directed towards the substrate. The growth rate of the process is very slow, the deposition conditions are very clean, and as a result four superior capabilities arise, when compared to other deposition methods:

- UHV conditions minimize the amount of oxidation/contamination of the growing structures

- Relatively low growth temperatures prevent interdiffusion (mixing) of nano-sized heterostructures

- Very thin-film analysis techniques can be used *in-situ* (during growth), such as reflection high energy electron diffraction (RHEED) to monitor the microstructure at the surface of the substrate as well as the chemical composition, using Auger electron spectroscopy.

Nanopore Sequencing

Nanopore sequencing is a method under development since 1995 for sequencing DNA (determining the order in which nucleotides occur on a strand of DNA).

A nanopore is simply a small hole with an internal diameter of the order of 1 nanometer. Certain porous transmembrane cellular proteins act as nanopores, and nanopores have also been made by etching a somewhat larger hole (several tens of nanometers) in a piece of silicon, and then gradually filling it in using ion-beam sculpting methods which results in a much smaller diameter hole: the nanopore. Graphene is also being explored as a synthetic substrate for solid-state nanopores.

The theory behind nanopore sequencing is that when a nanopore is immersed in a conducting fluid and a potential (voltage) is applied across it, an electric current due to conduction of ions through the nanopore can be observed. The amount of current is very sensitive to the size and shape of the nanopore. If single nucleotides (bases), strands of DNA or other molecules pass through or near the nanopore, this can create a characteristic change in the magnitude of the current through the nanopore.

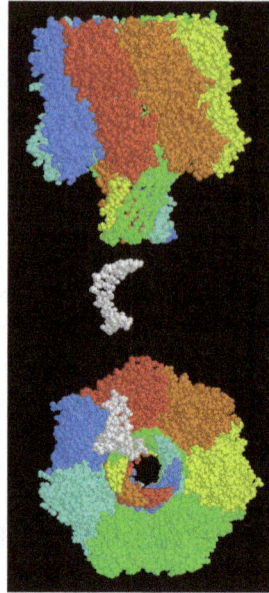

alpha-hemolysin pore (made up of 7 identical subunits in 7 colors) and 12-mer single-stranded DNA (in white) on the same scale to illustrate DNA effects on conductance when moving through a nanopore. Below is an orthogonal view of the same molecules.

Background

DNA could be passed through the nanopore for various reasons. For example, electrophoresis might attract the DNA towards the nanopore, and it might eventually pass through it. Or, enzymes attached to the nanopore might guide DNA towards the nanopore. The scale of the nanopore means that the DNA may be forced through the hole as a long string, one base at a time, rather like threading through the eye of a needle. As it does so, each nucleotide on the DNA molecule may obstruct the nanopore to a different, characteristic degree. The amount of current which can pass through the nanopore at any given moment therefore varies depending on whether the nanopore is blocked by an A, a C, a G or a T or a section of DNA that includes more than one of these bases (kmer). The change in the current through the nanopore as the DNA molecule passes through the nanopore represents a direct reading of the DNA sequence. This is the 'strand sequencing' approach commercialised by Oxford Nanopore Technologies.

Alternatively, a nanopore might be used to identify individual DNA bases as they pass through the nanopore in the correct order - this approach was published but not commercially developed by Oxford Nanopore Technologies and Professor Hagan Bayley.

Using Nanopore sequencing, a single molecule of DNA can be sequenced directly using a nanopore, without the need for an intervening PCR amplification step or a chemical labelling step or the need for optical instrumentation to identify the chemical label. The versatility of the nanopore concept is underlined by the fact that it has also been proposed for detection of life on other planets, since it is not necessarily restricted to the detection of the genetic information carrier DNA, but in general can be applied to sequence chain-like genetic information carriers without knowing the exact structure of their building blocks. Nanopore-based DNA analysis techniques are being industrially developed by Oxford Nanopore Technologies (developing strand sequencing using protein nanopores, and solid-state sequencing through internal R&D and collaborations with academic institutions). The MinION was made commercially available in June 2015 after having been introduced in 2014 in an early access programme. Publications on the method outline its use in rapid identification of viral pathogens, monitoring ebola, environmental monitoring, food safety monitoring, monitoring of antibiotic resistance, haplotyping and other applications. Other companies have noted that they have Nanopore programmes, NabSys (using a library of DNA probes and using nanopores to detect where these probes have hybridized to single stranded DNA) and NobleGen (using nanopores in combination with fluorescent labels). IBM has noted research projects on computer simulations of translocation of a DNA strand through a solid-state nanopore, but not projects on identifying the DNA bases on that strand.

Types of Nanopores

Alpha Hemolysin

Alpha hemolysin (αHL), a nanopore from bacteria that causes lysis of red blood cells, has been studied for over 15 years. To this point, studies have shown that all four bases can be identified using ionic current measured across the αHL pore. The structure of αHL is advantageous to identify specific bases moving through the pore. The αHL pore is ~10 nm long, with two distinct 5 nm sections. The upper section consists of a larger, vestibule-like structure and the lower section consists of three possible recognition sites (R1, R2, R3), and is able to discriminate between each base.

Sequencing using αHL has been developed through basic study and structural mutations, moving towards the sequencing of very long reads. Protein mutation of αHL has improved the detection abilities of the pore. The next proposed step is to bind an exonuclease onto the αHL pore. The enzyme would periodically cleave single bases, enabling the pore to identify successive bases. Coupling an exonuclease to the biological pore would slow the translocation of the DNA through the pore, and increase the accuracy of data acquisition.

Notably, theorists have shown that sequencing via exonuclease enzymes as described here is not feasible. This is mainly due to diffusion related effects imposing a limit on the capture probability of each nucleotide as it is cleaved. This results in a significant probability that a nucleotide is either not captured before it diffuses into the bulk or captured out of order, and therefore is not properly sequenced by the nanopore, leading to insertion and deletion errors. Therefore, major changes are needed to this method before it can be considered a viable strategy.

A recent study has pointed to the ability of αHL to detect nucleotides at two separate sites in the lower half of the pore. The R1 and R2 sites enable each base to be monitored twice as it moves through the pore, creating 16 different measurable ionic current values instead of 4. This method improves upon the single read through the nanopore by doubling the sites that the sequence is read per nanopore.

MspA

Mycobacterium smegmatis porin A (MspA) is the second biological nanopore currently being investigated for DNA sequencing. The MspA pore has been identified as a potential improvement over αHL due to a more favorable structure. The pore is described as a goblet with a thick rim and a diameter of 1.2 nm at the bottom of the pore. A natural MspA, while favorable for DNA sequencing because of shape and diameter, has a negative core that prohibited single stranded DNA(ssDNA) translocation. The natural nanopore was modified to improve translocation by replacing three negatively charged aspartic acids with neutral asparagines.

The electric current detection of nucleotides across the membrane has been shown to be tenfold more specific than αHL for identifying bases. Utilizing this improved specificity, a group at the University of Washington has proposed using double stranded DNA (dsDNA) between each single stranded molecule to hold the base in the reading section of the pore. The dsDNA would halt the base in the correct section of the pore and enable identification of the nucleotide. A recent grant has been awarded to a collaboration from UC Santa Cruz, the University of Washington, and Northeastern University to improve the base recognition of MspA using phi29 polymerase in conjunction with the pore.

CsgG

In March 2016, the lab of Han Remaut (VIB/Vrije Universiteit Brussels) announced a collaboration with Oxford Nanopore Technologies using the protein nanopore CsgG for DNA sequencing. Oxford Nanopore simultaneously announced the release of this new chemistry for its MinION device. – designed to improve accuracy and yields. The R9 nanopore was launched in late May 2016 and early reports indicate higher performance levels. More information can be found at the Oxford Nanopore page.

Fluorescence

An effective technique to determine a DNA sequence has been developed using solid state nanopores and fluorescence. This fluorescence sequencing method converts each base into a characteristic representation of multiple nucleotides which bind to a fluorescent probe strand-forming dsDNA. With the two color system proposed, each base is identified by two separate fluorescences, and will therefore be converted into two specific sequences. Probes consist of a fluorophore and quencher at the start and end of each sequence, respectively. Each fluorophore will be extinguished by the quencher at the end of the preceding sequence. When the dsDNA is translocating through a solid state nanopore, the probe strand will be stripped off, and the upstream fluorophore will fluoresce.

This sequencing method has a capacity of 50-250 bases per second per pore, and a four color fluorophore system (each base could be converted to one sequence instead of two), will sequence over 500 bases per second. Advantages of this method are based on the clear sequencing readout—using a camera instead of noisy current methods. However, the method does require sample preparation to convert each base into an expanded binary code before sequencing. Instead of one base being identified as it translocates through the pore, ~12 bases are required to find the sequence of one base.

Electron Tunneling

Measurement of electron tunneling through bases as ssDNA translocates through the nanopore is an improved solid state nanopore sequencing method. Most research has focused on proving bases could be determined using electron tunneling. These studies were conducted using a scanning probe microscope as the sensing electrode, and have proved that bases can be identified by specific tunneling currents. After the proof of principle research, a functional system must be created to couple the solid state pore and sensing devices.

Researchers at the Harvard Nanopore group have engineered solid state pores with single walled carbon nanotubes across the diameter of the pore. Arrays of pores are created and chemical vapor deposition is used to create nanotubes that grow across the array. Once a nanotube has grown across a pore, the diameter of the pore is adjusted to the desired size. Successful creation of a nanotube coupled with a pore is an important step towards identifying bases as the ssDNA translocates through the solid state pore.

Another method is the use of nanoelectrodes on either side of a pore. The electrodes are specifically created to enable a solid state nanopore's formation between the two electrodes. This technology could be used to not only sense the bases but to help control base translocation speed and orientation.

Challenges

One challenge for the 'strand sequencing' method was in refining the method to improve its resolution to be able to detect single bases. In the early papers methods, a nucleotide needed to be repeated in a sequence about 100 times successively in order to produce a measurable characteristic change. This low resolution is because the DNA strand moves rapidly at the rate of 1 to 5μs per base through the nanopore. This makes recording difficult and prone to background noise, failing in obtaining single-nucleotide resolution. The problem is being tackled by either improving the recording technology or by controlling the speed of DNA strand by various protein engineering strategies and

Oxford Nanopore employs a 'kmer approach', analyzing more than one base at any one time so that stretches of DNA are subject to repeat interrogation as the strand moves through the nanopore one base at a time. Various techniques including algorithmic have been used to improve the performance of the MinION technology since it was first made available to users. More recently effects of single bases due to secondary structure or released mononucleotides have been shown. Professor Hagan Bayley, founder of Oxford Nanopore, proposed in 2010 that creating two recognition sites within an alpha hemolysin pore may confer advantages in base recognition.

One challenge for the 'exonuclease approach', initially developed but not progressed by Oxford Nanopore Technologies, where a processive enzyme feeds individual bases, in the correct order, into the nanopore, is to integrate the exonuclease and the nanopore detection systems. In particular, the problem is that when an exonuclease hydrolyzes the phosphodiester bonds between nucleotides in DNA, the subsequently released nucleotide is not necessarily guaranteed to directly move in to, say, a nearby alpha-hemolysin nanopore. One idea is to attach the exonuclease to the nanopore, perhaps through biotinylation to the beta barrel hemolsyin. The central pore of the protein may be lined with charged residues arranged so that the positive and negative charges appear on opposite sides of the pore. However, this mechanism is primarily discriminatory and does not constitute a mechanism to guide nucleotides down some particular path.

Commercialization

Agilent Laboratories was the first to license and develop nanopores but does not have any current disclosed research in the area.

The company Oxford Nanopore Technologies now sells the MinION portable DNA sequencing device and is introducing the PromethION high-throughput device into an early access programme in 2016. In 2016 it announced that the first mobile-phone sequencer, SmidgION, was in development. In 2008 licensed technology from Harvard, UCSC and other universities and is developing protein and solid state nanopore technology with the aim of analysing a range of biological molecules including DNA, RNA and proteins. They revealed the idea of their first working device in February 2012. The Company introduced its pocket-sized sensing device MinION to an early-access community in early 2014 and made it commercially available in May 2015. A variety of applications for the device have now been developed including environmental or pathogen monitoring, genome research, investigation of fetal DNA.

Sequenom licensed nanopore technology from Harvard in 2007 using an approach that combines nanopores and fluorescent labels. This technology was subsequently licensed to Noblegen.

NABsys was spun out of Brown University and is researching nanopores as a method of identifying areas of single stranded DNA that have been hybridized with specific DNA probes.

Genia applied for a patent on a method of nanopore sequencing which uses a series of RNA "speed-bump" molecules to infer the sequence of the molecule based on the pattern of delayed progression.

Molecular Self-assembly

Molecular self-assembly is the process by which molecules adopt a defined arrangement without guidance or management from an outside source. There are two types of self-assembly. These are intramolecular self-assembly and intermolecular self-assembly. Commonly, the term molecular self-assembly refers to intermolecular self-assembly, while the intramolecular analog is more commonly called folding.

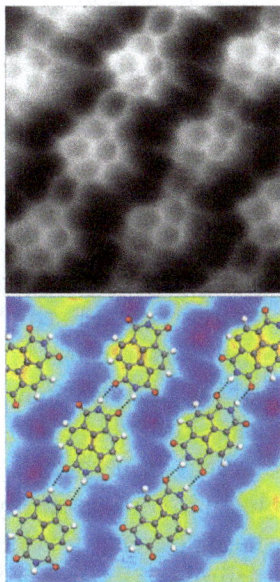

AFM image of napthalenetetracarboxylic diimide molecules on silver interacting via hydrogen bonding (77 K).

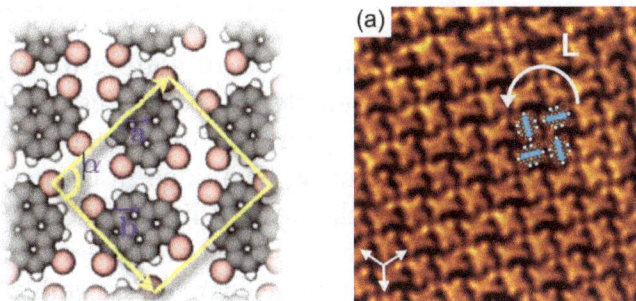

STM image of self-assembled Br_4-pyrene molecules on Au(111) surface (top) and its model (bottom; pink spheres are Br atoms).

Supramolecular Systems

Molecular self-assembly is a key concept in supramolecular chemistry. This is because assembly of molecules in such systems is directed through noncovalent interactions (e.g., hydrogen bonding, metal coordination, hydrophobic forces, van der Waals forces, π-π interactions, and/or electrostatic) as well as electromagnetic interactions. Common examples include the formation of micelles, vesicles, liquid crystal phases, and Langmuir monolayers by surfactant molecules. Further examples of supramolecular assemblies demonstrate that a variety of different shapes and sizes can be obtained using molecular self-assembly.

Molecular self-assembly allows the construction of challenging molecular topologies. One example is Borromean rings, interlocking rings wherein removal of one ring unlocks each of the other rings. DNA has been used to prepare a molecular analog of Borromean rings. More recently, a similar structure has been prepared using non-biological building blocks. While a mechanistic understanding of how supramolecular self-assembly occurs remains largely unknown, both experimental and theoretic work has been pursued on this topic.

Biological Systems

Molecular self-assembly underlies the construction of biologic macromolecular assemblies in living organisms, and so is crucial to the function of cells. It is exhibited in the self-assembly of lipids to form the membrane, the formation of double helical DNA through hydrogen bonding of the individual strands, and the assembly of proteins to form quaternary structures. Molecular self-assembly of incorrectly folded proteins into insoluble amyloid fibers is responsible for infectious prion-related neurodegenerative diseases. Molecular self-assembly of nanoscale structures plays a role in the growth of the remarkable β-keratin lamellae/setae/spatulae structures used to give geckos the ability to climb walls and adhere to ceilings and rock overhangs.

Nanotechnology

Molecular self-assembly is an important aspect of bottom-up approaches to nanotechnology. Using molecular self-assembly the final (desired) structure is programmed in the shape and functional groups of the molecules. Self-assembly is referred to as a 'bottom-up' manufacturing technique in contrast to a 'top-down' technique such as lithography where the desired final structure is carved from a larger block of matter. In the speculative vision of molecular nanotechnology, microchips of the future might be made by molecular self-assembly. An advantage to constructing nanostructure using molecular self-assembly for biological materials is that they will degrade back into individual molecules that can be broken down by the body.

DNA nanotechnology

DNA nanotechnology is an area of current research that uses the bottom-up, self-assembly approach for nanotechnological goals. DNA nanotechnology uses the unique molecular recognition properties of DNA and other nucleic acids to create self-assembling branched DNA complexes with useful properties. DNA is thus used as a structural material rather than as a carrier of biological information, to make structures such as two-dimensional periodic lattices (both tile-based as well as using the "DNA origami" method) and three-dimensional structures in the shapes of polyhedra. These DNA structures have also been used as templates in the assembly of other molecules such as gold nanoparticles and streptavidin proteins.

Two-dimensional Monolayers

The spontaneous assembly of a single layer of molecules at interfaces is usually referred to as two-dimensional self-assembly. Early direct proofs showing that molecules can assembly into higher-order architectures at solid interfaces came with the development of scanning tunneling microscopy and shortly thereafter. Eventually two strategies became popular for the self-assembly of 2D architectures, namely self-assembly following ultra-high-vacuum deposition and annealing and self-assembly at the solid-liquid interface. The design of molecules and conditions leading to the formation of highly-crystalline architectures is considered today a form of 2D crystal engineering at the nanoscopic scale.

DNA Nanotechnology

DNA nanotechnology is the design and manufacture of artificial nucleic acid structures for technological uses. In this field, nucleic acids are used as non-biological engineering materials for nanotechnology rather than as the carriers of genetic information in living cells. Researchers in the field have created static structures such as two- and three-dimensional crystal lattices, nanotubes, polyhedra, and arbitrary shapes, and functional devices such as molecular machines and DNA computers. The field is beginning to be used as a tool to solve basic science problems in structural biology and biophysics, including applications in X-ray crystallography and nuclear magnetic resonance spectroscopy of proteins to determine structures. Potential applications in molecular scale electronics and nanomedicine are also being investigated.

The conceptual foundation for DNA nanotechnology was first laid out by Nadrian Seeman in the early 1980s, and the field began to attract widespread interest in the mid-2000s. This use of nucleic acids is enabled by their strict base pairing rules, which cause only portions of strands with complementary base sequences to bind together to form strong, rigid double helix structures. This allows for the rational design of base sequences that will selectively assemble to form complex target structures with precisely controlled nanoscale features. Several assembly methods are used to make these structures, including tile-based structures that assemble from smaller structures, folding structures using the DNA origami method, and dynamically reconfigurable structures using strand displacement methods. While the field's name specifically references DNA, the same principles have been used with other types of nucleic acids as well, leading to the occasional use of the alternative name *nucleic acid nanotechnology*.

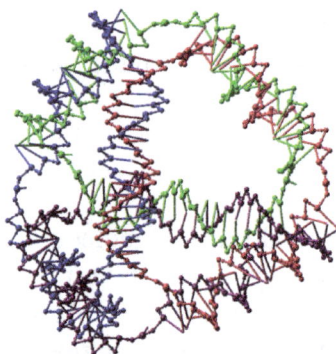

DNA nanotechnology involves forming artificial, designed nanostructures out of nucleic acids, such as this DNA tetrahedron. Each edge of the tetrahedron is a 20 base pair DNA double helix, and each vertex is a three-arm junction. The 4 DNA strands that form the 4 tetrahedral faces are color-coded.

Fundamental Concepts

These four strands associate into a DNA four-arm junction because this structure maximizes the number of correct base pairs, with A matched to T and C matched to G. See this image for a more realistic model of the four-arm junction showing its tertiary structure.

This double-crossover (DX) supramolecular complex consists of five DNA single strands that form two double-helical domains, on the top and the bottom in this image. There are two crossover points where the strands cross from one domain into the other.

Properties of Nucleic Acids

Nanotechnology is often defined as the study of materials and devices with features on a scale below 100 nanometers. DNA nanotechnology, specifically, is an example of bottom-up molecular self-assembly, in which molecular components spontaneously organize into stable structures; the particular form of these structures is induced by the physical and chemical properties of the components selected by the designers. In DNA nanotechnology, the component materials are strands of nucleic acids such as DNA; these strands are often synthetic and are almost always used outside the context of a living cell. DNA is well-suited to nanoscale construction because the binding between two nucleic acid strands depends on simple base pairing rules which are well understood, and form the specific nanoscale structure of the nucleic acid double helix. These qualities make the assembly of nucleic acid structures easy to control through nucleic acid design. This property is absent in other materials used in nanotechnology, including proteins, for which protein design is very difficult, and nanoparticles, which lack the capability for specific assembly on their own.

The structure of a nucleic acid molecule consists of a sequence of nucleotides distinguished by which nucleobase they contain. In DNA, the four bases present are adenine (A), cytosine (C), guanine (G), and thymine (T). Nucleic acids have the property that two molecules will only bind to each other to form a double helix if the two sequences are complementary, meaning that they form matching sequences of base pairs, with A only binding to T, and C only binding to G. Because the formation of correctly matched base pairs is energetically favorable, nucleic acid strands are expected in most cases to bind to each other in the conformation that maximizes the number of correctly paired bases. The sequences of bases in a system of strands thus determine the pattern of binding and the overall structure in an easily controllable way. In DNA nanotechnology, the base sequences of strands are rationally designed by researchers so that the base pairing interactions cause the strands to assemble in the desired conformation. While DNA is the dominant material used, structures incorporating other nucleic acids such as RNA and peptide nucleic acid (PNA) have also been constructed.

Subfields

DNA nanotechnology is sometimes divided into two overlapping subfields: structural DNA nanotechnology and dynamic DNA nanotechnology. Structural DNA nanotechnology, sometimes abbreviated as SDN, focuses on synthesizing and characterizing nucleic acid complexes and materials that assemble into a static, equilibrium end state. On the other hand, dynamic DNA nanotechnology focuses on complexes with useful non-equilibrium behavior such as the ability to reconfigure based on a chemical or physical stimulus. Some complexes, such as nucleic acid nanomechanical devices, combine features of both the structural and dynamic subfields.

The complexes constructed in structural DNA nanotechnology use topologically branched nucleic acid structures containing junctions. (In contrast, most biological DNA exists as an unbranched double helix.) One of the simplest branched structures is a four-arm junction that consists of four individual DNA strands, portions of which are complementary in a specific pattern. Unlike in natural Holliday junctions, each arm in the artificial immobile four-arm junction has a different base sequence, causing the junction point to be fixed at a certain position. Multiple junctions can be combined in the same complex, such as in the widely used double-crossover (DX) structural motif, which contains two parallel double helical domains with individual strands crossing between the domains at two crossover points. Each crossover point is, topologically, a four-arm junction, but is constrained to one orientation, in contrast to the flexible single four-arm junction, providing a rigidity that makes the DX motif suitable as a structural building block for larger DNA complexes.

Dynamic DNA nanotechnology uses a mechanism called toehold-mediated strand displacement to allow the nucleic acid complexes to reconfigure in response to the addition of a new nucleic acid strand. In this reaction, the incoming strand binds to a single-stranded toehold region of a double-stranded complex, and then displaces one of the strands bound in the original complex through a branch migration process. The overall effect is that one of the strands in the complex is replaced with another one. In addition, reconfigurable structures and devices can be made using functional nucleic acids such as deoxyribozymes and ribozymes, which can perform chemical reactions, and aptamers, which can bind to specific proteins or small molecules.

Structural DNA Nanotechnology

Structural DNA nanotechnology, sometimes abbreviated as SDN, focuses on synthesizing and characterizing nucleic acid complexes and materials where the assembly has a static, equilibrium endpoint. The nucleic acid double helix has a robust, defined three-dimensional geometry that makes it possible to predict and design the structures of more complicated nucleic acid complexes. Many such structures have been created, including two- and three-dimensional structures, and periodic, aperiodic, and discrete structures.

Extended Lattices

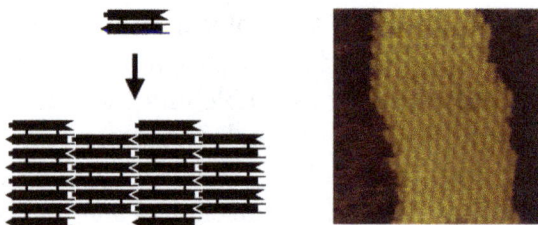

The assembly of a DX array. *Left*, schematic diagram. Each bar represents a double-helical domain of DNA, with the shapes representing complementary sticky ends. The DX complex at top will combine with other DX complexes into the two-dimensional array shown at bottom. *Right*, an atomic force microscopy image of the assembled array. The individual DX tiles are clearly visible within the assembled structure. The field is 150 nm across.

Small nucleic acid complexes can be equipped with sticky ends and combined into larger two-dimensional periodic lattices containing a specific tessellated pattern of the individual molecular tiles. The earliest example of this used double-crossover (DX) complexes as the basic tiles, each containing four sticky ends designed with sequences that caused the DX units to combine into periodic two-dimensional flat sheets that are essentially rigid two-dimensional crystals of DNA. Two-dimensional arrays have been made from other motifs as well, including the Holliday junction rhombus lattice, and various DX-based arrays making use of a double-cohesion scheme. The top two images at right show examples of tile-based periodic lattices.

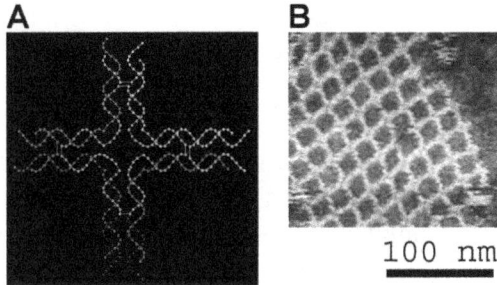

Left, a model of a DNA tile used to make another two-dimensional periodic lattice. Right, an atomic force micrograph of the assembled lattice.

Two-dimensional arrays can be made to exhibit aperiodic structures whose assembly implements a specific algorithm, exhibiting one form of DNA computing. The DX tiles can have their sticky end sequences chosen so that they act as Wang tiles, allowing them to perform computation. A DX array whose assembly encodes an XOR operation has been demonstrated; this allows the DNA array to implement a cellular automaton that generates a fractal known as the Sierpinski gasket. The third image at right shows this type of array. Another system has the function of a binary counter, displaying a representation of increasing binary numbers as it grows. These results show that computation can be incorporated into the assembly of DNA arrays.

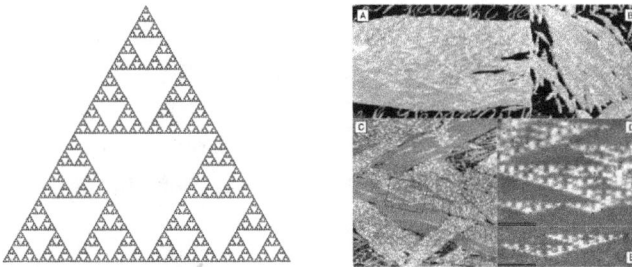

An example of an aperiodic two-dimensional lattice that assembles into a fractal pattern. Left, the Sierpinski gasket fractal. Right, DNA arrays that display a representation of the Sierpinski gasket on their surfaces

DX arrays have been made to form hollow nanotubes 4–20 nm in diameter, essentially two-dimensional lattices which curve back upon themselves. These DNA nanotubes are somewhat similar in size and shape to carbon nanotubes, and while they lack the electrical conductance of carbon nanotubes, DNA nanotubes are more easily modified and connected to other structures. One of many schemes for constructing DNA nanotubes uses a lattice of curved DX tiles that curls around itself and closes into a tube. In an alternative method that allows the circumference to be specified in a simple, modular fashion using single-stranded tiles, the rigidity of the tube is an emergent property.

Forming three-dimensional lattices of DNA was the earliest goal of DNA nanotechnology, but this proved to be one of the most difficult to realize. Success using a motif based on the concept of tensegrity, a balance between tension and compression forces, was finally reported in 2009.

Discrete Structures

Researchers have synthesized many three-dimensional DNA complexes that each have the connectivity of a polyhedron, such as a cube or octahedron, meaning that the DNA duplexes trace the edges of a polyhedron with a DNA junction at each vertex. The earliest demonstrations of DNA polyhedra were very work-intensive, requiring multiple ligations and solid-phase synthesis steps to create catenated polyhedra. Subsequent work yielded polyhedra whose synthesis was much easier. These include a DNA octahedron made from a long single strand designed to fold into the correct conformation, and a tetrahedron that can be produced from four DNA strands in one step, pictured at the top of this article.

Nanostructures of arbitrary, non-regular shapes are usually made using the DNA origami method. These structures consist of a long, natural virus strand as a "scaffold", which is made to fold into the desired shape by computationally designed short "staple" strands. This method has the advantages of being easy to design, as the base sequence is predetermined by the scaffold strand sequence, and not requiring high strand purity and accurate stoichiometry, as most other DNA nanotechnology methods do. DNA origami was first demonstrated for two-dimensional shapes, such as a smiley face and a coarse map of the Western Hemisphere. Solid three-dimensional structures can be made by using parallel DNA helices arranged in a honeycomb pattern, and structures with two-dimensional faces can be made to fold into a hollow overall three-dimensional shape, akin to a cardboard box. These can be programmed to open and reveal or release a molecular cargo in response to a stimulus, making them potentially useful as programmable molecular cages.

Templated Assembly

Nucleic acid structures can be made to incorporate molecules other than nucleic acids, sometimes called heteroelements, including proteins, metallic nanoparticles, quantum dots, and fullerenes. This allows the construction of materials and devices with a range of functionalities much greater than is possible with nucleic acids alone. The goal is to use the self-assembly of the nucleic acid structures to template the assembly of the nanoparticles hosted on them, controlling their position and in some cases orientation. Many of these schemes use a covalent attachment scheme, using oligonucleotides with amide or thiol functional groups as a chemical handle to bind the heteroelements. This covalent binding scheme has been used to arrange gold nanoparticles on a DX-based array, and to arrange streptavidin protein molecules into specific patterns on a DX array. A non-covalent hosting scheme using Dervan polyamides on a DX array was used to arrange streptavidin proteins in a specific pattern on a DX array. Carbon nanotubes have been hosted on DNA arrays in a pattern allowing the assembly to act as a molecular electronic device, a carbon nanotube field-effect transistor. In addition, there are nucleic acid metallization methods, in which the nucleic acid is replaced by a metal which assumes the general shape of the original nucleic acid structure, and schemes for using nucleic acid nanostructures as lithography masks, transferring their pattern into a solid surface.

Dynamic DNA Nanotechnology

Dynamic DNA nanotechnology focuses on forming nucleic acid systems with designed dynamic functionalities related to their overall structures, such as computation and mechanical motion. There is some overlap between structural and dynamic DNA nanotechnology, as structures can be formed through annealing and then reconfigured dynamically, or can be made to form dynamically in the first place.

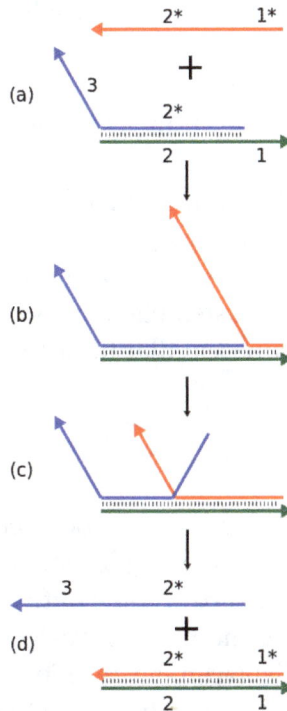

Dynamic DNA nanotechnology often makes use of toehold-mediated strand displacement reactions. In this example, the red strand binds to the single stranded toehold region on the green strand (region 1), and then in a branch migration process across region 2, the blue strand is displaced and freed from the complex. Reactions like these are used to dynamically reconfigure or assemble nucleic acid nanostructures. In addition, the red and blue strands can be used as signals in a molecular logic gate.

Nanomechanical Devices

DNA complexes have been made that change their conformation upon some stimulus, making them one form of nanorobotics. These structures are initially formed in the same way as the static structures made in structural DNA nanotechnology, but are designed so that dynamic reconfiguration is possible after the initial assembly. The earliest such device made use of the transition between the B-DNA and Z-DNA forms to respond to a change in buffer conditions by undergoing a twisting motion. This reliance on buffer conditions, however, caused all devices to change state at the same time. Subsequent systems could change states based upon the presence of control strands, allowing multiple devices to be independently operated in solution. Some examples of such systems are a "molecular tweezers" design that has an open and a closed state, a device that could switch from a paranemic-crossover (PX) conformation to a double-junction (JX2) conformation, undergoing rotational motion in the process, and a two-dimensional array that could dynamically expand and contract in response to control strands. Structures have also been made that dynamically open or close, potentially acting as a molecular cage to release or reveal a functional cargo upon opening.

DNA walkers are a class of nucleic acid nanomachines that exhibit directional motion along a linear track. A large number of schemes have been demonstrated. One strategy is to control the motion of the walker along the track using control strands that need to be manually added in sequence. Another approach is to make use of restriction enzymes or deoxyribozymes to cleave the strands and cause the walker to move forward, which has the advantage of running autonomously. A later system could walk upon a two-dimensional surface rather than a linear track, and demonstrated the ability to selectively pick up and move molecular cargo. Additionally, a linear walker has been demonstrated that performs DNA-templated synthesis as the walker advances along the track, allowing autonomous multistep chemical synthesis directed by the walker. The synthetic DNA walkers' function is similar to that of the proteins dynein and kinesin.

Strand Displacement Cascades

Cascades of strand displacement reactions can be used for either computational or structural purposes. An individual strand displacement reaction involves revealing a new sequence in response to the presence of some initiator strand. Many such reactions can be linked into a cascade where the newly revealed output sequence of one reaction can initiate another strand displacement reaction elsewhere. This in turn allows for the construction of chemical reaction networks with many components, exhibiting complex computational and information processing abilities. These cascades are made energetically favorable through the formation of new base pairs, and the entropy gain from disassembly reactions. Strand displacement cascades allow isothermal operation of the assembly or computational process, in contrast to traditional nucleic acid assembly's requirement for a thermal annealing step, where the temperature is raised and then slowly lowered to ensure proper formation of the desired structure. They can also support catalytic function of the initiator species, where less than one equivalent of the initiator can cause the reaction to go to completion.

Strand displacement complexes can be used to make molecular logic gates capable of complex computation. Unlike traditional electronic computers, which use electric current as inputs and outputs, molecular computers use the concentrations of specific chemical species as signals. In the case of nucleic acid strand displacement circuits, the signal is the presence of nucleic acid strands that are released or consumed by binding and unbinding events to other strands in displacement complexes. This approach has been used to make logic gates such as AND, OR, and NOT gates. More recently, a four-bit circuit was demonstrated that can compute the square root of the integers 0–15, using a system of gates containing 130 DNA strands.

Another use of strand displacement cascades is to make dynamically assembled structures. These use a hairpin structure for the reactants, so that when the input strand binds, the newly revealed sequence is on the same molecule rather than disassembling. This allows new opened hairpins to be added to a growing complex. This approach has been used to make simple structures such as three- and four-arm junctions and dendrimers.

Applications

DNA nanotechnology provides one of the few ways to form designed, complex structures with precise control over nanoscale features. The field is beginning to see application to solve basic science problems in structural biology and biophysics. The earliest such application envisaged for the field, and one still in development, is in crystallography, where molecules that are difficult to crystallize in isolation could be arranged within a three-dimensional nucleic acid lattice, allowing determination of their structure. Another application is the use of DNA origami rods to replace liquid crystals in residual dipolar coupling experiments in protein NMR spectroscopy; using DNA origami is advantageous because, unlike liquid crystals, they are tolerant of the detergents needed to suspend membrane proteins in solution. DNA walkers have been used as nanoscale assembly lines to move nanoparticles and direct chemical synthesis. Further, DNA origami structures have aided in the biophysical studies of enzyme function and protein folding.

DNA nanotechnology is moving toward potential real-world applications. The ability of nucleic acid arrays to arrange other molecules indicates its potential applications in molecular scale electronics. The assembly of a nucleic acid structure could be used to template the assembly of a molecular electronic elements such as molecular wires, providing a method for nanometer-scale control of the placement and overall architecture of the device analogous to a molecular breadboard. DNA nanotechnology has been compared to the concept of programmable matter because of the coupling of computation to its material properties.

In a study conducted by a group of scientists from iNANO center and CDNA Center in Aarhus university (Aarhus), researchers were able to construct a small multi-switchable 3D DNA Box Origami. The proposed nanoparticle was characterized by atomic force microscopy (AFM), transmission electron microscopy (TEM) and Förster resonance energy transfer (FRET). The constructed box was shown to have a unique reclosing mechanism, which enabled it to repeatedly open and close in response to a unique set of DNA or RNA keys. The authors proposed that this "DNA device can potentially be used for a broad range of applications such as controlling the function of single molecules, controlled drug delivery, and molecular computing.".

There are potential applications for DNA nanotechnology in nanomedicine, making use of its ability to perform computation in a biocompatible format to make "smart drugs" for targeted drug delivery. One such system being investigated uses a hollow DNA box containing proteins that induce apoptosis, or cell death, that will only open when in proximity to a cancer cell. There has additionally been interest in expressing these artificial structures in engineered living bacterial cells, most likely using the transcribed RNA for the assembly, although it is unknown whether these complex structures are able to efficiently fold or assemble in the cell's cytoplasm. If successful, this could enable directed evolution of nucleic acid nanostructures. Scientists at Oxford University reported the self-assembly of four short strands of synthetic DNA into a cage which can enter cells and survive for at least 48 hours. The fluorescently labeled DNA tetrahedra were found to remain intact in the laboratory cultured human kidney cells despite the attack by cellular enzymes after two days. This experiment showed the potential of drug delivery inside the living cells using the DNA 'cage'. A DNA tetrahedron was used to deliver RNA Interference (RNAi) in a mouse model, reported a team of researchers in MIT. Delivery of the interfering RNA for treatment has showed some success using polymer or lipid, but there are limits of safety and imprecise targeting, in addition to short shelf life in the blood stream. The DNA nanostructure created by the team consists of six strands of DNA to form a tetrahedron, with one strand of RNA affixed to each of the six edges. The tetrahedron is further equipped with targeting protein, three folate molecules, which lead the DNA nanoparticles to the abundant folate receptors found on some tumors. The result showed that the gene expression targeted by the RNAi, luciferase, dropped by more than half. This study shows promise in using DNA nanotechnology as an effective tool to deliver treatment using the emerging RNA Interference technology.

Design

DNA nanostructures must be rationally designed so that individual nucleic acid strands will assemble into the desired structures. This process usually begins with specification of a desired target structure or function. Then, the overall secondary structure of the target complex is determined, specifying the arrangement of nucleic acid strands within the structure, and which portions of those strands should be bound to each other.

The last step is the primary structure design, which is the specification of the actual base sequences of each nucleic acid strand.

Structural Design

The first step in designing a nucleic acid nanostructure is to decide how a given structure should be represented by a specific arrangement of nucleic acid strands. This design step determines the secondary structure, or the positions of the base pairs that hold the individual strands together in the desired shape. Several approaches have been demonstrated:

- Tile-based structures. This approach breaks the target structure into smaller units with strong binding between the strands contained in each unit, and weaker interactions between the units. It is often used to make periodic lattices, but can also be used to implement algorithmic self-assembly, making them a platform for DNA computing. This was the dominant design strategy used from the mid-1990s until the mid-2000s, when the DNA origami methodology was developed.

- Folding structures. An alternative to the tile-based approach, folding approaches make the nanostructure from one long strand, which can either have a designed sequence that folds due to its interactions with itself, or it can be folded into the desired shape by using shorter, "staple" strands. This latter method is called DNA origami, which allows forming nanoscale two- and three-dimensional shapes.

- Dynamic assembly. This approach directly controls the kinetics of DNA self-assembly, specifying all of the intermediate steps in the reaction mechanism in addition to the final product. This is done using starting materials which adopt a hairpin structure; these then assemble into the final conformation in a cascade reaction, in a specific order. This approach has the advantage of proceeding isothermally, at a constant temperature. This is in contrast to the thermodynamic approaches, which require a thermal annealing step where a temperature change is required to trigger the assembly and favor proper formation of the desired structure.

Sequence Design

After any of the above approaches are used to design the secondary structure of a target complex, an actual sequence of nucleotides that will form into the desired structure must be devised. Nucleic acid design is the process of assigning a specific nucleic acid base sequence to each of a structure's constituent strands so that they will associate into a desired conformation. Most methods have the goal of designing sequences so that the target structure has the lowest energy, and is thus the most thermodynamically favorable, while incorrectly assembled structures have higher energies and are thus disfavored. This is done either through simple, faster heuristic methods such as sequence symmetry minimization, or by using a full nearest-neighbor thermodynamic model, which is more accurate but slower and more computationally intensive. Geometric models are used to examine tertiary structure of the nanostructures and to ensure that the complexes are not overly strained.

Nucleic acid design has similar goals to protein design. In both, the sequence of monomers is designed to favor the desired target structure and to disfavor other structures. Nucleic acid design has the advantage of being much computationally easier than protein design, because the simple base pairing rules are sufficient to predict a structure's energetic favorability, and detailed information about the overall three-dimensional folding of the structure is not required. This allows the use of simple heuristic methods that yield experimentally robust designs. However, nucleic acid structures are less versatile than proteins in their function because of proteins' increased ability to fold into complex structures, and the limited chemical diversity of the four nucleotides as compared to the twenty proteinogenic amino acids.

Materials and Methods

The sequences of the DNA strands making up a target structure are designed computationally, using molecular modeling and thermodynamic modeling software. The nucleic acids themselves are then synthesized using standard oligonucleotide synthesis methods, usually automated in an oligonucleotide synthesizer, and strands of custom sequences are commercially available. Strands can be purified by denaturing gel electrophoresis if needed, and precise concentrations determined via any of several nucleic acid quantitation methods using ultraviolet absorbance spectroscopy.

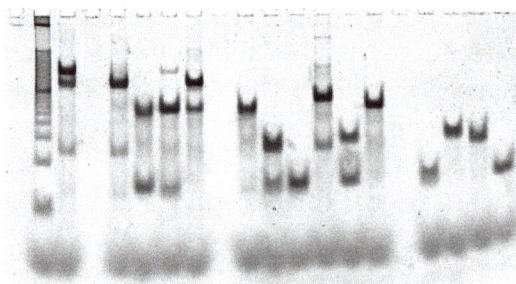

Gel electrophoresis methods, such as this formation assay on a DX complex, are used to ascertain whether the desired structures are forming properly. Each vertical lane contains a series of bands, where each band is characteristic of a particular reaction intermediate.

The fully formed target structures can be verified using native gel electrophoresis, which gives size and shape information for the nucleic acid complexes. An electrophoretic mobility shift assay can assess whether a structure incorporates all desired strands. Fluorescent labeling and Förster resonance energy transfer (FRET) are sometimes used to characterize the structure of the complexes.

Nucleic acid structures can be directly imaged by atomic force microscopy, which is well suited to extended two-dimensional structures, but less useful for discrete three-dimensional structures because of the microscope tip's interaction with the fragile nucleic acid structure; transmission electron microscopy and cryo-electron microscopy are often used in this case. Extended three-dimensional lattices are analyzed by X-ray crystallography.

History

The conceptual foundation for DNA nanotechnology was first laid out by Nadrian Seeman in the early 1980s. Seeman's original motivation was to create a three-dimensional DNA lattice for orienting other large molecules, which would simplify their crystallographic study by eliminating the difficult process of obtaining pure crystals. This idea had reportedly come to him in late 1980, after realizing the similarity between the woodcut *Depth* by M. C. Escher and an array of DNA six-arm junctions. Several natural branched DNA structures were known at the time, including the DNA replication fork and the mobile Holliday junction, but Seeman's insight was that immobile nucleic acid junctions could be created by properly designing the strand sequences to remove symmetry in the assembled molecule, and that these immobile junctions could in principle be combined into rigid crystalline lattices. The first theoretical paper proposing this scheme was published in 1982, and the first experimental demonstration of an immobile DNA junction was published the following year.

The woodcut *Depth* (pictured) by M. C. Escher reportedly inspired Nadrian Seeman to consider using three-dimensional lattices of DNA to orient hard-to-crystallize molecules. This led to the beginning of the field of DNA nanotechnology.

In 1991, Seeman's laboratory published a report on the synthesis of a cube made of DNA, the first synthetic three-dimensional nucleic acid nanostructure, for which he received the 1995 Feynman Prize in Nanotechnology. This was followed by a DNA truncated octahedron. However, it soon became clear that these structures, polygonal shapes with flexible junctions as their vertices, were not rigid enough to form extended three-dimensional lattices. Seeman developed the more rigid double-crossover (DX) structural motif, and in 1998, in collaboration with Erik Winfree, published the creation of two-dimensional lattices of DX tiles. These tile-based structures had the advantage that they provided the ability to implement DNA computing, which was demonstrated by Winfree and Paul Rothemund in their 2004 paper on the algorithmic self-assembly of a Sierpinski gasket structure, and for which they shared the 2006 Feynman Prize in Nanotechnology. Winfree's key insight was that the DX tiles could be used as Wang tiles, meaning that their assembly could perform computation. The synthesis of a three-dimensional lattice was finally published by Seeman in 2009, nearly thirty years after he had set out to achieve it.

New abilities continued to be discovered for designed DNA structures throughout the 2000s. The first DNA nanomachine—a motif that changes its structure in response to an input—was demonstrated in 1999 by Seeman. An improved system, which was the first nucleic acid device to make use of toehold-mediated strand displacement, was demonstrated by Bernard Yurke the following year. The next advance was to translate this into mechanical motion, and in 2004 and 2005, several DNA walker systems were demonstrated by the groups of Seeman, Niles Pierce, Andrew Turberfield, and Chengde Mao. The idea of using DNA arrays to template the assembly of other molecules such as nanoparticles and proteins, first suggested by Bruche Robinson and Seeman in 1987, was demonstrated in 2002 by Seeman, Kiehl et al. and subsequently by many other groups.

In 2006, Rothemund first demonstrated the DNA origami method for easily and robustly forming folded DNA structures of arbitrary shape. Rothemund had conceived of this method as being conceptually intermediate between Seeman's DX lattices, which used many short strands, and William Shih's DNA octahedron, which consisted mostly of one very long strand. Rothemund's DNA origami contains a long strand which folding is assisted by several short strands. This method allowed forming much larger structures than formerly possible, and which are less technically demanding to design and synthesize. DNA origami was the cover story of *Nature* on March 15, 2006. Rothemund's research demonstrating two-dimensional DNA origami structures was followed by the demonstration of solid three-dimensional DNA origami by Douglas *et al.* in 2009, while the labs of Jørgen Kjems and Yan demonstrated hollow three-dimensional structures made out of two-dimensional faces.

DNA nanotechnology was initially met with some skepticism due to the unusual non-biological use of nucleic acids as materials for building structures and doing computation, and the preponderance of proof of principle experiments that extended the abilities of the field but were far from actual applications. Seeman's 1991 paper on the synthesis of the DNA cube was rejected by the journal *Science* after one reviewer praised its originality while another criticized it for its lack of biological relevance. By the early

2010s, however, the field was considered to have increased its abilities to the point that applications for basic science research were beginning to be realized, and practical applications in medicine and other fields were beginning to be considered feasible. The field had grown from very few active laboratories in 2001 to at least 60 in 2010, which increased the talent pool and thus the number of scientific advances in the field during that decade.

Self-assembled Monolayer

Self-assembled monolayers (SAM) of organic molecules are molecular assemblies formed spontaneously on surfaces by adsorption and are organized into more or less large ordered domains. In some cases molecules that form the monolayer do not interact strongly with the substrate. This is the case for instance of the two-dimensional supramolecular networks of e.g. Perylene-tetracarboxylicacid-dianhydride (PTCDA) on gold or of e.g. porphyrins on highly oriented pyrolitic graphite (HOPG). In other cases the molecules possess a head group that has a strong affinity to the substrate and anchors the molecule to it. Such a SAM consisting of a head group, tail and functional end group is depicted in Figure 1. Common head groups include thiols, silanes, phosphonates, etc.

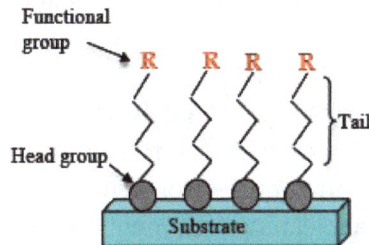

Figure 1. Representation of a SAM structure

SAMs are created by the chemisorption of "head groups" onto a substrate from either the vapor or liquid phase followed by a slow organization of "tail groups". Initially, at small molecular density on the surface, adsorbate molecules form either a disordered mass of molecules or form an ordered two-dimensional "lying down phase", and at higher molecular coverage, over a period of minutes to hours, begin to form three-dimensional crystalline or semicrystalline structures on the substrate surface. The "head groups" assemble together on the substrate, while the tail groups assemble far from the substrate. Areas of close-packed molecules nucleate and grow until the surface of the substrate is covered in a single monolayer.

Adsorbate molecules adsorb readily because they lower the surface free-energy of the substrate and are stable due to the strong chemisorption of the "head groups." These bonds create monolayers that are more stable than the physisorbed bonds of Langmuir–Blodgett films. A Trichlorosilane based "head group", for example in a FDTS molecule reacts with an hydroxyl group on a substrate, and forms very stable, covalent bond [R-Si-O-substrate] with an energy of 452 kJ/mol. Thiol-metal bonds, that are on the order of 100 kJ/mol, making the bond a fairly stable in a variety of temperature, solvents, and potentials. The monolayer packs tightly due to van der Waals interactions, thereby reducing its own free energy. The adsorption can be described by the Langmuir adsorption isotherm if lateral interactions are neglected. If they cannot be neglected, the adsorption is better described by the Frumkin isotherm.

Types

Selecting the type of head group depends on the application of the SAM. Typically, head groups are connected to a molecular chain in which the terminal end can be functionalized (i.e. adding –OH, –NH2, –COOH, or –SH groups) to vary the wetting and interfacial properties. An appropriate substrate is chosen to react with the head group. Substrates can be planar surfaces, such as silicon and metals, or curved surfaces, such as nanoparticles. Alkanethiols are the most commonly used molecules for SAMs. Alkanethiols are molecules with an alkyl chain, $(C-C)^n$ chain, as the back bone, a tail group, and a S-H head group. Other types of interesting molecules include aromatic thiols, of interest in molecular electronics, in which the alkane chain is (partly) replaced by aromatic rings. An example is the dithiol 1,4-Benzenedimethanethiol ($SHCH_2C_6H_4CH_2SH$)). Interest in such dithiols stems from the possibility of linking the two sulfur ends to metallic contacts, which was first used in molecular conduction measurements. Thiols are frequently used on noble metal substrates because of the strong affinity of sulfur for these metals. The sulfur gold interaction is semi-covalent and has a strength of approximately 45kcal/mol. In addition, gold is an inert and biocompatible material that is easy to acquire. It is also easy to pattern via lithography, a useful feature for applications in nanoelectromechanical systems (NEMS). Additionally, it can withstand harsh chemical cleaning treatments. Recently other chalcogenide SAMs: selenides and tellurides have attracted attention in a search for different bonding characteristics to substrates affecting the SAM characteristics and which could be of interest in some applications such as molecular electronics. Silanes are generally used on nonmetallic oxide surfaces; however monolayers formed from covalent bonds between silicon and carbon or oxygen cannot be considered self assembled because they do not form reversibly. Self-assembled monolayers of thiolates on noble metals are a special case because the metal-metal bonds become reversible after the formation of the thiolate-metal complex. This reversibility is what gives rise to vacancy islands and it is why SAMs of alkanethiolates can be thermally desorbed and undergo exchange with free thiols.

Preparation

Metal substrates for use in SAMs can be produced through physical vapor deposition techniques, electrodeposition or electroless deposition. Thiol or selenium SAMs produced by adsorption from solution are typically made by immersing a substrate into a dilute solution of alkane thiol in ethanol, though many different solvents can be used besides use of pure liquids. While SAMs are often allowed to form over 12 to 72 hours at room temperature, SAMs of alkanethiolates form within minutes. Special attention is essential in some cases, such as that of dithiol SAMs to avoid problems due to oxidation or photoinduced processes, which can affect terminal groups and lead to disorder and multilayer formation. In this case appropriate choice of solvents, their degassing by inert gasses and preparation in the absence of light is crucial and allows formation of "standing up" SAMs with free –SH groups. Self-assembled monolayers can also be adsorbed from the vapor phase. In some cases when obtaining an ordered assembly is difficult or when different density phases need to be obtained substitutional self-assembly is used. Here one first forms the SAM of a given type of molecules, which give rise to ordered assembly and then a second assembly phase is performed (e.g. by immersion into a different solution). This method has also been used to give information on relative binding strengths of SAMs with different head groups and more generally on self-assembly characteristics.

Characterization

The thicknesses of SAMs can be measured using ellipsometry and X-ray photoelectron spectroscopy (XPS), which also give information on interfacial properties. The order in the SAM and orientation of molecules can be probed by Near Edge Xray Absorption Fine Structure (NEXAFS) and Fourier Transform Infrared Spectroscopy in Reflection Absorption Infrared Spectroscopy (RAIRS) studies. Numerous other spectroscopic techniques are used such as Second-harmonic generation (SHG), Sum-frequency generation (SFG), Surface-enhanced Raman scattering (SERS), as well as High-resolution electron energy loss spectroscopy (HREELS). The structures of SAMs are commonly determined using scanning probe microscopy techniques such as atomic force microscopy (AFM) and scanning tunneling microscopy (STM). STM has been able to help understand the mechanisms of SAM formation as well as determine the important structural features that lend SAMs their integrity as surface-stable entities. In particular STM can image the shape, spatial distribution, terminal groups and their packing structure. AFM offers an equally powerful tool without the requirement of the SAM being conducting or semi-conducting. AFM has been used to determine chemical functionality, conductance, magnetic properties, surface charge, and frictional forces of SAMs. More recently, however, diffractive methods have also been used. The structure can be used to characterize the kinetics and defects found on the monolayer surface. These techniques have also shown physical differences between SAMs with planar substrates and nanoparticle substrates. An alternative characterisation instrument for measuring the self-assembly in real time is dual polarisation interferometry where the refractive index, thickness, mass and birefringence of the self assembled layer are quantified at high resolution. Contact angle measurements can be used to determine the surface free-energy which reflects the average composition of the surface of the SAM and can be used to probe the kinetics and thermodynamics of the formation of SAMs. The kinetics of adsorption and temperature induced desorption as well as information on structure can also be obtained in real time by ion scattering techniques such as low energy ion scattering (LEIS) and time of flight direct recoil spectroscopy (TOFDRS).

Defects

Defects due to both external and intrinsic factors may appear. External factors include the cleanliness of the substrate, method of preparation, and purity of the adsorbates. SAMs intrinsically form defects due to the thermodynamics of formation, e.g. thiol SAMs on gold typically exhibit etch pits (monatomic vacancy islands) likely due to extraction of adatoms from the substrate and formation of adatom-adsorbate moieties. Recently, a new type of fluorosurfactants have found that can form nearly perfect monolayer on gold substrate due to the increase of mobility of gold surface atoms.

Nanoparticle Properties

The structure of SAMs is also dependent on the curvature of the substrate. SAMs on nanoparticles, including colloids and nanocrystals, "stabilize the reactive surface of the particle and present organic functional groups at the particle-solvent interface". These organic functional groups are useful for applications, such as immunoassays, that are dependent on chemical composition of the surface.

Kinetics

There is evidence that SAM formation occurs in two steps: an initial fast step of adsorption and a second slower step of monolayer organization. Adsorption occurs at the liquid–liquid, liquid–vapor, and liquid-solid interfaces. The transport of molecules to the surface occurs due to a combination of diffusion and convective transport. According to the Langmuir or Avrami kinetic model the rate of deposition onto the surface is proportional to the free space of the surface.

$$k(1-\theta) = \frac{d\theta}{dt}.$$

Where θ is the proportional amount of area deposited and k is the rate constant. Although this model is robust it is only used for approximations because it fails to take into account intermediate processes. Dual polarisation interferometry being a real time technique with ~10 Hz resolution can measure the kinetics of monolayer self-assembly directly.

Once the molecules are at the surface the self-organization occurs in three phases:

1. A low-density phase with random dispersion of molecules on the surface.

2. An intermediate-density phase with conformational disordered molecules or

molecules lying flat on the surface.

3. A high-density phase with close-packed order and molecules standing normal to the substrate's surface.

The phase transitions in which a SAM forms depends on the temperature of the environment relative to the triple point temperature, the temperature in which the tip of the low-density phase intersects with the intermediate-phase region. At temperatures below the triple point the growth goes from phase 1 to phase 2 where many islands form with the final SAM structure, but are surrounded by random molecules. Similar to nucleation in metals, as these islands grow larger they intersect forming boundaries until they end up in phase 3.

At temperatures above the triple point the growth is more complex and can take two paths. In the first path the heads of the SAM organize to their near final locations with the tail groups loosely formed on top. Then as they transit to phase 3, the tail groups become ordered and straighten out. In the second path the molecules start in a lying down position along the surface. These then form into islands of ordered SAMs, where they grow into phase 3.

The nature in which the tail groups organize themselves into a straight ordered monolayer is dependent on the inter-molecular attraction, or Van der Waals forces, between the tail groups. To minimize the free energy of the organic layer the molecules adopt conformations that allow high degree of Van der Waals forces with some hydrogen bonding. The small size of the SAM molecules are important here because Van der Waals forces arise from the dipoles of molecules and are thus much weaker than the surrounding surface forces at larger scales. The assembly process begins with a small group of molecules, usually two, getting close enough that the Van der Waals forces overcome the surrounding force. The forces between the molecules orient them so they are in their straight, optimal, configuration. Then as other molecules come close by they interact with these already organized molecules in the same fashion and become a part of the conformed group. When this occurs across a large area the molecules support each other into forming their SAM shape seen in Figure 1. The orientation of the molecules can be described with two parameters: α and β. α is the angle of tilt of the backbone from the surface normal. In typical applications α varies from 0 to 60 degrees depending on the substrate and type of SAM molecule. β is the angle of rotation along the long axis of tee molecule. β is usually between 30 and 40 degrees. In some cases existence of kinetic traps hindering the final ordered orientation has been pointed out. Thus in case of dithiols formation of a "lying down" phase was considered an impediment to formation of "standing up" phase, however various recent studies indicate this is not the case.

Many of the SAM properties, such as thickness, are determined in the first few minutes. However, it may take hours for defects to be eliminated via annealing and for final SAM properties to be determined. The exact kinetics of SAM formation depends on the adsorbate, solvent and substrate properties. In general, however, the kinetics are dependent on both preparations conditions and material properties of the solvent, adsorbate and substrate. Specifically, kinetics for adsorption from a liquid solution are dependent on:

- Temperature – room-temperature preparation improves kinetics and reduces defects.

- Concentration of adsorbate in the solution – low concentrations require longer immersion times and often create highly crystalline domains.

- Purity of the adsorbate – impurities can affect the final physical properties of the SAM

- Dirt or contamination on the substrate – imperfections can cause defects in the SAM

The final structure of the SAM is also dependent on the chain length and the structure of both the adsorbate and the substrate. Steric hindrance and metal substrate properties, for example, can affect the packing density of the film, while chain length affects SAM thickness. Longer chain length also increases the thermodynamic stability.

Patterning

1. Locally Attract

This first strategy involves locally depositing self-assembled monolayers on the surface only where the nanostructure will later be located. This strategy is advantageous because it involves high throughput methods that generally involve fewer steps than the other two strategies. The major techniques that use this strategy are:

- Micro-contact printing

  Micro-contact printing or soft lithography is analogous to printing ink with a rubber stamp. The SAM molecules are inked onto a pre-shaped elastomeric stamp with a solvent and transferred to the substrate surface by stamping. The SAM solution is applied to the entire stamp but only areas that make contact with the surface allow transfer of the SAMs. The transfer of the SAMs is a complex diffusion process that depends on the type of molecule, concentration, duration of contact, and pressure applied. Typical stamps use PDMS because its elastomeric properties, $E = 1.8$ MPa, allow it to fit the countour of micro surfaces and its low surface energy, $\gamma = 21.6$ dyn/cm^2. This is a parallel process and can thus place nanoscale objects over a large area in a short time.

- Dip-pen nanolithography

Dip-pen nanolithography is a process that uses an atomic force microscope to transfer molecules on the tip to a substrate. Initially the tip is dipped into a reservoir with an ink. The ink on the tip evaporates and leaves the desired molecules attached to the tip. When the tip is brought into contact with the surface a water meniscus forms between the tip and the surface resulting in the diffusion of molecules from the tip to the surface. These tips can have radii in the tens of nanometers, and thus SAM molecules can be very precisely deposited onto a specific location of the surface. This process was discovered by Chad Mirkin and co-workers at Northwestern University.

2. Locally Remove

The locally remove strategy begins with covering the entire surface with a SAM. Then individual SAM molecules are removed from locations where the deposition of nanostructures is not desired. The end result is the same as in the locally attract strategy, the difference being in the way this is achieved. The major techniques that use this strategy are:

- Scanning tunneling microscope

  The scanning tunneling microscope can remove SAM molecules in many different ways. The first is to remove them mechanically by dragging the tip across the substrate surface. This is not the most desired technique as these tips are expensive and dragging them causes a lot of wear and reduction of the tip quality. The second way is to degrade or desorb the SAM molecules by shooting them with an electron beam. The scanning tunneling microscope can also remove SAMs by field desorption and field enhanced surface diffusion.

- Atomic force microscope

  The most common use of this technique is to remove the SAM molecules in a process called shaving, where the atomic force microscope tip is dragged along the surface mechanically removing the molecules. An atomic force microscope can also remove SAM molecules by local oxidation nanolithography.

- Ultraviolet irradiation

  In this process, UV light is projected onto the surface with a SAM through a pattern of apperatures in a chromium film. This leads to photo oxidation of the SAM molecules. These can then be washed away in a polar solvent. This process has 100 nm resolutions and requires exposure time of 15–20 minutes.

3. Modify Tail Groups

The final strategy focuses not on the deposition or removal of SAMS, but the modification of terminal groups. In the first case the terminal group can be modified to remove functionality so that SAM molecule will be inert. In the same regards the terminal group can be modified to add functionality so it can accept different materials or have different properties than the original SAM terminal group. The major techniques that use this strategy are:

- Focused electron beam and ultraviolet irradiation

  Exposure to electron beams and UV light changes the terminal group chemistry. Some of the changes that can occur include the cleavage of bonds, the forming of double carbon bonds, cross-linking of adjacent molecules, fragmentation of molecules, and confromational disorder.

- Atomic force microscope

  A conductive AFM tip can create an electrochemical reaction that can change the terminal group.

Applications

Thin-film SAMs

SAMs are an inexpensive and versatile surface coating for applications including control of wetting and adhesion, chemical resistance, bio compatibility, sensitization, and molecular recognition for sensors and nano fabrication. Areas of application for SAMs include biology, electrochemistry and electronics, nanoelectromechanical systems (NEMS) and microelectromechanical systems (MEMS), and everyday household goods. SAMs can serve as models for studying membrane properties of cells and organelles and cell attachment on surfaces. SAMs can also be used to modify the surface properties of electrodes for electrochemistry, general electronics, and various NEMS and MEMS. For example, the properties of SAMs can be used to control electron transfer in electrochemistry. They can serve to protect metals from harsh chemicals and etchants. SAMs can also reduce sticking of NEMS and MEMS components in humid environments. In the same way, SAMs can alter the properties of glass. A common household product, Rain-X, utilizes SAMs to create a hydrophobic monolayer on car windshields to keep them clear of rain. Another application is an anti-adhesion coating on nanoimprint lithography (NIL) tools and stamps. One can also coat injection molding tools for polymer replication with a Perfluordecyltrichlorosilane SAM.

Thin film SAMs can also be placed on nanostructures. In this way they functionalize the nanostructure. This is advantageous because the nanostructure can now selectively attach itself to other molecules or SAMs. This technique is useful in biosensors or other MEMS devices that need to separate one type of molecule from its environment. One example is the use of magnetic nanoparticles to remove a fungus from a blood stream. The nanoparticle is coated with a SAM that binds to the fungus. As the contaminated blood is filtered through a MEMS device the magnetic nanoparticles are inserted into the blood where they bind to the fungus and are then magnetically driven out of the blood stream into a nearby laminar waste stream.

Patterned SAMs

SAMs are also useful in depositing nanostructures, because each adsorbate molecule can be tailored to attract two different materials. Current techniques utilize the head to attract to a surface, like a plate of gold. The terminal group is then modified to attract a specific material like a particular nanoparticle, wire, ribbon, or other nanostructure. In this way, wherever the a SAM is patterned to a surface there will be nanostructures attached to the tail groups. One example is the use of two types of SAMs to align single wall carbon nanotubes, SWNTs. Dip pen nanolithography was used to pattern a 16-mercaptohexadecanoic acid (MHA)SAM and the rest of the surface was passivated with 1-octadecanethiol (ODT) SAM. The polar solvent that is carrying the SWNTs is attracted to the hydrophilic MHA; as the solvent evaporates, the SWNTs are close enough to the MHA SAM to attach to it due to Van der Waals forces. The nanotubes thus line up with the MHA-ODT boundary. Using this technique Chad Mirkin, Schatz and their co-workers were able to make complex two-dimensional shapes, a representation of a shape created is shown to the right. Another application of patterned SAMs is the functionalization of biosensors. The tail groups can be modified so they have an affinity for cells, proteins, or molecules. The SAM can then be placed onto a biosensor so that binding of these molecules can be detected. The ability to pattern these SAMs allows them to be placed in configurations that increase sensitivity and do not damage or interfere with other components of the biosensor.

Metal Organic Superlattices

There has been considerable interest in use of SAMs for new materials e.g. via formation of two- or three-dimensional metal organic superlattices by assembly of SAM capped nanoparticles or layer by layer SAM-nanoparticle arrays using dithiols.

Supramolecular Assembly

A supramolecular assembly or "supermolecule" is a well defined complex of molecules held together by noncovalent bonds. While a supramolecular assembly can be simply composed of two molecules (e.g., a DNA double helix or an inclusion compound), it is more often used to denote larger complexes of molecules that form sphere-, rod-, or sheet-like species. Micelles, liposomes and biological membranes are examples of supramolecular assemblies. The dimensions of supramolecular assemblies can range from nanometers to micrometers. Thus they allow access to nanoscale objects using a bottom-up approach in far fewer steps than a single molecule of similar dimensions.

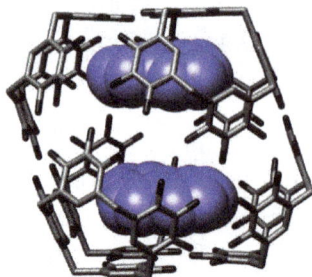

An example of a supramolecular assembly reported by Atwood and coworkers in Science 2005, 309, 2037.

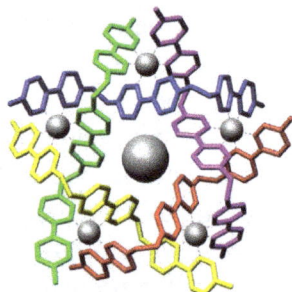

An example of a supramolecular assembly reported by Jean-Marie Lehn and coworkers in Angew. Chem. Int. Ed. Engl. 1996, 35, 1838-1840.

The process by which a supramolecular assembly forms is called molecular self-assembly. Some try to distinguish self-assembly as the process by which individual molecules form the defined aggregate. Self-organization, then, is the process by which those aggregates create higher-order structures. This can become useful when talking about liquid crystals and block copolymers.

Templating Reactions

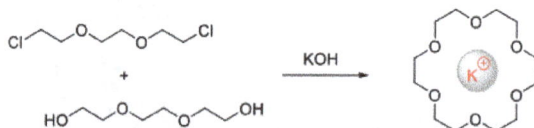

18-crown-6 can be synthesized from using potassium ion as the template cation

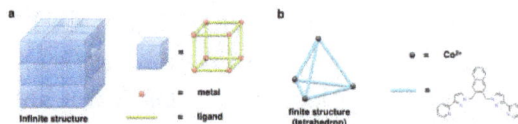

Illustrations of a. metal-organic frameworks and b. supramolecular coordination complexes

As studied in coordination chemistry, metals (usually transition metals) exist in solution bound to ligands, In many cases, the coordination sphere defines geometries conducive to reactions either between ligands or involving ligands and other external reagents.

A well known metal-templating was described by Charles Pederson in his synthesis of various crown ethers using metal cations as template. For example, 18-crown-6 strongly coordinates potassium ion thus can be prepared through the Williamson ether synthesis using potassium ion as the template metal.

Metals are frequently used for assembly of large supramolecular structures. Metal organic frameworks (MOFs) are one example. MOFs are infinite structures where metal serve as nodes to connect organic ligands together. SCCs are discrete systems where selected metals and ligands undergo self-assembly to form finite supramolecular complexes, usually the size and structure of the complex formed can be determined by the angularity of chosen metal-ligand bonds.

Hydrogen Bond Assisted Supramolecular Assembly

Hydrogen bond-assisted supramolecular assembly is the process of assembling small organic molecules to form large supramolecular structures by non-covalent hydrogen bonding interactions. The directionality, reversibility, and strong bonding nature of hydrogen bond make it an attractive and useful approach in supramolecular assembly. Functional groups such as carboxylic acids, ureas, amines, and amides are commonly used to assemble higher order structures upon hydrogen bonding.

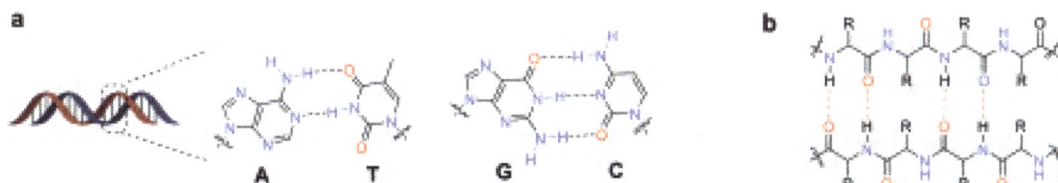

Hydrogen bonds in (a) DNA duplex formation and (b) protein β-sheet structure
(a) Representative hydrogen bond patterns in supramolecular assembly. (b) Hydrogen bond network in cyanuric acid-melamine crystals.

Hydrogen bond play an essential role in the assembly of secondary and tertiary structures of large biomolecules. DNA double helix is formed by hydrogen bonding between nucleobases: adenine and thymine forms two hydrogen bonds, while guanine and cytosine forms three hydrogen bonds (Figure "Hydrogen bonds in (a) DNA duplex formation"). Another prominent example of hydrogen bond-assisted assembly in nature is the formation of protein secondary structures. Both the α-helix and β-sheet are formed through hydrogen bonding between the amide hydrogen and the amide carbonyl oxygen (Figure "Hydrogen bonds in (b) protein β-sheet structure").

In supramolecular chemistry, hydrogen bonds have been broadly applied to crystal engineering, molecular recognition, and catalysis. Hydrogen bonds are among the mostly used synthons in bottom-up approach to engineering molecular interactions in crystals. Representative hydrogen bond patterns for supramolecular assembly is shown in Figure "Representative hydrogen bond patterns in supramolecular assembly". A

1: 1 mixture of cyanuric acid and melamine forms crystal with a highly dense hydrogen-bonding network. This supramolecular aggregates has been used as templates to engineering other crystal structures.

Applications

Supramolecular assemblies have no specific applications but are the subject of many intriguing reactions. A supramolecular assembly of peptide amphiphiles in the form of nanofibers has been shown to promote the growth of neurons. An advantage to this supramolecular approach is that the nanofibers will degrade back into the individual peptide molecules that can be broken down by the body. By self-assembling of dendritic dipeptides, hollow cylinders can be produced. The cylindrical assemblies possess internal helical order and self-organize into columnar liquid crystalline lattices. When inserted into vesicular membranes, the porous cylindrical assemblies mediate transport of protons across the membrane. Self-assembly of dendrons generates arrays of nanowires. Electron donor-acceptor complexes comprise the core of the cylindrical supramolecular assemblies, which further self-organize into two-dimensional columnar liquid crystaline lattices. Each cylindrical supramolecular assembly functions as an individual wire. High charge carrier mobilities for holes and electrons were obtained.

References

- Lu, Yicheng; Zhong, Jian (2004). Todd Steiner, ed. Semiconductor Nanostructures for Optoelectronic Applications. Norwood, MA: Artech House, Inc. pp. 191–192. ISBN 978-1-58053-751-3.

- Wang, Ji-Tao (2002). Nonequilibrium Nondissipative Thermodynamics: With Application to Low-pressure Diamond Synthesis. Berlin: Springer Verlag. p. 65. ISBN 978-3-540-42802-2.

- Rosen, Milton J. (2004). Surfactants and interfacial phenomena. Hoboken, NJ: Wiley-Interscience. ISBN 978-0-471-47818-8.

- Background: Pelesko, John A. (2007). Self-assembly: the science of things that put themselves together. New York: Chapman & Hall/CRC. pp. 5, 7. ISBN 978-1-58488-687-7.

- Applications: Rietman, Edward A. (2001). Molecular engineering of nanosystems. Springer. pp. 209–212. ISBN 978-0-387-98988-4. Retrieved 17 April 2011.

- Methods: Gallagher, S. R.; Desjardins, P. (1 July 2011). "Quantitation of nucleic acids and proteins". Current Protocols Essential Laboratory Techniques. doi:10.1002/9780470089941. et0202s5. ISBN 0470089938.

- History: Pelesko, John A. (2007). Self-assembly: the science of things that put themselves together. New York: Chapman & Hall/CRC. pp. 201, 242, 259. ISBN 978-1-58488-687-7.

- Ariga, Katsuhiko; Kunitake, Toyoki (2006). Supramolecular chemistry - fundamentals and applications (English ed.). Berlin [u.a.]: Springer. ISBN 3540012982. Retrieved 4 April 2016.

Nanomaterials and Nanoparticles

Nanomaterials vary between 1-100 nanometres. Nanomaterials have become commercialized in the past few years. Several nanometres, very small in number form the quantum dots. Quantum dots is very central to nanotechnology. Hybrid material, metamaterial, metallogels, nanoparticles, silver nanoparticles and platinum nanoparticles are some of the topics that have been elucidated in this text.

Nanomaterials

Nanomaterials describe, in principle, materials of which a single unit is sized (in at least one dimension) between 1 and 1000 nanometres (10^{-9} meter) but is usually 1—100 nm (the usual definition of nanoscale).

Nanomaterials research takes a materials science-based approach to nanotechnology, leveraging advances in materials metrology and synthesis which have been developed in support of microfabrication research. Materials with structure at the nanoscale often have unique optical, electronic, or mechanical properties.

Nanomaterials are slowly becoming commercialized and beginning to emerge as commodities.

Types

Natural Nanomaterials

Biological systems often feature natural, functional nanomaterials. The structure of foraminifera (mainly chalk) and viruses (protein, capsid), the wax crystals covering a lotus or nasturtium leaf, spider and spider-mite silk, the blue hue of tarantulas, the "spatulae" on the bottom of gecko feet, some butterfly wing scales, natural colloids (milk, blood), horny materials (skin, claws, beaks, feathers, horns, hair), paper, cotton, nacre, corals, and even our own bone matrix are all natural *organic* nanomaterials.

Natural *inorganic* nanomaterials occur through crystal growth in the diverse chemical conditions of the Earth's crust. For example, clays display complex nanostructures due to anisotropy of their underlying crystal structure, and volcanic activity can give rise to opals, which are an instance of a naturally occurring photonic crystals due to their nanoscale structure. Fires represent particularly complex reactions and can produce pigments, cement, fumed silica etc.

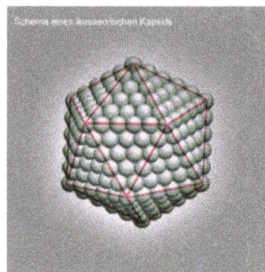

"Lotus effect", hydrophobic effect
with self-cleaning ability

Viral capsid

Fullerenes

The fullerenes are a class of allotropes of carbon which conceptually are graphene sheets rolled into tubes or spheres. These include the carbon nanotubes (or silicon nanotubes) which are of interest both because of their mechanical strength and also because of their electrical properties.

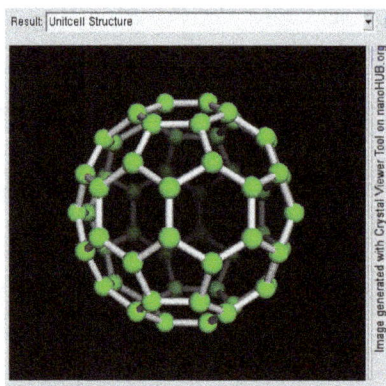

Rotating view of C_{60}, one kind of fullerene.

The first fullerene molecule to be discovered, and the family's namesake, buckminster-fullerene (C_{60}), was prepared in 1985 by Richard Smalley, Robert Curl, James Heath, Sean O'Brien, and Harold Kroto at Rice University. The name was a homage to Buckminster Fuller, whose geodesic domes it resembles. Fullerenes have since been found to occur in nature. More recently, fullerenes have been detected in outer space.

For the past decade, the chemical and physical properties of fullerenes have been a hot topic in the field of research and development, and are likely to continue to be for a long time. In April 2003, fullerenes were under study for potential medicinal use: binding specific antibiotics to the structure of resistant bacteria and even target certain types of cancer cells such as melanoma. The October 2005 issue of Chemistry and Biology contains an article describing the use of fullerenes as light-activated antimicrobial agents. In the field of nanotechnology, heat resistance and superconductivity are among the properties attracting intense research.

A common method used to produce fullerenes is to send a large current between two nearby graphite electrodes in an inert atmosphere. The resulting carbon plasma arc between the electrodes cools into sooty residue from which many fullerenes can be isolated.

There are many calculations that have been done using ab-initio Quantum Methods applied to fullerenes. By DFT and TDDFT methods one can obtain IR, Raman and UV spectra. Results of such calculations can be compared with experimental results.

Graphene Nanostructures

2D materials are crystalline materials consisting of a two-dimensional single layer of atoms. The most important representative graphene was discovered in 2004. Other 2D materials based on other elements have since been reported.

Box-shaped graphene (BSG) nanostructure is an example of 3D nanomaterial. BSG nanostructure has appeared after mechanical cleavage of pyrolytic graphite. This nanostructure is a multilayer system of parallel hollow nanochannels located along the surface and having quadrangular cross-section. The thickness of the channel walls is approximately equal to 1 nm. The typical width of channel facets makes about 25 nm.

Nanoparticles

Inorganic nanomaterials, (e.g. quantum dots, nanowires and nanorods) because of their interesting optical and electrical properties, could be used in optoelectronics. Furthermore, the optical and electronic properties of nanomaterials which depend on their size and shape can be tuned via synthetic techniques. There are the possibilities to use those materials in organic material based optoelectronic devices such as Organic solar cells, OLEDs etc. The operating principles of such devices are governed by photoinduced processes like electron transfer and energy transfer. The performance of the devices depends on the efficiency of the photoinduced process responsible for their functioning. Therefore, better understanding of those photoinduced processes in organic/inorganic nanomaterial composite systems is necessary in order to use them in organic optoelectronic devices.

Nanoparticles or nanocrystals made of metals, semiconductors, or oxides are of particular interest for their mechanical, electrical, magnetic, optical, chemical and other properties. Nanoparticles have been used as quantum dots and as chemical catalysts such as nanomaterial-based catalysts. Recently, a range of nanoparticles are extensively investigated for biomedical applications including tissue engineering, drug delivery, biosensor.

Nanoparticles are of great scientific interest as they are effectively a bridge between bulk materials and atomic or molecular structures. A bulk material should have constant physical properties regardless of its size, but at the nano-scale this is often not the case. Size-dependent properties are observed such as quantum confinement in semi-conductor particles, surface plasmon resonance in some metal particles and superparamagnetism in magnetic materials.

Nanoparticles exhibit a number of special properties relative to bulk material. For example, the bending of bulk copper (wire, ribbon, etc.) occurs with movement of copper atoms/clusters at about the 50 nm scale. Copper nanoparticles smaller than 50 nm are considered super hard materials that do not exhibit the same malleability and ductility as bulk copper. The change in properties is not always desirable. Ferroelectric materials smaller than 10 nm can switch their magnetisation direction using room temperature thermal energy, thus making them useless for memory storage. Suspensions of nanoparticles are possible because the interaction of the particle surface with the solvent is strong enough to overcome differences in density, which usually result in a material either sinking or floating in a liquid. Nanoparticles often have unexpected visual properties because they are small enough to confine their electrons and produce quantum effects. For example, gold nanoparticles appear deep red to black in solution.

The often very high surface area to volume ratio of nanoparticles provides a tremendous driving force for diffusion, especially at elevated temperatures. Sintering is possible at lower temperatures and over shorter durations than for larger particles. This theoretically does not affect the density of the final product, though flow difficulties and the tendency of nanoparticles to agglomerate do complicate matters. The surface effects of nanoparticles also reduces the incipient melting temperature.

Nanozymes

Nanozymes are nanomaterials with enzyme-like characteristics. They are an emerging type of artificial enzyme, which have been used for wide applications in such as biosensing, bioimaging, tumor diagnosis and therapy, antibiofouling, etc.

Synthesis

The goal of any synthetic method for nanomaterials is to yield a material that exhibits properties that are a result of their characteristic length scale being in the nanometer range (~1 – 100 nm). Accordingly, the synthetic method should exhibit control of size in this range so that one property or another can be attained. Often the methods are divided into two main types "Bottom Up" and "Top Down."

Bottom Up Methods

Bottom up methods involve the assembly of atoms or molecules into nanostructured arrays. In these methods the raw material sources can be in the form of gases, liquids or solids. The latter requiring some sort of disassembly prior to their incorporation onto a nanostructure. Bottom methods generally fall into two categories: chaotic and controlled.

Chaotic Processes

Chaotic processes involve elevating the constituent atoms or molecules to a chaotic state and then suddenly changing the conditions so as to make that state unstable. Through the clever manipulation of any number of parameters, products form largely as a result of the insuring kinetics. The collapse from the chaotic state can be difficult or impossible to control and so ensemble statistics often govern the resulting size distribution and average size. Accordingly, nanoparticle formation is controlled through manipulation of the end state of the products.

Examples of Chaotic Processes are: Laser ablation, Exploding wire, Arc, Flame pyrolysis, Combustion, Precipitation synthesis techniques.

Controlled Processes

Controlled Processes involve the controlled delivery of the constituent atoms or molecules to the site(s) of nanoparticle formation such that the nanoparticle can grow to a prescribed sizes in a controlled manner. Generally the state of the constituent atoms or molecules are never far from that needed for nanoparticle formation. Accordingly, nanoparticle formation is controlled through the control of the state of the reactants.

Examples of controlled processes are self-limiting growth solution, self-limited chemical vapor deposition, shaped pulse femtosecond laser techniques, and molecular beam epitaxy.

Characterization

Novel effects can occur in materials when structures are formed with sizes comparable to any one of many possible length scales, such as the de Broglie wavelength of electrons, or the optical wavelengths of high energy photons. In these cases quantum mechanical effects can dominate material properties. One example is quantum confinement where the electronic properties of solids are altered with great reductions in particle size. The optical properties of nanoparticles, e.g. fluorescence, also become a function of the particle diameter. This effect does not come into play by going from macrosocopic to micrometer dimensions, but becomes pronounced when the nanometer scale is reached.

In addition to optical and electronic properties, the novel mechanical properties of many nanomaterials is the subject of nanomechanics research. When added to a bulk material, nanoparticles can strongly influence the mechanical properties of the material, such as the stiffness or elasticity. For example, traditional polymers can be reinforced by nanoparticles (such as carbon nanotubes) resulting in novel materials which can be used as lightweight replacements for metals. Such composite materials may enable a weight reduction accompanied by an increase in stability and improved functionality.

Finally, nanostructured materials with small particle size such as zeolites, and asbestos, are used as catalysts in a wide range of critical industrial chemical reactions. The further development of such catalysts can form the basis of more efficient, environmentally friendly chemical processes.

The first observations and size measurements of nano-particles were made during the first decade of the 20th century. Zsigmondy made detailed studies of gold sols and other nanomaterials with sizes down to 10 nm and less. He published a book in 1914. He used an ultramicroscope that employs a *dark field* method for seeing particles with sizes much less than light wavelength.

There are traditional techniques developed during 20th century in Interface and Colloid Science for characterizing nanomaterials. These are widely used for *first generation* passive nanomaterials specified in the next section.

These methods include several different techniques for characterizing particle size distribution. This characterization is imperative because many materials that are expected to be nano-sized are actually aggregated in solutions. Some of methods are based on light scattering. Others apply ultrasound, such as ultrasound attenuation spectroscopy for testing concentrated nano-dispersions and microemulsions.

There is also a group of traditional techniques for characterizing surface charge or zeta potential of nano-particles in solutions. This information is required for proper system stabilzation, preventing its aggregation or flocculation. These methods include microelectrophoresis, electrophoretic light scattering and electroacoustics. The last one, for instance colloid vibration current method is suitable for characterizing concentrated systems.

Uniformity

The chemical processing and synthesis of high performance technological components for the private, industrial and military sectors requires the use of high purity ceramics, polymers, glass-ceramics and material composites. In condensed bodies formed from fine powders, the irregular sizes and shapes of nanoparticles in a typical powder often lead to non-uniform packing morphologies that result in packing density variations in the powder compact.

Uncontrolled agglomeration of powders due to attractive van der Waals forces can also give rise to in microstructural inhomogeneities. Differential stresses that develop as a result of non-uniform drying shrinkage are directly related to the rate at which the solvent can be removed, and thus highly dependent upon the distribution of porosity. Such stresses have been associated with a plastic-to-brittle transition in consolidated bodies, and can yield to crack propagation in the unfired body if not relieved.

In addition, any fluctuations in packing density in the compact as it is prepared for the kiln are often amplified during the sintering process, yielding inhomogeneous densification. Some pores and other structural defects associated with density variations have been shown to play a detrimental role in the sintering process by growing and thus limiting end-point densities. Differential stresses arising from inhomogeneous densification have also been shown to result in the propagation of internal cracks, thus becoming the strength-controlling flaws.

It would therefore appear desirable to process a material in such a way that it is physically uniform with regard to the distribution of components and porosity, rather than using particle size distributions which will maximize the green density. The containment of a uniformly dispersed assembly of strongly interacting particles in suspension requires total control over particle-particle interactions. It should be noted here that a number of dispersants such as ammonium citrate (aqueous) and imidazoline or oleyl alcohol (nonaqueous) are promising solutions as possible additives for enhanced dispersion and deagglomeration. Monodisperse nanoparticles and colloids provide this potential.

Monodisperse powders of colloidal silica, for example, may therefore be stabilized sufficiently to ensure a high degree of order in the colloidal crystal or polycrystalline colloidal solid which results from aggregation. The degree of order appears to be limited by the time and space allowed for longer-range correlations to be established. Such defective polycrystalline colloidal structures would appear to be the basic elements of sub-micrometer colloidal materials science, and, therefore, provide the first step in developing a more rigorous understanding of the mechanisms involved in microstructural evolution in high performance materials and components.

Legal Definition

On 18 October 2011, the European Commission adopted the following definition of a nanomaterial: "A natural, incidental or manufactured material containing particles, in an unbound state or as an aggregate or as an agglomerate and where, for 50% or more of the particles in the number size distribution, one or more external dimensions is in the size range 1 nm – 100 nm. In specific cases and where warranted by concerns for the environment, health, safety or competitiveness the number size distribution threshold of 50% may be replaced by a threshold between 1 and 50%."

However, this differs from the definition adopted by the International Organization for Standardization (ISO), which is: "Material with any external dimension in the nanoscale or having internal structure in the nanoscale."

"Nanoscale" is, in turn, defined as: "Size range from approximately 1 nm to 100 nm."

It is not currently known which of these, if any, will prevail in courts of law.

Safety

Health Impact

While nanomaterials and nanotechnologies are expected to yield numerous health and health care advances, such as more targeted methods of delivering drugs, new cancer therapies, and methods of early detection of diseases, they also may have unwanted effects. Increased rate of absorption is the main concern associated with manufactured nanoparticles.

When materials are made into nanoparticles, their surface area to volume ratio increases. The greater specific surface area (surface area per unit weight) may lead to increased rate of absorption through the skin, lungs, or digestive tract and may cause unwanted effects to the lungs as well as other organs. However, the particles must be absorbed in sufficient quantities in order to pose health risks.

Nanoparticles created adventitiously (e.g., through the rubbing of prostheses) have long been known to be a health hazard, but as the use of nanomaterials increases worldwide, concerns for worker and user safety are mounting. To address such concerns, the Swedish Karolinska Institute conducted a study in which various nanoparticles were introduced to human lung epithelial cells. The results, released in 2008, showed that iron oxide nanoparticles caused little DNA damage and were non-toxic. Zinc oxide nanoparticles were slightly worse. Titanium dioxide caused only DNA damage. Carbon nanotubes caused DNA damage at low levels. Copper oxide was found to be the worst offender, and was the only nanomaterial identified by the researchers as a clear health risk. Though nanomaterials are not confirmed as a health risk to workers who produce them, NIOSH recommends that exposure precautions and personal protective equipment be used to protect workers until the risks of nanomaterial manufacture are better understood.

Occupational Safety

Beyond normal chemical safety practices such as using personal protective equipment, The United States National Institute for Occupational Safety and Health recommends the use of HEPA filtration on local exhaust ventilation such as fume hoods. General laboratory ventilation is often not sufficient to effectively clear nanomaterials released into the general room air over a 30-minute period; therefore, researchers should leave the hood fan on even after synthesis is complete. Nanomaterials in dry powder form have the most risk of airborne dust generation and dermal contact, followed by liquid suspensions and then by nanomaterials embedded in a solid matrix.

Nanoparticles behave differently than other similarly sized particles. It is therefore necessary to develop specialized approaches to testing and monitoring their effects on human health and on the environment. The OECD Chemicals Committee has established the Working Party on Manufactured Nanomaterials to address this issue and to study the practices of OECD member countries in regards to nanomaterial safety.

Quantum Dot

Quantum dots (QD) are very small semiconductor particles, only several nanometres in size, so small that their optical and electronic properties differ from those of larger particles. They are a central theme in nanotechnology. Many types of quantum dot will emit light of specific frequencies if electricity or light is applied to them, and these frequencies can be precisely tuned by changing the dots' size, shape and material, giving rise to many applications.

Colloidal quantum dots irradiated with a UV light. Different sized quantum dots emit different color light due to quantum confinement.

In the language of materials science, nanoscale semiconductor materials tightly confine either electrons or electron holes. Quantum dots are also sometimes referred to as artificial atoms, a term that emphasizes that a quantum dot is a single object with bound, discrete electronic states, as is the case with naturally occurring atoms or molecules.

Quantum dots exhibit properties that are intermediate between those of bulk semiconductors and those of discrete molecules. Their optoelectronic properties change as a function of both size and shape. Larger QDs (radius of 5–6 nm, for example) emit longer wavelengths resulting in emission colors such as orange or red. Smaller QDs (radius of 2–3 nm, for example) emit shorter wavelengths resulting in colors like blue and green, although the specific colors and sizes vary depending on the exact composition of the QD.

Because of their highly tunable properties, QDs are of wide interest. Potential applications include transistors, solar cells, LEDs, diode lasers and second-harmonic generation, quantum computing, and medical imaging. Additionally, their small size allows for QDs to be suspended in solution which leads to possible uses in inkjet printing and spin-coating. These processing techniques result in less-expensive and less time consuming methods of semiconductor fabrication.

Production

Quantum Dots with gradually stepping emission from violet to deep red are being produced in a kg scale at PlasmaChem GmbH

There are several ways to prepare quantum dots, the principal ones involving colloids.

Colloidal Synthesis

Colloidal semiconductor nanocrystals are synthesized from solutions, much like traditional chemical processes. The main difference is the product neither precipitates as a bulk solid nor remains dissolved. Heating the solution at high temperature, the precursors decompose forming monomers which then nucleate and generate nanocrystals. Temperature is a critical factor in determining optimal conditions for the nanocrystal growth. It must be high enough to allow for rearrangement and annealing of atoms during the synthesis process while being low enough to promote crystal growth. The concentration of monomers is another critical factor that has to be stringently controlled during nanocrystal growth. The growth process of nanocrystals can occur in two different regimes, "focusing" and "defocusing". At high monomer concentrations, the critical size (the size where nanocrystals neither grow nor shrink) is relatively small, resulting in growth of nearly all particles. In this regime, smaller particles grow faster than large ones (since larger crystals need more atoms to grow than small crystals) resulting in "focusing" of the size distribution to yield nearly monodisperse particles. The size focusing is optimal when the monomer concentration is kept such that the average nanocrystal size present is always slightly larger than the critical size. Over time, the monomer concentration diminishes, the critical size becomes larger than the average size present, and the distribution "defocuses".

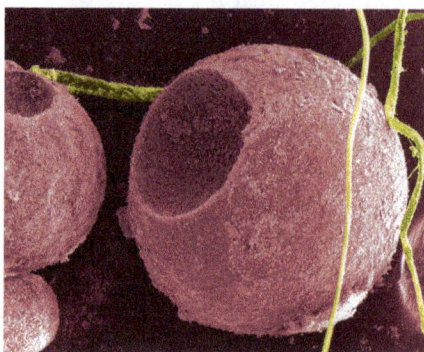

Cadmium sulfide quantum dots on cells

There are colloidal methods to produce many different semiconductors. Typical dots are made of binary compounds such as lead sulfide, lead selenide, cadmium selenide, cadmium sulfide, indium arsenide, and indium phosphide. Dots may also be made from ternary compounds such as cadmium selenide sulfide. These quantum dots can contain as few as 100 to 100,000 atoms within the quantum dot volume, with a diameter of ~ 10 to 50 atoms. This corresponds to about 2 to 10 nanometers, and at 10 nm in diameter, nearly 3 million quantum dots could be lined up end to end and fit within the width of a human thumb.

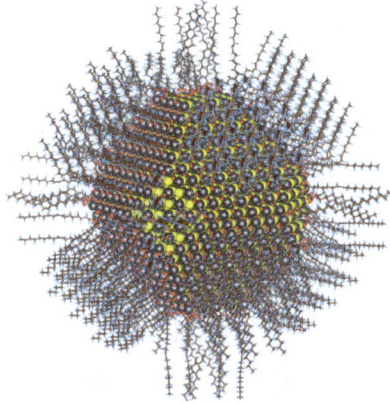

Ideallized image of colloidal nanoparticle of lead sulfide (selenide) with complete passivation by oleic acid, oleyl and hydroxyl (size ~5nm)

Large batches of quantum dots may be synthesized via colloidal synthesis. Due to this scalability and the convenience of benchtop conditions, colloidal synthetic methods are promising for commercial applications. It is acknowledged to be the least toxic of all the different forms of synthesis.

Plasma Synthesis

Plasma synthesis has evolved to be one of the most popular gas-phase approaches for the production of quantum dots, especially those with covalent bonds. For example, silicon (Si) and germanium (Ge) quantum dots have been synthesized by using non-thermal plasma. The size, shape, surface and composition of quantum dots can all be controlled in nonthermal plasma. Doping that seems quite challenging for quantum dots has also been realized in plasma synthesis. Quantum dots synthesized by plasma are usually in the form of powder, for which surface modification may be carried out. This can lead to excellent dispersion of quantum dots in either organic solvents or water (i. e., colloidal quantum dots).

Fabrication

- Self-assembled quantum dots are typically between 5 and 50 nm in size. Quantum dots defined by lithographically patterned gate electrodes, or by etching on two-dimensional electron gases in semiconductor heterostructures can have lateral dimensions between 20 and 100 nm.

- Some quantum dots are small regions of one material buried in another with a larger band gap. These can be so-called core–shell structures, e.g., with CdSe in the core and ZnS in the shell or from special forms of silica called ormosil.

- Quantum dots sometimes occur spontaneously in quantum well structures due to monolayer fluctuations in the well's thickness.

- Self-assembled quantum dots nucleate spontaneously under certain conditions during molecular beam epitaxy (MBE) and metallorganic vapor phase epitaxy (MOVPE), when a material is grown on a substrate to which it is not lattice matched. The resulting strain produces coherently strained islands on top of a two-dimensional wetting layer. This growth mode is known as Stranski–Krastanov growth. The islands can be subsequently buried to form the quantum dot. This fabrication method has potential for applications in quantum cryptography (i.e. single photon sources) and quantum computation. The main limitations of this method are the cost of fabrication and the lack of control over positioning of individual dots.

- Individual quantum dots can be created from two-dimensional electron or hole gases present in remotely doped quantum wells or semiconductor heterostructures called lateral quantum dots. The sample surface is coated with a thin layer of resist. A lateral pattern is then defined in the resist by electron beam lithography. This pattern can then be transferred to the electron or hole gas by etching, or by depositing metal electrodes (lift-off process) that allow the application of external voltages between the electron gas and the electrodes. Such quantum dots are mainly of interest for experiments and applications involving electron or hole transport, i.e., an electrical current.

- The energy spectrum of a quantum dot can be engineered by controlling the geometrical size, shape, and the strength of the confinement potential. Also, in contrast to atoms, it is relatively easy to connect quantum dots by tunnel barriers to conducting leads, which allows the application of the techniques of tunneling spectroscopy for their investigation.

The quantum dot absorption features correspond to transitions between discrete,-three-dimensional particle in a box states of the electron and the hole, both confined to the same nanometer-size box.These discrete transitions are reminiscent of atomic spectra and have resulted in quantum dots also being called *artificial atoms*.

- Confinement in quantum dots can also arise from electrostatic potentials (generated by external electrodes, doping, strain, or impurities).

- CMOS technology can be employed to fabricate silicon quantum dots. Ultra small (L=20 nm, W=20 nm) CMOS transistors behave as single electron quantum dots when operated at cryogenic temperature over a range of −269 °C (4 K) to about −258 °C (15 K). The transistor displays Coulomb blockade due to progressive charging of electrons one by one. The number of electrons confined in the channel is driven by the gate voltage, starting from an occupation of zero electrons, and it can be set to 1 or many.

Viral Assembly

genetically engineered M13 bacteriophage viruses allow preparation of quantum dot biocomposite structures. It had previously been shown that genetically engineered viruses can recognize specific semiconductor surfaces through the method of selection by combinatorial phage display. Additionally, it is known that liquid crystalline structures of wild-type viruses (Fd, M13, and TMV) are adjustable by controlling the solution concentrations, solution ionic strength, and the external magnetic field applied to the solutions. Consequently, the specific recognition properties of the virus can be used to organize inorganic nanocrystals, forming ordered arrays over the length scale defined by liquid crystal formation. Using this information, Lee et al. (2000) were able to create self-assembled, highly oriented, self-supporting films from a phage and ZnS precursor solution. This system allowed them to vary both the length of bacteriophage and the type of inorganic material through genetic modification and selection.

Electrochemical Assembly

Highly ordered arrays of quantum dots may also be self-assembled by electrochemical techniques. A template is created by causing an ionic reaction at an electrolyte-metal interface which results in the spontaneous assembly of nanostructures, including quantum dots, onto the metal which is then used as a mask for mesa-etching these nanostructures on a chosen substrate.

Bulk-manufacture

Quantum dot manufacturing relies on a process called "high temperature dual injection" which has been scaled by multiple companies for commercial applications that require large quantities (hundreds of kilograms to tonnes) of quantum dots. This reproducible production method can be applied to a wide range of quantum dot sizes and compositions.

The bonding in certain cadmium-free quantum dots, such as III-V-based quantum dots, is more covalent than that in II-VI materials, therefore it is more difficult to separate nanoparticle nucleation and growth via a high temperature dual injection synthesis. An alternative method of quantum dot synthesis, the "molecular seeding" process, provides a reproducible route to the production of high quality quantum dots in large volumes. The process utilises identical molecules of a molecular cluster compound as the nucleation sites for nanoparticle growth, thus avoiding the need for a high temperature injection step. Particle growth is maintained by the periodic addition of precursors at moderate temperatures until the desired particle size is reached. The molecular seeding process is not limited to the production of cadmium-free quantum dots; for example, the process can be used to synthesise kilogram batches of high quality II-VI quantum dots in just a few hours.

Another approach for the mass production of colloidal quantum dots can be seen in the transfer of the well-known hot-injection methodology for the synthesis to a technical continuous flow system. The batch-to-batch variations arising from the needs during the mentioned methodology can be overcome by utilizing technical components for mixing and growth as well as transport and temperature adjustments. For the production of CdSe based semiconductor nanoparticles this method has been investigated and tuned to production amounts of kg per month. Since the use of technical components allows for easy interchange in regards of maximum through-put and size, it can be further enhanced to tens or even hundreds of kilograms.

In 2011 a consortium of U.S. and Dutch companies reported a "milestone" in high volume quantum dot manufacturing by applying the traditional high temperature dual injection method to a flow system.

On January 23, 2013 Dow entered into an exclusive licensing agreement with UK-based Nanoco for the use of their low-temperature molecular seeding method for bulk manufacture of cadmium-free quantum dots for electronic displays, and on September 24, 2014 Dow commenced work on the production facility in South Korea capable of producing sufficient quantum dots for "millions of cadmium-free televisions and other devices, such as tablets". Mass production is due to commence in mid-2015. On 24 March 2015 Dow announced a partnership deal with LG Electronics to develop the use of cadmium free quantum dots in displays.

Heavy Metal-free Quantum Dots

In many regions of the world there is now a restriction or ban on the use of heavy metals in many household goods, which means that most cadmium based quantum dots are unusable for consumer-goods applications.

For commercial viability, a range of restricted, heavy metal-free quantum dots has been developed showing bright emissions in the visible and near infra-red region of the spectrum and have similar optical properties to those of CdSe quantum dots. Among these systems are InP/ZnS and CuInS/ZnS, for example.

Peptides are being researched as potential quantum dot material. Since peptides occur naturally in all organisms, such dots would likely be nontoxic and easily biodegraded.

Safety

Toxicity

Some quantum dots pose risks to human health and the environment under certain conditions. Notably, the studies on quantum dots toxicity are focused on cadmium containing particles and has yet to be demonstrated in animal models after physiologically relevant dosing. In vitro studies, based on cell cultures, on quantum dots (QD) toxicity suggests that their toxicity may derive from multiple factors including its physicochemical characteristics (size, shape, composition, surface functional groups, and surface charges) and environment. Assessing their potential toxicity is complex as these factors include properties such as QD size, charge, concentration, chemical composition, capping ligands, and also on their oxidative, mechanical and photolytic stability.

Many studies have focused on the mechanism of QD cytotoxicity using model cell cultures. It has been demonstrated that after exposure to ultraviolet radiation or oxidized by air CdSe QDs release free cadmium ions causing cell death. Group II-VI QDs also have been reported to induce the formation of reactive oxygen species after exposure to light, which in turn can damage cellular components such as proteins, lipids and DNA. Some studies have also demonstrated that addition of a ZnS shell inhibit the process of reactive oxygen species in CdSe QDs. Another aspect of QD toxicity is the process of their size dependent intracellular pathways that concentrate these particles in cellular organelles that are inaccessible by metal ions, which may result in unique patterns of cytotoxicity compared to their constituent metal ions. The reports of QD localization in the cell nucleus present additional modes of toxicity because they may induce DNA mutation, which in turn will propagate through future generation of cells causing diseases.

Although concentration of QDs in certain organelles have been reported in in vivo studies using animal models, interestingly, no alterations in animal behavior, weight, hematological markers or organ damage has been found through either histological or biochemical analysis. These finding have led scientists to believe that intracellular dose is the most important deterring factor for QD toxicity. Therefore, factors determining the QD endocytosis that determine the effective intracellular concentration, such as QD size, shape and surface chemistry determine their toxicity. Excretion of QDs through urine in animal models also have demonstrated via injecting radio-labeled ZnS capped CdSe QDs where the ligand shell was labelled with ^{99m}Tc. Though multiple other studies have concluded retention of QDs in cellular levels, exocytosis of QDs is still poorly studied in the literature.

While significant research efforts have broadened the understanding of toxicity of QDs, there are large discrepancies in the literature and questions still remains to be answered. Diversity of this class material as compared to normal chemical substances makes the assessment of their toxicity very challenging. As their toxicity may also be dynamic depending on the environmental factors such as pH level, light exposure and cell type, traditional methods of assessing toxicity of chemicals such as LD_{50} are not applicable for QDs. Therefore, researchers are focusing on introducing novel approaches and adapting existing methods to include this unique class of materials. Furthermore, novel strategies to engineer safer QDs are still under exploration by the scientific community. A recent novelty in the field is the discovery of carbon quantum dots, a new generation of optically-active nanoparticles potentially capable of replacing semiconductor QDs, but with the advantage of much lower toxicity.

Optical Properties

In semiconductors, light absorption generally leads to an electron being excited from the valence to the conduction band, leaving behind a hole. The electron and the hole can bind to each other to form an exciton. When this exciton recombines (i.e. the electron resumes its ground state), the exciton's energy can be emitted as light. This is called Fluorescence. In a simplified model, the energy of the emitted photon can be understood as the sum of the band gap energy between the highest occupied level and the lowest unoccupied energy level, the confinement energies of the hole and the excited electron, and the bound energy of the exciton (the electron-hole pair):

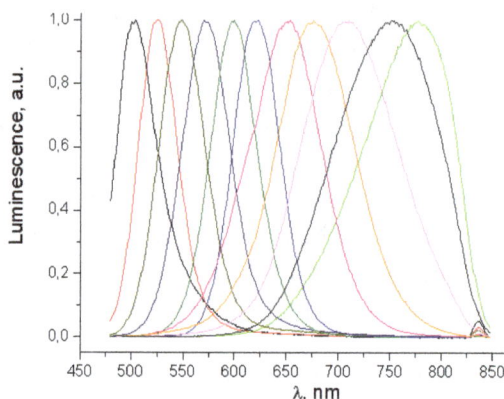

Fluorescence spectra of CdTe quantum dots of various sizes. Different sized quantum dots emit different color light due to quantum confinement.

As the confinement energy depends on the quantum dot's size, both absorption onset and fluorescence emission can be tuned by changing the size of the quantum dot during its synthesis. The larger the dot, the redder (lower energy) its absorption onset and fluorescence spectrum. Conversely, smaller dots absorb and emit bluer (higher energy) light. Recent articles in *Nanotechnology* and in other journals have begun to suggest that the shape of the quantum dot may be a factor in the coloration as well, but as yet not enough information is available. Furthermore, it was shown that the lifetime of fluorescence is determined by the size of the quantum dot. Larger dots have more closely spaced energy levels in which the electron-hole pair can be trapped. Therefore, electron-hole pairs in larger dots live longer causing larger dots to show a longer lifetime.

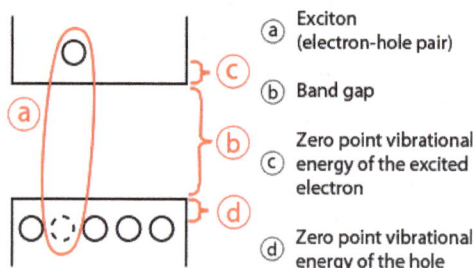

To improve fluorescence quantum yield, quantum dots can be made with "shells" of a larger bandgap semiconductor material around them. The improvement is suggested to be due to the reduced access of electron and hole to non-radiative surface recombination pathways in some cases, but also due to reduced auger recombination in others.

Potential Applications

Quantum dots are particularly promising for optical applications due to their high extinction coefficient. They operate like a single electron transistor and show the Coulomb blockade effect. Quantum dots have also been suggested as implementations of qubits for quantum information processing.

Tuning the size of quantum dots is attractive for many potential applications. For instance, larger quantum dots have a greater spectrum-shift towards red compared to smaller dots, and exhibit less pronounced quantum properties. Conversely, the smaller particles allow one to take advantage of more subtle quantum effects.

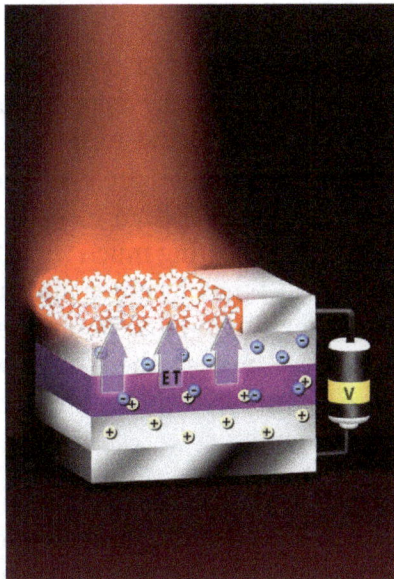

A device that produces visible light, through energy transfer from thin layers of quantum wells to crystals above the layers.

Being zero-dimensional, quantum dots have a sharper density of states than higher-dimensional structures. As a result, they have superior transport and optical properties. They have potential uses in diode lasers, amplifiers, and biological sensors. Quantum dots may be excited within a locally enhanced electromagnetic field produced by gold nanoparticles, which can then be observed from the surface plasmon resonance in the photoluminescent excitation spectrum of (CdSe)ZnS nanocrystals. High-quality quantum dots are well suited for optical encoding and multiplexing applications due to their broad excitation profiles and narrow/symmetric emission spectra. The new generations of quantum dots have far-reaching potential for the study of intracellular processes at the single-molecule level, high-resolution cellular imaging, long-term in vivo observation of cell trafficking, tumor targeting, and diagnostics.

CdSe nanocrystals are efficient triplet photosensitizers. Laser excitation of small CdSe nanoparticles enables the extraction of the excited state energy from the Quantum Dots into bulk solution, thus opening the door to a wide range of potential applications such as photodynamic therapy, photovoltaic devices, molecular electronics, and catalysis.

Computing

Quantum dot technology is potentially relevant to solid-state quantum computation. By applying small voltages to the leads, current through the quantum dot can be controlled and thereby precise measurements of the spin and other properties therein can be made. With several entangled quantum dots, or qubits, plus a way of performing operations, quantum calculations and the computers that would perform them might be possible.

Biology

In modern biological analysis, various kinds of organic dyes are used. However, as technology advances, greater flexibility in these dyes is sought. To this end, quantum dots have quickly filled in the role, being found to be superior to traditional organic dyes on several counts, one of the most immediately obvious being brightness (owing to the high extinction coefficient combined with a comparable quantum yield to fluorescent dyes) as well as their stability (allowing much less photobleaching). It has been estimated that quantum dots are 20 times brighter and 100 times more stable than traditional fluorescent reporters. For single-particle tracking, the irregular blinking of quantum dots is a minor drawback. However, there have been groups which have developed quantum dots which are essentially nonblinking and demonstrated their utility in single molecule tracking experiments.

The use of quantum dots for highly sensitive cellular imaging has seen major advances. The improved photostability of quantum dots, for example, allows the acquisition of many consecutive focal-plane images that can be reconstructed into a high-resolution three-dimensional image. Another application that takes advantage of the extraordinary photostability of quantum dot probes is the real-time tracking of molecules and cells over extended periods of time. Antibodies, streptavidin, peptides, DNA, nucleic acid aptamers, or small-molecule ligands can be used to target quantum dots to specific proteins on cells. Researchers were able to observe quantum dots in lymph nodes of mice for more than 4 months.

Semiconductor quantum dots have also been employed for in vitro imaging of pre-labeled cells. The ability to image single-cell migration in real time is expected to be important to several research areas such as embryogenesis, cancer metastasis, stem cell therapeutics, and lymphocyte immunology.

One application of quantum dots in biology is as donor fluorophores in Förster resonance energy transfer, where the large extinction coefficient and spectral purity of these fluorophores make them superior to molecular fluorophores It is also worth noting that the broad absorbance of QDs allows selective excitation of the QD donor and a minimum excitation of a dye acceptor in FRET-based studies. The applicability of the FRET model, which assumes that the Quantum Dot can be approximated as a point dipole, has recently been demonstrated

The use of quantum dots for tumor targeting under in vivo conditions employ two targeting schemes: active targeting and passive targeting. In the case of active targeting, quantum dots are functionalized with tumor-specific binding sites to selectively bind to tumor cells. Passive targeting uses the enhanced permeation and retention of tumor cells for the delivery of quantum dot probes. Fast-growing tumor cells typically have more permeable membranes than healthy cells, allowing the leakage of small nanoparticles into the cell body. Moreover, tumor cells lack an effective lymphatic drainage system, which leads to subsequent nanoparticle-accumulation.

Quantum dot probes exhibit in vivo toxicity. For example, CdSe nanocrystals are highly toxic to cultured cells under UV illumination, because the particles dissolve, in a process known as photolysis, to release toxic cadmium ions into the culture medium. In the absence of UV irradiation, however, quantum dots with a stable polymer coating have been found to be essentially nontoxic. Hydrogel encapsulation of quantum dots allows for quantum dots to be introduced into a stable aqueous solution, reducing the possibility of cadmium leakage.Then again, only little is known about the excretion process of quantum dots from living organisms.

In another potential application, quantum dots are being investigated as the inorganic fluorophore for intra-operative detection of tumors using fluorescence spectroscopy.

Delivery of undamaged quantum dots to the cell cytoplasm has been a challenge with existing techniques. Vector-based methods have resulted in aggregation and endosomal sequestration of quantum dots while electroporation can damage the semi-conducting particles and aggregate delivered dots in the cytosol. Via cell squeezing, quantum dots can be efficiently delivered without inducing aggregation, trapping material in endosomes, or significant loss of cell viability. Moreover, it has shown that individual quantum dots delivered by this approach are detectable in the cell cytosol, thus illustrating the potential of this technique for single molecule tracking studies.

Photovoltaic Devices

The tunable absorption spectrum and high extinction coefficients of quantum dots make them attractive for light harvesting technologies such as photovoltaics. Quantum dots may be able to increase the efficiency and reduce the cost of today's typical silicon photovoltaic cells. According to an experimental proof from 2004, quantum dots of lead selenide can produce more than one exciton from one high energy photon via the process of carrier multiplication or multiple exciton generation (MEG). This compares favorably to today's photovoltaic cells which can only manage one exciton per high-energy photon, with high kinetic energy carriers losing their energy as heat. Quantum dot photovoltaics would theoretically be cheaper to manufacture, as they can be made "using simple chemical reactions."

Quantum Dot Only Solar Cells

Aromatic self-assembled monolayers (SAMs) (e.g. 4-nitrobenzoic acid) can be used to improve the band alignment at electrodes for better efficiencies. This technique has provided a record power conversion efficiency (PCE) of 10.7%. The SAM is positioned between ZnO-PbS colloidal quantum dot (CQD) film junction to modify band alignment via the dipole moment of the constituent SAM molecule, and the band tuning may be modified via the density, dipole and the orientation of the SAM molecule.

Quantum Dot in Hybrid Solar Cells

Colloidal quantum dots are also used in inorganic/organic hybrid solar cells. These solar cells are attractive because of potentially their low-cost fabrication and relatively high efficiency. Incorporation of metal oxides, such as ZnO, TiO_2, and Nb_2O_5 nanomaterials into organic photovoltaics have been commercialized using full roll-to-roll processing. A 13.2% power conversion efficiency is claimed in Si nanowire/PEDOT:PSS hybrid solar cells.

Quantum Dot with Nanowire in Solar Cells

Another potential use involves capped single-crystal ZnO nanowires with CdSe quantum dots, immersed in mercaptopropionic acid as hole transport medium in order to obtain a QD-sensitized solar cell. The morphology of the nanowires allowed the electrons to have a direct pathway to the photoanode. This form of solar cell exhibits 50-60% internal quantum efficiencies.

Nanowires with quantum dot coatings on silicon nanowires (SiNW) and carbon quantum dots. The use of SiNWs instead of planar silicon enhances the antiflection properties of Si. The SiNW exhibits a light-trapping effect due to light trapping in the SiNW. This use of SiNWs in conjunction with carbon quantum dots resulted in a solar cell that reached 9.10% PCE.

Graphene quantum dots have also been blended with organic electronic materials to improve efficiency and lower cost in photovoltaic devices and organic light emitting diodes (OLEDs) in compared to graphene sheets. These graphene quantum dots were functionalized with organic ligands that experience photoluminescence from UV-Vis absorption.

Light Emitting Devices

Several methods are proposed for using quantum dots to improve existing light-emitting diode (LED) design, including "Quantum Dot Light Emitting Diode" (QD-LED) displays and "Quantum Dot White Light Emitting Diode" (QD-WLED) displays. Because Quantum dots naturally produce monochromatic light, they can be more efficient than light sources which must be color filtered. QD-LEDs can be fabricated on a silicon substrate, which allows them to be integrated onto standard silicon-based integrated circuits or microelectromechanical systems. Quantum dots are valued for displays, because they emit light in very specific gaussian distributions. This can result in a display with visibly more accurate colors. A conventional color liquid crystal display (LCD) is usually backlit by fluorescent lamps (CCFLs) or conventional white LEDs that are color filtered to produce red, green, and blue pixels. An improvement is using conventional blue-emitting LEDs as the light sources and converting part of the emitted light into *pure* green and red light by the appropriate quantum dots placed in front of the blue LED or using a quantum dot infused diffuser sheet in the backlight optical stack. This type of white light as the backlight of an LCD panel allows for the best color gamut at lower cost than a RGB LED combination using three LEDs.

The ability of QDs to precisely convert and tune a spectrum makes them attractive for LCD displays. Previous LCD displays can waste energy converting red-green poor, blue-yellow rich white light into a more balanced lighting. By using QDs, only the necessary colors for ideal images are contained in the screen. The result is a screen that is brighter, clearer, and more energy-efficient. The first commercial application of quantum dots was the Sony XBR X900A series of flat panel televisions released in 2013.

In June 2006, QD Vision announced technical success in making a proof-of-concept quantum dot display and show a bright emission in the visible and near infra-red region of the spectrum. A QD-LED integrated at a scanning microscopy tip was used to demonstrate fluorescence near-field scanning optical microscopy (NSOM) imaging.

Photodetector Devices

Quantum dot photodetectors (QDPs) can be fabricated either via solution-processing, or from conventional single-crystalline semiconductors. Conventional single-crystalline semiconductor QDPs are precluded from integration with flexible organic electronics due to the incompatibility of their growth conditions with the process windows required by organic semiconductors. On the other hand, solution-processed QDPs can be readily integrated with an almost infinite variety of substrates, and also postprocessed atop other integrated circuits. Such colloidal QDPs have potential applications in surveillance, machine vision, industrial inspection, spectroscopy, and fluorescent biomedical imaging.

Photocatalysts

Quantum dots also function as photocatalysts for the light driven chemical conversion of water into hydrogen as a pathway to solar fuel. In photocatalysis, electron hole pairs formed in the dot under band gap excitation drive redox reactions in the surrounding liquid. Generally, the photocatalytic activity of the dots is related to the particle size and its degree of quantum confinement. This is because the band gap determines the chemical energy that is stored in the dot in the excited state. An obstacle for the use of quantum dots in photocatalysis is the presence of surfactants on the surface of the dots. These surfactants (or ligands) interfere with the chemical reactivity of the dots by slowing down mass transfer and electron transfer processes. Also, quantum dots made of metal chalcogenides are chemically unstable under oxidizing conditions and undergo photo corrosion reactions.

Theory

Quantum dots are theoretically described as a point like, or a zero dimensional (0D) entity. Most of their properties depend on the dimensions, shape and materials of which QDs are made. Generally QDs present different thermodynamic properties from the bulk materials of which they are made. One of these effects is the Melting-point depression. Optical properties of spherical metallic QDs are well described by the Mie scattering theory.

Quantum Confinement in Semiconductors

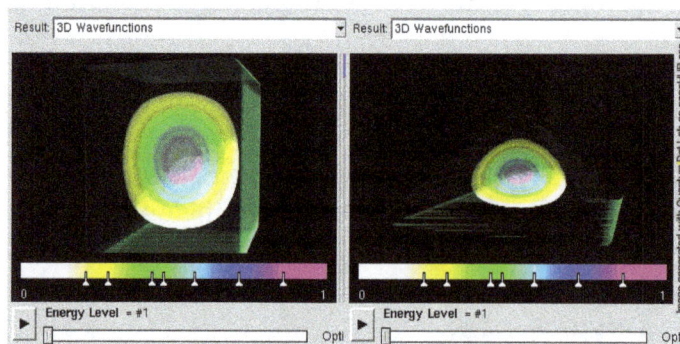

3D confined electron wave functions in a quantum dot. Here, rectangular and triangular-shaped quantum dots are shown. Energy states in rectangular dots are more s-type and p-type. However, in a triangular dot the wave functions are mixed due to confinement symmetry. (Click for animation)

In a semiconductor crystallite whose size is smaller than twice the size of its exciton Bohr radius, the excitons are squeezed, leading to quantum confinement. The energy levels can then be predicted using the particle in a box model in which the energies of states depends on the length of the box. Comparing the quantum dots size to the Bohr radius of the electron and hole wave functions, 3 regimes can be defined. A 'strong confinement regime' is defined as the quantum dots radius being smaller than both electron and hole Bohr radius, 'weak confinement' is given when the quantum dot is larger than both. For semiconductors in which electron and hole radii are markedly different, an 'intermediate confinement regime' exists, where the quantum dot's radius is larger than the Bohr radius of one charge carrier (typically the hole), but not the other charge carrier.

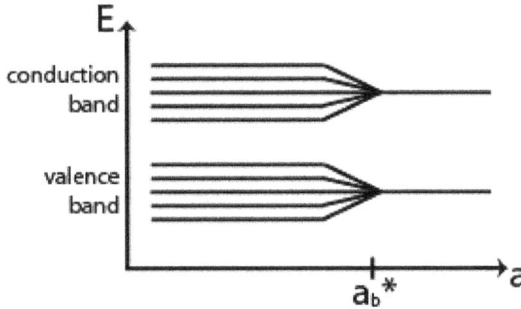

Splitting of energy levels for small quantum dots due to the quantum confinement effect. The horizontal axis is the radius, or the size, of the quantum dots and a_b^* is the Exciton Bohr radius.

Band gap energy

The band gap can become smaller in the strong confinement regime as the energy levels split up. The Exciton Bohr radius can be expressed as:

where a_b is the Bohr radius=0.053 nm, m is the mass, μ is the reduced mass, and ε_r is the size-dependent dielectric constant (Relative permittivity). This results in the increase in the total emission energy (the sum of the energy levels in the smaller band gaps in the strong confinement regime is larger than the energy levels in the band gaps of the original levels in the weak confinement regime) and the emission at various wavelengths. If the size distribution of QDs is not enough peaked, the convolution of multiple emission wavelenghts is observed as a continuous spectra.

Confinement energy

The exciton entity can be modeled using the particle in the box. The electron and the hole can be seen as hydrogen in the Bohr model with the hydrogen nucleus replaced by the hole of positive charge and negative electron mass. Then the energy levels of the exciton can be represented as the solution to the particle in a box at the ground level (n = 1) with the mass replaced by the reduced mass. Thus by varying the size of the quantum dot, the confinement energy of the exciton can be controlled.

Bound exciton energy

There is Coulomb attraction between the negatively charged electron and the positively charged hole. The negative energy involved in the attraction is proportional to Rydberg's energy and inversely proportional to square of the size-dependent dielectric constant of the semiconductor. When the size of the semiconductor crystal is smaller than the Exciton Bohr radius, the Coulomb interaction must be modified to fit the situation.

Therefore, the sum of these energies can be represented as:

$$E_{confinement} = \frac{\hbar^2 \pi^2}{2a^2}\left(\frac{1}{m_e}+\frac{1}{m_h}\right) = \frac{\hbar^2 \pi^2}{2\mu a^2} E_{exciton} \qquad m \qquad R \qquad R\,E \quad E_{bandgap} \quad E_{confinement}$$

$$E_{exciton} = -\frac{1}{\epsilon_r^2}\frac{\mu}{m_e}R_y = -R_y^*.$$

$$E = E_{bandgap} + E_{confinement} + E_{exciton}$$

$$= E_{bandgap} + \frac{\hbar^2 \pi^2}{2\mu a^2} - R_y^*$$

where μ is the reduced mass, a is the radius, m_e is the free electron mass, m_h is the hole mass, and ε_r is the size-dependent dielectric constant.

Although the above equations were derived using simplifying assumptions, they imply that the electronic transitions of the quantum dots will depend on their size. These quantum confinement effects are apparent only below the critical size. Larger particles do not exhibit this effect. This effect of quantum confinement on the quantum dots has been repeatedly verified experimentally and is a key feature of many emerging electronic structures.

The Coulomb interaction between confined carriers can also be studied by numerical means when results unconstrained by asymptotic approximations are pursued.

Besides confinement in all three dimensions (i.e., a quantum dot), other quantum confined semiconductors include:

- Quantum wires, which confine electrons or holes in two spatial dimensions and allow free propagation in the third.

- Quantum wells, which confine electrons or holes in one dimension and allow free propagation in two dimensions.

Models

A variety of theoretical frameworks exist to model optical, electronic, and structural properties of quantum dots. These may be broadly divided into quantum mechanical, semiclassical, and classical.

Quantum Mechanics

Quantum mechanical models and simulations of quantum dots often involve the interaction of electrons with a pseudopotential or random matrix.

Semiclassical

Semiclassical models of quantum dots frequently incorporate a chemical potential. For example, The thermodynamic chemical potential of an N-particle system is given by

$$\mu(N) = E(N) - E(N-1)$$

whose energy terms may be obtained as solutions of the Schrödinger equation. The definition of capacitance,

$$\frac{1}{C} \equiv \frac{\Delta V}{\Delta Q},,$$

with the potential difference

$$\Delta V = \frac{\Delta \mu}{e} = \frac{\mu(N + \Delta N) - \mu(N)}{e}$$

may be applied to a quantum dot with the addition or removal of individual electrons,

$\Delta N = 1$ and $\Delta Q = e$.

Then

$$C(N) = \frac{e^2}{\mu(N+1) - \mu(N)} = \frac{e^2}{I(N) - A(N)}$$

is the "quantum capacitance" of a quantum dot, where we denoted by $I(N)$ the ionization potential and by $A(N)$ the electron affinity of the N-particle system.

Classical Mechanics

Classical models of electrostatic properties of electrons in quantum dots are similar in nature to the Thomson problem of optimally distributing electrons on a unit sphere.

The classical electrostatic treatment of electrons confined to spherical quantum dots is similar to their treatment in the Thomson, or plum pudding model, of the atom.

The classical treatment of both two-dimensional and three-dimensional quantum dots exhibit electron shell-filling behavior. A "periodic table of classical artificial atoms" has been described for two-dimensional quantum dots. As well, several connections have been reported between the three-dimensional Thomson problem and electron shell-filling patterns found in naturally-occurring atoms found throughout the periodic table. This latter work originated in classical electrostatic modeling of electrons in a spherical quantum dot represented by an ideal dielectric sphere.

Hybrid Material

Hybrid materials are composites consisting of two constituents at the nanometer or molecular level. Commonly one of these compounds is inorganic and the other one organic in nature. Thus, they differ from traditional composites where the constituents are at the macroscopic (micrometer to millimeter) level. Mixing at the microscopic scale leads to a more homogeneous material that either show characteristics in between the two original phases or even new properties.

Introduction

Hybrid Materials in Nature

Many natural materials consist of inorganic and organic building blocks distributed on the nanoscale. In most cases the inorganic part provides mechanical strength and an overall structure to the natural objects while the organic part delivers bonding between the inorganic building blocks and/or the soft tissue. Typical examples of such materials are bone, or nacre.

Development of Hybrid Materials

The first hybrid materials were the paints made from inorganic and organic components that were used thousands of years ago. Rubber is an example of the use of inorganic materials as fillers for organic polymers. The sol–gel process developed in the 1930s was one of the major driving forces what has become the broad field of inorganic–organic hybrid materials.

Classification

Hybrid materials can be classified based on the possible interactions connecting the inorganic and organic species. *Class I* hybrid materials are those that show weak interactions between the two phases, such as van der Waals, hydrogen bonding or weak electrostatic interactions. *Class II* hybrid materials are those that show strong chemical interactions between the components such as covalent bonds.

Structural properties can also be used to distinguish between various hybrid materials. An organic moiety containing a functional group that allows the attachment to an inorganic network, e.g. a trialkoxysilane group, can act as a *network modifier* because in the final structure the inorganic network is only modified by the organic group. Phenyltrialkoxysilanes are an example for such compounds; they modify the silica network in the sol–gel process via the reaction of the trialkoxysilane group without supplying additional functional groups intended to undergo further chemical reactions to the material formed. If a reactive functional group is incorporated the system is called a *network functionalizer*. The situation is different if two or three of such anchor groups modify an organic segment; this leads to materials in which the inorganic group is afterwards an integral part of the hybrid network. The latter type of system is known as *network builder*

Blends are formed if no strong chemical interactions exist between the inorganic and organic building blocks. One example for such a material is the combination of inorganic clusters or particles with organic polymers lacking a strong (e.g. covalent) interaction between the components. In this case a material is formed that consists for example of an organic polymer with entrapped discrete inorganic moieties in which, depending on the functionalities of the components, for example weak crosslinking occurs by the entrapped inorganic units through physical interactions or the inorganic components are entrapped in a crosslinked polymer matrix. If an inorganic and an organic network interpenetrate each other without strong chemical interactions, so called interpenetrating networks (IPNs) are formed, which is for example the case if a sol–gel material is formed in presence of an organic polymer or vice versa. Both materials described belong to class I hybrids. Class II hybrids are formed when the discrete inorganic building blocks, e.g. clusters, are covalently bonded to the organic polymers or inorganic and organic polymers are covalently connected with each other.

Distinction Between Nanocomposites and Hybrid Materials

The term nanocomposite is used if the combination of organic and inorganic structural units yield a material with composite properties. That is to say that the original properties of the separate organic and inorganic components are still present in the composite and are unchanged by mixing these materials. However, if a new property emerges from the intimate mixture, then the material becomes a hybrid. A macroscopic example is the mule, which is more suited for hard work than either of its parents, the horse and the donkey. The size of the individual components and the nature of their interaction (covalent, electrostatic, etc.) do not enter into the definition of a hybrid material.

Advantages of Hybrid Materials Over Traditional Composites

- Inorganic clusters or nanoparticles with specific optical, electronic or magnetic properties can be incorporated in organic polymer matrices.

- Contrary to pure solid state inorganic materials that often require a high temperature treatment for their processing, hybrid materials show a more polymer-like handling, either because of their large organic content or because of the formation of crosslinked inorganic networks from small molecular precursors just like in polymerization reactions.

- Light scattering in homogeneous hybrid material can be avoided and therefore optical transparency of the resulting hybrid materials and nanocomposites can be achieved.

Synthesis

Two different approaches can be used for the formation of hybrid materials: Either well-defined preformed building blocks are applied that react with each other to form the final hybrid material in which the precursors still at least partially keep their original integrity or one or both structural units are formed from the precursors that are transformed into a new (network) structure. It is important that the interface between the inorganic and organic materials which has to be tailored to overcome serious problems in the preparation of hybrid materials. Different building blocks and approaches can be used for their preparation and these have to be adapted to bridge the differences of inorganic and organic materials.

Building Block Approach

Building blocks at least partially keep their molecular integrity throughout the material formation, which means that structural units that are present in these sources for materials formation can also be found in the final material. At the same time typical properties of these building blocks usually survive the matrix formation, which is not the case if material precursors are transferred into novel materials. Representative examples of such well-defined building blocks are modified inorganic clusters or nanoparticles with attached reactive organic groups.

Cluster compounds often consist of at least one functional group that allows an interaction with an organic matrix, for example by copolymerization. Depending on the number of groups that can interact, these building blocks are able to modify an organic matrix (one functional group) or form partially or fully crosslinked materials (more than one group). For instance, two reactive groups can lead to the formation of chain structures. If the building blocks contain at least three reactive groups they can be used without additional molecules for the formation of a crosslinked material.

Beside the molecular building blocks mentioned, nanosized building blocks, such as particles or nanorods, can also be used to form nanocomposites. The building block approach has one large advantage compared with the in situ formation of the inorganic or organic entities: because at least one structural unit (the building block) is well-defined and usually does not undergo significant structural changes during the matrix formation, better structure–property predictions are possible. Furthermore, the building blocks can be designed in such a way to give the best performance in the materials' formation, for example good solubility of inorganic compounds in organic monomers by surface groups showing a similar polarity as the monomers.

In recent years many building blocks have been synthesized and used for the preparation of hybrid materials. Chemists can design these compounds on a molecular scale with highly sophisticated methods and the resulting systems are used for the formation of functional hybrid materials. Many future applications, in particular in nanotechnology, focus on a bottom-up approach in which complex structures are hierarchically formed by these small building blocks. This idea is also one of the driving forces of the building block approach in hybrid materials.

In Situ Formation of The Components

The in situ formation of the hybrid materials is based on the chemical transformation of the precursors used throughout materials' preparation. Typically this is the case if organic polymers are formed but also if the sol–gel process is applied to produce the inorganic component. In these cases well-defined discrete molecules are transformed to multidimensional structures, which often show totally different properties from the original precursors. Generally simple, commercially available molecules are applied and the internal structure of the final material is determined by the composition of these precursors but also by the reaction conditions. Therefore control over the latter is a crucial step in this process. Changing one parameter can often lead to two very different materials. If, for example, the inorganic species is a silica derivative formed by the sol–gel process, the change from base to acid catalysis makes a large difference because base catalysis leads to a more particle-like microstructure while acid catalysis leads to a polymer-like microstructure. Hence, the final performance of the derived materials is strongly dependent on their processing and its optimization.

In Situ Formation of Inorganic Materials

Many of the classical inorganic solid state materials are formed using solid precursors and high temperature processes, which are often not compatible with the presence of organic groups because they are decomposed at elevated temperatures. Hence, these high temperature processes are not suitable for the in situ formation of hybrid materials. Reactions that are employed should have more the character of classical covalent bond formation in solutions. One of the most prominent processes which fulfill these demands is the sol–gel process. However, such rather low temperature processes often do not lead to the thermodynamically most stable structure but to kinetic products, which has some implications for the structures obtained. For example low temperature derived inorganic materials are often amorphous or crystallinity is only observed on a very small length scale, i.e. the nanometer range. An example of the latter is the formation of metal nanoparticles in organic or inorganic matrices by reduction of metal salts or organometallic precursors.

Some methods of in situ formation of inorganic materials are:

- Sol-gel process
- Nonhydrolytic sol–gel process
- Sol–gel reactions of non-silicates

Formation of Organic Polymers in Presence of Preformed Inorganic Materials

If the organic polymerization occurs in the presence of an inorganic material to form the hybrid material one has to distinguish between several possibilities to overcome the incompatibility of the two species. The inorganic material can either have no surface functionalization but the bare material surface; it can be modified with nonreactive organic groups (e.g. alkyl chains); or it can contain reactive surface groups such as polymerizable functionalities. Depending on these prerequisites the material can be pretreated, for example a pure inorganic surface can be treated with surfactants or silane coupling agents to make it compatible with the organic monomers, or functional monomers can be added that react with the surface of the inorganic material. If the inorganic component has nonreactive organic groups attached to its surface and it can be dissolved in a monomer which is subsequently polymerized, the resulting material after the organic polymerization, is a blend. In this case the inorganic component interact only weakly or not at all with the organic polymer; hence, a class I material is formed. Homogeneous materials are only obtained in this case if agglomeration of the inorganic components in the organic environment is prevented. This can be achieved if the interactions between the inorganic components and the monomers are better or at least the same as between the inorganic components. However, if no strong chemical interactions are formed, the long-term stability of a once homogeneous material is questionable because of diffusion effects in the resulting hybrid material. The stronger the respective interaction between the components, the more stable is the final material. The strongest interaction is achieved if class II materials are formed, for example with covalent interactions.

Hybrid Materials by Simultaneous Formation of Both Components

Simultaneous formation of the inorganic and organic polymers can result in the most homogeneous type of interpenetrating networks. Usually the precursors for the sol–gel process are mixed with monomers for the organic polymerization and both processes are carried out at the same time with or without solvent. Applying this method, three processes are competing with each other:

(a) the kinetics of the hydrolysis and condensation forming the inorganic phase, (b) the kinetics of the polymerization of the organic phase, and (c) the thermodynamics of the phase separation between the two phases.

By tailoring the kinetics of the two polymerizations in such a way that they occur simultaneously and rapidly enough, phase separation is avoided or minimized. Additional parameters such as attractive interactions between the two moieties, as described above can also be used to avoid phase separation.

One problem that also arises from the simultaneous formation of both networks is the sensitivity of many organic polymerization processes for sol–gel conditions or the composition of the materials formed. Ionic polymerizations, for example, often interact with the precursors or intermediates formed in the sol–gel process. Therefore, they are not usually applied in these reactions.

Applications

- Decorative coatings obtained by the embedding of organic dyes in hybrid coatings.

- Scratch-resistant coatings with hydrophobic or anti-fogging properties.

- Nanocomposite based devices for electronic and optoelectronic applications including light-emitting diodes, photodiodes, solar cells, gas sensors and field effect transistors.

- Fire retardant materials for construction industry.

- Nanocomposite based dental filling materials.

- Composite electrolyte materials for applications such as solid-state lithium batteries or supercapacitors.

- Proton conducting membranes used in fuel cells.

- Antistatic / anti-reflection coatings

- Corrosion protection

- Porous hybrid materials

Metamaterial

A metamaterial is a material engineered to have a property that is not found in nature. They are made from assemblies of multiple elements fashioned from composite materials such as metals or plastics. The materials are usually arranged in repeating patterns, at scales that are smaller than the wavelengths of the phenomena they influence. Metamaterials derive their properties not from the properties of the base materials, but from their newly designed structures. Their precise shape, geometry, size, orientation and arrangement gives them their smart properties capable of manipulating electromagnetic waves: by blocking, absorbing, enhancing, or bending waves, to achieve benefits that go beyond what is possible with conventional materials.

Appropriately designed metamaterials can affect waves of electromagnetic radiation or sound in a manner not observed in bulk materials. Those that exhibit a negative index of refraction for particular wavelengths have attracted significant research. These materials are known as negative-index metamaterials.

Negative-index metamaterial array configuration, which was constructed of copper split-ring resonators and wires mounted on interlocking sheets of fiberglass circuit board. The total array consists of 3 by 20×20 unit cells with overall dimensions of 10 mm × 100 mm × 100 mm (0.39 in × 3.94 in × 3.94 in).

Potential applications of metamaterials are diverse and include optical filters, medical devices, remote aerospace applications, sensor detection and infrastructure monitoring, smart solar power management, crowd control, radomes, high-frequency battlefield communication and lenses for high-gain antennas, improving ultrasonic sensors, and even shielding structures from earthquakes. Metamaterials offer the potential to create superlenses. Such a lens could allow imaging below the diffraction limit that is the minimum resolution that can be achieved by conventional glass lenses. A form of 'invisibility' was demonstrated using gradient-index materials. Acoustic and seismic metamaterials are also research areas.

Metamaterial research is interdisciplinary and involves such fields as electrical engineering, electromagnetics, classical optics, solid state physics, microwave and antennae engineering, optoelectronics, material sciences, nanoscience and semiconductor engineering.

History

Explorations of artificial materials for manipulating electromagnetic waves began at the end of the 19th century. Some of the earliest structures that may be considered metamaterials were studied by Jagadish Chandra Bose, who in 1898 researched substances with chiral properties. Karl Ferdinand Lindman studied wave interaction with metallic helices as artificial chiral media in the early twentieth century.

Winston E. Kock developed materials that had similar characteristics to metamaterials in the late 1940s. In the 1950s and 1960s, artificial dielectrics were studied for lightweight microwave antennas. Microwave radar absorbers were researched in the 1980s and 1990s as applications for artificial chiral media.

Negative-index materials were first described theoretically by Victor Veselago in 1967.

He proved that such materials could transmit light. He showed that the phase velocity could be made anti-parallel to the direction of Poynting vector. This is contrary to wave propagation in naturally-occurring materials.

John Pendry was the first to identify a practical way to make a left-handed metamaterial, a material in which the right-hand rule is not followed. Such a material allows an electromagnetic wave to convey energy (have a group velocity) against its phase velocity. Pendry's idea was that metallic wires aligned along the direction of a wave could provide negative permittivity (dielectric function $\varepsilon < 0$). Natural materials (such as ferroelectrics) display negative permittivity; the challenge was achieving negative permeability ($\mu < 0$). In 1999 Pendry demonstrated that a split ring (C shape) with its axis placed along the direction of wave propagation could do so. In the same paper, he showed that a periodic array of wires and rings could give rise to a negative refractive index. Pendry also proposed a related negative-permeability design, the Swiss roll.

In 2000, Smith et al. reported the experimental demonstration of functioning electromagnetic metamaterials by horizontally stacking, periodically, split-ring resonators and thin wire structures. A method was provided in 2002 to realize negative-index metamaterials using artificial lumped-element loaded transmission lines in microstrip technology. In 2003, complex (both real and imaginary parts of) negative refractive index and imaging by flat lens using left handed metamaterials were demonstrated. By 2007, experiments that involved negative refractive index had been conducted by many groups. At microwave frequencies, the first, imperfect invisibility cloak was realized in 2006.

Electromagnetic Metamaterials

An electromagnetic metamaterial affects electromagnetic waves that impinge on or interact with its structural features, which are smaller than the wavelength. To behave as a homogeneous material accurately described by an effective refractive index, its features must be much smaller than the wavelength.

For microwave radiation, the features are on the order of millimeters. Microwave frequency metamaterials are usually constructed as arrays of electrically conductive elements (such as loops of wire) that have suitable inductive and capacitive characteristics. One microwave metamaterial uses the split-ring resonator.

Photonic metamaterials, nanometer scale, manipulate light at optical frequencies. To date, subwavelength structures have shown only a few, questionable, results at visible wavelengths. Photonic crystals and frequency-selective surfaces such as diffraction gratings, dielectric mirrors and optical coatings exhibit similarities to subwavelength structured metamaterials. However, these are usually considered distinct from subwavelength structures, as their features are structured for the wavelength at which they function and thus cannot be approximated as a homogeneous material. However, material structures such as photonic crystals are effective in the visible light spectrum. The middle of the visible spectrum has a wavelength of approximately 560 nm (for sunlight). Photonic crystal structures are generally half this size or smaller, that is <280 nm.

Plasmonic metamaterials utilize surface plasmons, which are packets of electrical charge that collectively oscillate at the surfaces of metals at optical frequencies.

Frequency selective surfaces (FSS) can exhibit subwavelength characteristics and are known variously as artificial magnetic conductors (AMC) or High Impedance Surfaces (HIS). FSS display inductive and capacitive characteristics that are directly related to their subwavelength structure.

Negative Refractive Index

Almost all materials encountered in optics, such as glass or water, have positive values for both permittivity ε and permeability μ. However, metals such as silver and gold have negative permittivity at shorter wavelengths. A material such as a surface plasmon that has either (but not both) ε or μ negative is often opaque to electromagnetic radiation. However, anisotropic materials with only negative permittivity can produce negative refraction due to chirality.

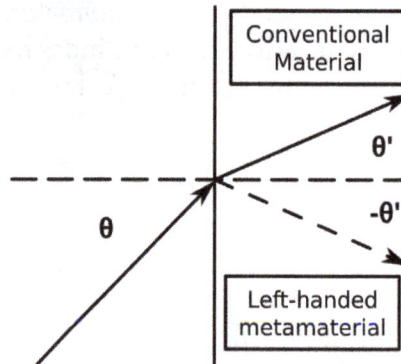

A comparison of refraction in a left-handed metamaterial to that in a normal material

Although the optical properties of a transparent material are fully specified by the parameters εr and μr, refractive index n is often used in practice, which can be determined from $n = \pm\sqrt{\epsilon_r \mu_r}..$ All known non-metamaterial transparent materials possess positive εr and μr. By convention the positive square root is used for n.

However, some engineered metamaterials have $\varepsilon r < 0$ and $\mu r < 0$. Because the product $\varepsilon r \mu r$ is positive, n is real. Under such circumstances, it is necessary to take the negative square root for n.

Video representing negative refraction of light at uniform planar interface.

The foregoing considerations are simplistic for actual materials, which must have complex-valued εr and μr. The real parts of both εr and μr do not have to be negative for a passive material to display negative refraction. Metamaterials with negative n have numerous interesting properties:

- Snell's law ($n_1\sin\theta_1 = n_2\sin\theta_2$), but as n_2 is negative, the rays are refracted on the *same* side of the normal on entering the material.

- Cherenkov radiation points the other way.

- The time-averaged Poynting vector is antiparallel to phase velocity. However, for waves (energy) to propagate, a $-\mu$ must be paired with a $-\varepsilon$ in order to satisfy the wave number dependence on the material parameters $kc = \omega\sqrt{\mu\epsilon}$.

Negative index of refraction derives mathematically from the vector triplet E, H and k.

For plane waves propagating in electromagnetic metamaterials, the electric field, magnetic field and wave vector follow a left-hand rule, the reverse of the behavior of conventional optical materials.

Classification

Electromagnetic metamaterials are divided into different classes, as follows:

Negative Index

In negative-index metamaterials (NIM), both permittivity and permeability are negative, resulting in a negative index of refraction. These are also known as double negative metamaterials or double negative materials (DNG). Other terms for NIMs include "left-handed media", "media with a negative refractive index", and "backward-wave media".

In optical materials, if both permittivity ε and permeability μ are positive, wave propagation travels in the forward direction. If both ε and μ are negative, a backward wave is produced. If ε and μ have different polarities, waves do not propagate.

Mathematically, quadrant II and quadrant IV have coordinates (0,0) in a coordinate plane where ε is the horizontal axis, and μ is the vertical axis.

To date, only metamaterials exhibit a negative index of refraction.

Single Negative

Single negative (SNG) metamaterials have either negative relative permittivity (εr) or negative relative permeability (μr), but not both. They act as metamaterials when combined with a different, complementary SNG, jointly acting as a DNG.

Epsilon negative media (ENG) display a negative εr while μr is positive. Many plasmas exhibit this characteristic. For example, noble metals such as gold or silver are ENG in the infrared and visible spectrums.

Mu-negative media (MNG) display a positive εr and negative μr. Gyrotropic or gyro-magnetic materials exhibit this characteristic. A gyrotropic material is one that has been altered by the presence of a quasistatic magnetic field, enabling a magneto-optic effect. A magneto-optic effect is a phenomenon in which an electromagnetic wave propagates through such a medium. In such a material, left- and right-rotating elliptical polarizations can propagate at different speeds. When light is transmitted through a layer of magneto-optic material, the result is called the Faraday effect: the polarization plane can be rotated, forming a Faraday rotator. The results of such a reflection are known as the magneto-optic Kerr effect (not to be confused with the nonlinear Kerr effect). Two gyrotropic materials with reversed rotation directions of the two principal polarizations are called optical isomers.

Joining a slab of ENG material and slab of MNG material resulted in properties such as resonances, anomalous tunneling, transparency and zero reflection. Like negative-index materials, SNGs are innately dispersive, so their εr, μr and refraction index n, are a function of frequency.

Bandgap

Electromagnetic bandgap metamaterials (EBM) control light propagation. This is accomplished either with photonic crystals (PC) or left-handed materials (LHM). PCs can prohibit light propagation altogether. Both classes can allow light to propagate in specific, designed directions and both can be designed with bandgaps at desired frequencies. The period size of EBGs is an appreciable fraction of the wavelength, creating constructive and destructive interference.

PC are distinguished from sub-wavelength structures, such as tunable metamaterials, because the PC derives its properties from its bandgap characteristics. PCs are sized to match the wavelength of light, versus other metamaterials that expose sub-wavelength structure. Furthermore, PCs function by diffracting light. In contrast, metamaterial does not use diffraction.

PCs have periodic inclusions that inhibit wave propagation due to the inclusions' destructive interference from scattering. The photonic bandgap property of PCs makes them the electromagnetic analog of electronic semi-conductor crystals.

EBGs have the goal of creating high quality, low loss, periodic, dielectric structures. An EBG affects photons in the same way semiconductor materials affect electrons. PCs are the perfect bandgap material, because they allow no light propagation. Each unit of the prescribed periodic structure acts like one atom, albeit of a much larger size.

EBGs are designed to prevent the propagation of an allocated bandwidth of frequencies, for certain arrival angles and polarizations. Various geometries and structures have been proposed to fabricate EBG's special properties. In practice it is impossible to build a flawless EBG device.

EBGs have been manufactured for frequencies ranging from a few gigahertz (GHz) to a few terahertz (THz), radio, microwave and mid-infrared frequency regions. EBG application developments include a transmission line, woodpiles made of square dielectric bars and several different types of low gain antennas.

Double Positive Medium

Double positive mediums (DPS) do occur in nature, such as naturally occurring dielectrics. Permittivity and magnetic permeability are both positive and wave propagation is in the forward direction. Artificial materials have been fabricated which combine DPS, ENG and MNG properties.

Bi-isotropic and Bianisotropic

Categorizing metamaterials into double or single negative, or double positive, normally assumes that the metamaterial has independent electric and magnetic responses described by ε and μ. However, in many cases, the electric field causes magnetic polarization, while the magnetic field induces electrical polarization, known as magnetoelectric coupling. Such media are denoted as bi-isotropic. Media that exhibit magnetoelectric coupling and that are anisotropic (which is the case for many metamaterial structures), are referred to as bi-anisotropic.

Four material parameters are intrinsic to magnetoelectric coupling of bi-isotropic media. They are the electric (E) and magnetic (H) field strengths, and electric (D) and magnetic (B) flux densities. These parameters are ε, μ, κ and χ or permittivity, permeability, strength of chirality, and the Tellegen parameter respectively. In this type of media, material parameters do not vary with changes along a rotated coordinate system of measurements. In this sense they are invariant or scalar.

The intrinsic magnetoelectric parameters, κ and χ, affect the phase of the wave. The effect of the chirality parameter is to split the refractive index. In isotropic media this results in wave propagation only if ε and μ have the same sign. In bi-isotropic media with χ assumed to be zero, and κ a non-zero value, different results appear. Either a backward wave or a forward wave can occur. Alternatively, two forward waves or two backward waves can occur, depending on the strength of the chirality parameter.

In the general case, the constitutive relations for bi-anisotropic materials read $\mathbf{D} = \varepsilon\mathbf{E} + \xi\mathbf{H}$, $\mathbf{B} = \zeta\mathbf{E} + \mu\mathbf{H}$, where ϵ and μ are the permittivity and the permeability tensors, respectively, whereas   and ζ are the two magneto-electric tensors. If the medium is reciprocal, permittivity and permeability are symmetric tensors, and $\xi = -\zeta^T = -i\kappa^T$, where κ is the chiral tensor describing chiral electromagnetic and reciprocal magneto-electric response. The chiral tensor can be expressed as $\kappa = \frac{1}{3}Tr(\kappa)I + N + J$, where $Tr(\kappa)$ is the trace of κ, I is the identity matrix, N is a symmetric trace-free tensor, and J is an antisymmetric tensor. Such decomposition allows us to classify the reciprocal bianisotropic response and we can identify the following three main classes: (i) chiral media ($Tr(\kappa) \neq 0, N \neq 0, J = 0$), (ii) pseudochiral media ($Tr(\kappa) = 0, N \neq 0, J = 0$),), (iii) omega media ($Tr(\kappa) = 0, N = 0, J \neq 0$). Generally the chiral and/or bianisotropic electromagnetic response is a consequence of 3D geometrical chirality: 3D chiral meta-materials are composed by embedding 3D chiral structures in a host medium and they show chirality-related polarization effects such as optical activity and circular dichro-ism. The concept of 2D chirality also exists and a planar object is said to be chiral if it cannot be superposed onto its mirror image unless it is lifted from the plane. On the other hand, bianisotropic response can arise from geometrical achiral structures pos-sessing neither 2D nor 3D intrinsic chirality. Plum *et al.* investigated extrinsic chiral metamaterials where the magneto-electric coupling results from the geometric chiral-ity of the whole structure and the effect is driven by the radiation wave vector contrib-uting to the overall chiral asymmetry (extrinsic electromagnetic chiralilty). Rizza *et al.* suggested 1D chiral metamaterials where the effective chiral tensor is not vanishing if the system is geometrically one-dimensional chiral (the mirror image of the entire structure cannot be superposed onto it by using translations without rotations).

Chiral

Chiral metamaterials are constructed from chiral materials in which the effective parameter k is non-zero. This is a potential source of confusion as the metamaterial literature includes two conflicting uses of the terms *left-* and *right-handed*. The first refers to one of the two circularly polarized waves that are the propagating modes in chiral media. The second relates to the triplet of electric field, magnetic field and Poynting vector that arise in negative refractive index media, which in most cases are not chiral.

Wave propagation properties in chiral metamaterials demonstrate that negative refrac-tion can be realized in metamaterials with a strong chirality and positive ε and μ. This is because the refractive index has distinct values for left and right, given by

$$n = \sqrt{\epsilon_r \mu_r} \pm \kappa$$

It can be seen that a negative index will occur for one polarization if $\kappa > \sqrt{\epsilon r \mu r}$. In this case, it is not necessary that either or both εr and μr be negative for backward wave propagation.

FSS based

Frequency selective surface-based metamaterials block signals in one waveband and pass those at another waveband. They have become an alternative to fixed frequen-cy metamaterials. They allow for optional changes of frequencies in a single medium, rather than the restrictive limitations of a fixed frequency response.

Other Types

Elastic

These metamaterials use different parameters to achieve a negative index of refraction in materials that are not electromagnetic. Furthermore, "a new design for elastic metamaterials that can behave either as liquids or solids over a limited frequency range may enable new applications based on the control of acoustic, elastic and seismic waves." They are also called mechanical metamaterials.

Acoustic

Acoustic metamaterials control, direct and manipulate sound in the form of sonic, infrasonic or ultrasonic waves in gases, liquids and solids. As with electromagnetic waves, sonic waves can exhibit negative refraction.

Control of sound waves is mostly accomplished through the bulk modulus β, mass density ρ and chirality. The bulk modulus and density are analogs of permittivity and permeability in electromagnetic metamaterials. Related to this is the mechanics of sound wave propagation in a lattice structure. Also materials have mass and intrinsic degrees of stiffness. Together, these form a resonant system and the mechanical (sonic) resonance may be excited by appropriate sonic frequencies (for example audible pulses).

Structural

Structural metamaterials provide properties such as crushability and light weight. Using projection micro-stereolithography, microlattices can be created using forms much like trusses and girders. Materials four orders of magnitude stiffer than conventional aerogel, but with the same density have been created. Such materials can withstand a load of at least 160,000 times their own weight by over-constraining the materials.

A ceramic nanotruss metamaterial can be flattened and revert to its original state.

Nonlinear

Metamaterials may be fabricated that include some form of nonlinear media, whose properties change with the power of the incident wave. Nonlinear media are essential for nonlinear optics. Most optical materials have a relatively weak response, meaning that their properties change by only a small amount for large changes in the intensity of the electromagnetic field. The local electromagnetic fields of the inclusions in nonlinear metamaterials can be much larger than the average value of the field. Besides, remarkable nonlinear effects have been predicted and observed if the metamaterial effective dielectric permittivity is very small (epsilon-near-zero media). In addition, exotic properties such as a negative refractive index, create opportunities to tailor the phase matching conditions that must be satisfied in any nonlinear optical structure.

Frequency Bands

Terahertz

Terahertz metamaterials interact at terahertz frequencies, usually defined as 0.1 to 10 THz. Terahertz radiation lies at the far end of the infrared band, just after the end of the microwave band. This corresponds to millimeter and submillimeter wavelengths between the 3 mm (EHF band) and 0.03 mm (long-wavelength edge of far-infrared light).

Photonic

Photonic metamaterial interact with optical frequencies (mid-infrared). The sub-wavelength period distinguishes them from photonic band gap structures.

Tunable

Tunable metamaterials allow arbitrary adjustments to frequency changes in the refractive index. A tunable metamaterial expands beyond the bandwidth limitations in left-handed materials by constructing various types of metamaterials.

Plasmonic

Plasmonic metamaterials exploit surface plasmons, which are produced from the interaction of light with metal-dielectrics. Under specific conditions, the incident light couples with the surface plasmons to create self-sustaining, propagating electromagnetic waves known as surface plasmon polaritons.

Applications

Metamaterials are under consideration for many applications. Metamaterial antennas are commercially available.

In 2007, one researcher stated that for metamaterial applications to be realized, energy loss must be reduced, materials must be extended into three-dimensional isotropic materials and production techniques must be industrialized.

Antennas

Metamaterial antennas are a class of antennas that use metamaterials to improve performance. Demonstrations showed that metamaterials could enhance an antenna's radiated power. Materials that can attain negative permeability allow for properties such as small antenna size, high directivity and tunable frequency.

Absorber

A metamaterial absorber manipulates the loss components of metamaterials' permittivity and magnetic permeability, to absorb large amounts of electromagnetic radiation. This is a useful feature for photodetection and solar photovoltaic applications. Loss components are also relevant in applications of negative refractive index (photonic metamaterials, antenna systems) or transformation optics (metamaterial cloaking, celestial mechanics), but often are not utilized in these applications.

Skin depth engineering can be used in metamaterial absorbers in photovoltaic applications as well as other optoelectronic devices, where optimizing the device performance demands minimizing resistive losses and power consumption, such as photodetectors, laser diodes, and light emitting diodes.

Superlens

A *superlens* is a two or three-dimensional device that uses metamaterials, usually with negative refraction properties, to achieve resolution beyond the diffraction limit (ideally, infinite resolution). Such a behaviour is enabled by the capability of double-negative materials to yield negative phase velocity. The diffraction limit is inherent in conventional optical devices or lenses.

Cloaking Devices

Metamaterials are a potential basis for a practical cloaking device. The proof of principle was demonstrated on October 19, 2006. No practical cloaks are publicly known to exist.

Seismic Protection

Seismic metamaterials counteract the adverse effects of seismic waves on man-made structures.

Sound Filtering

Metamaterials textured with nanoscale wrinkles could control sound or light signals, such as changing a material's color or improving ultrasound resolution. Uses include nondestructive material testing, medical diagnostics and sound suppression. The materials can be made through a high-precision, multi-layer deposition process. The thickness of each layer can be controlled within a fraction of a wavelength. The material is then compressed, creating precise wrinkles whose spacing can cause scattering of selected frequencies.

Theoretical Models

All materials are made of atoms, which are dipoles. These dipoles modify light velocity by a factor n (the refractive index). In a split ring resonator the ring and wire units act as atomic dipoles: the wire acts as a ferroelectric atom, while the ring acts as an inductor L, while the open section acts as a capacitor C. The ring as a whole acts as an LC circuit. When the electromagnetic field passes through the ring, an induced current is created. The generated field is perpendicular to the light's magnetic field. The magnetic resonance results in a negative permeability; the refraction index is negative as well. (The lens is not truly flat, since the structure's capacitance imposes a slope for the electric induction.)

Several (mathematical) material models frequency response in DNGs. One of these is the Lorentz model, which describes electron motion in terms of a driven-damped, harmonic oscillator. The Debye relaxation model applies when the acceleration component of the Lorentz mathematical model is small compared to the other components of the equation. The Drude model applies when the restoring force component is negligible and the coupling coefficient is generally the plasma frequency. Other component distinctions call for the use of one of these models, depending on its polarity or purpose.

Three-dimensional composites of metal/non-metallic inclusions periodically/randomly embedded in a low permittivity matrix are usually modeled by analytical methods, including mixing formulas and scattering-matrix based methods. The particle is modeled by either an electric dipole parallel to the electric field or a pair of crossed electric and magnetic dipoles parallel to the electric and magnetic fields, respectively, of the applied wave. These dipoles are the leading terms in the multipole series. They are the only existing ones for a homogeneous sphere, whose polarizability can be easily obtained from the Mie scattering coefficients. In general, this procedure is known as the "point-dipole approximation", which is a good approximation for metamaterials consisting of composites of electrically small spheres. Merits of these methods include low calculation cost and mathematical simplicity.

Other first principles techniques for analyzing regular 3D lattices of scatterers may be found in Computing photonic band structure

Institutional Networks

MURI

The Multidisciplinary University Research Initiative (MURI) encompasses dozens of Universities and a few government organizations. Participating universities include UC Berkeley, UC Los Angeles, UC San Diego, Massachusetts Institute of Technology, and Imperial College in London, UK. The sponsors are Office of Naval Research and the Defense Advanced Research Project Agency.

MURI supports research that intersects more than one traditional science and engineering discipline to accelerate both research and translation to applications. As of 2009, 69 academic institutions were expected to participate in 41 research efforts.

Metamorphose

The Virtual Institute for Artificial Electromagnetic Materials and Metamaterials "Metamorphose VI AISBL" is an international association to promote artificial electromagnetic materials and metamaterials. It organizes scientific conferences, supports specialized journals, creates and manages research programs, provides training programs (including PhD and training programs for industrial partners); and technology transfer to European Industry.

Metallogels

Metallogels are one-dimensional nanostructured materials, which constitute a growing class in the Supramolecular chemistry field. Non-covalent interactions, such as hydrophobic interactions, π-π interactions, and hydrogen bonding, are among the responsible forces for the formation of those gels from small molecules. However, the main driving force for the formation of a metallogel is the metal-ligand coordination. Once the structure has been established, it resists gravitational force when inverted.

Synthesis Method

Since the properties of gels depend on the type of non-covalent interactions involved, the metal-ligand interaction provide not only thermodynamic stability, but also kinetic liability. The general method for synthesizing gels is to heat the solution, which contains the metal ion being used and investigated, along with the ligand that will form the metallogel around it, as well as any other compounds used to create the appropriate conditions for the reaction to proceed well, until all added solids (depending on the type of gel prepared) are dissolved in the solvent used, and then cooling it down until the gels are self-assembled and properly formed. However, this method has not shown favorable results with the additions of several transition metals, along with the use of lanthanides, in a acetonitrile nitrile solution of the ligand. In these studies, the ligand used is a 2,6-Bis(1'-alkylbenzimidazolyl)pyridine, due to its commercial abundance and the wide variety of synthetic pathways that allow the functionalization of this ligand, and therefore the chemical tuning of the metallogel. Therefore, under controlled heating and cooling conditions, the addition of a source of transition metals to the solution containing the ligand, along with lanthanide ions yielded stable gels, who have passed the inversion test. Self-assembling occurs from the influence of non-covalent interactions, as depicted in figure1. These linear, self-assembling compounds can continue to self-assemble forming columnar, helical structures that further aggregate to form bundles of fibers.

Another approach to form gels as functional nanomaterials, is the bottom up method used in subcomponent self-assembly. This method aims to save resources, shorten the time of synthesis, and offer a wider range of gels by the quick exchange of one of the reaction components.

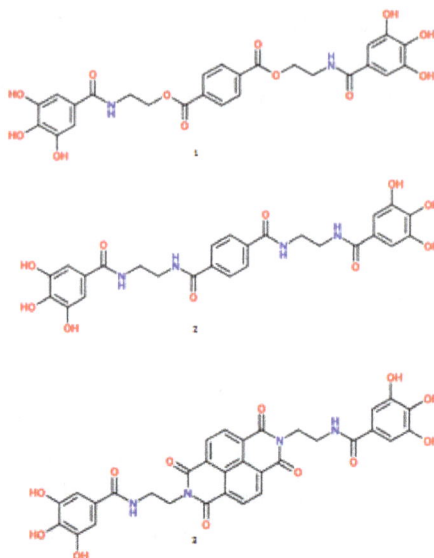

Examples

Although the synthesis method generally is the same, relying on the self-assembly of small molecules in the appropriate conditions, the metallogels differ mainly in the metal ion used, which directly influences their functions, chemical, optic, and electronic properties. Among numerous metal ions that are used, gold ions have been investigated for their wide variety of foreseen applications, as discussed in the applications section. They are further divided into two categories, based on the type of solvent used during the synthesis process. Gold Organometallogelators which are formed by Au(I) in trinuclear gold(I) pyrazolate complexes with long akyl chains, which appears as a red-luminescent organogel. Gold Hydrometallogelators are made of glutathione and Au (III), which appear as a transparent gel.

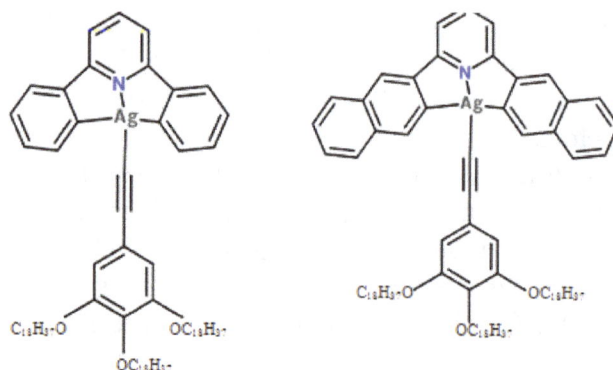

Examples of Silver Organometallogelators

Silver metal ions also show properties of self-assembly, since they have high affinity to bind nitrogen, which can act as the driving force to form stable supramolecular structures.

However, copper ions have a promiscuous nature that allows them to bind to a variety of ligands, which readily form stable metallogels with tunable properties, widening the scope of their applications. Bipyridines were among the most important ligands, since the formation of those metallogels can lead to research about the coordination of copper ions to DNA base pairs. Oxalic acid dihydrate is another important ligand, that easily forms stable structures when copper salts are added, which can be used as proton conductors. Furthermore, a bile acid–picolinic acid conjugate can form gels in solvents that are 30%-50% organic. The increased water content renders this gel more bio-compatible, offering room for further investigation. One of the most recent studies in copper hydrogels involved adding glyme to the polymer to observe the resultant effect on the network. It has formed a stronger amorphous polymer that is anticipated to have self-healing properties.

Palladium ions were among the transition metals used to form catalytic and irreversible metallogels, as well.

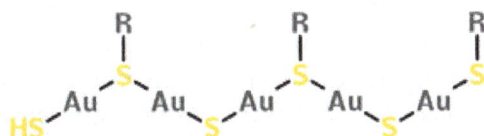

Gold Hydrometallogelators

Applications

Metallogels obtain multi-responsive property to widely environmental responses. In particular, metallogels which are made of transition metal and lanthanoid are thermo-responsive, mechano-responsive, chemo-responsive, and photo-responsive. A metallogel system of Co/La shows inverse gel-sol transition when being heated to 100°C. When being heated, the orange color of solution remains unchanged suggesting the reaction of La/ligands only due to the heat. Such behavior is classified as thermo-response. Metallogels are also mechano-responsive. A system of Zn/La shows the formation of gel-like material upon addition of CH_3CN as solvent followed by a gently shake. However, this metarial turns to transparent liquid after sitted for 20 seconds. As an example of a chemo-response, adding a small amount of formic acid to Zn/Eu will cause the breakdown of gel-like material as well as its mechanical stability and light-emission. Different system of metal and lanthanoids show different emission bands on photoluminescent spectra. Co/Eu emits no band on the spectra due to presence of low energy metal of the system. Zn/La shows signal of metal-bound ligand-based at 397 nm while Zn/Eu shows signals of lanthanide metal at 581, 594, 616, 652 nm and signals of ligand at 397 nm indicating that the ligand is sensitive to metal biding.

In addition to the multi-responsive properties, gold metallogels can prove to be useful in cosmetics, food processing, and lubrication. Those gels are used in drug delivery, to trap active enzymes and bacteria inside them. Furthermore, the basis of certain technologies, producing valves, clutches, and dampers, rely on the multi-responsive nature of metallogels to electric and magnetic stimuli.

A recent study on metal organic gels involving Cadmium and Zinc ion shows promising results to absorb dyes, which emulates the ability of natural systems to get rid of toxic material that are difficult to decompose.

Nanoparticle

Nanoparticles are particles between 1 and 100 nanometers in size. In nanotechnology, a particle is defined as a small object that behaves as a whole unit with respect to its transport and properties. Particles are further classified according to diameter. Ultrafine particles are the same as nanoparticles and between 1 and 100 nanometers in size, fine particles are sized between 100 and 2,500 nanometers, and coarse particles cover a range between 2,500 and 10,000 nanometers. Scientific research on nanoparticles is intense as they have many potential applications in medicine, optics, and electronics. The U.S. National Nanotechnology Initiative offers government funding focused on nanoparticle research.

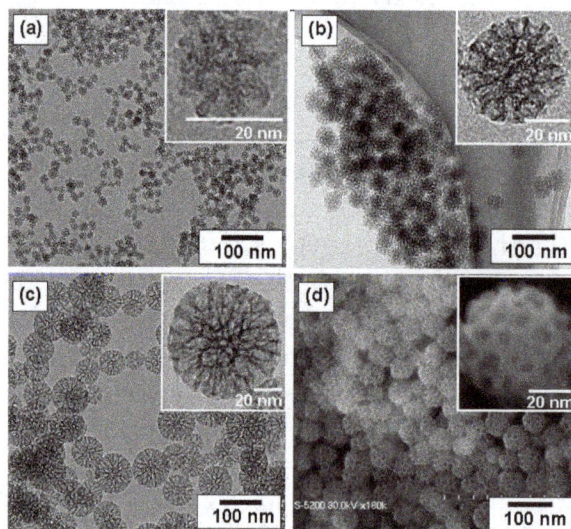

TEM (a, b, and c) images of prepared mesoporous silica nanoparticles with mean outer diameter: (a) 20nm, (b) 45nm, and (c) 80nm. SEM (d) image corresponding to (b). The insets are a high magnification of mesoporous silica particle.

Definition

IUPAC definition

Particle of any shape with dimensions in the 1×10^{-9} and 1×10^{-7} m range.

Note 1: Modified from definitions of nanoparticle and nanogel in [refs.,].

Note 2: The basis of the 100-nm limit is the fact that novel properties that differentiate particles from the bulk *material* typically develop at a criticallength scale of under 100 nm.

Note 3: Because other phenomena (transparency or turbidity, ultrafiltration, stable dispersion, etc.) that extend the upper limit are occasionally considered, the use of the prefix *nano* is accepted for dimensions smaller than 500 nm, provided reference to the definition is indicated.

Note 4: Tubes and fibers with only two dimensions below 100 nm are also nanoparticles.

The term "nanoparticle" is not usually applied to individual molecules; it usually refers to inorganic materials.

The reason for the synonymous definition of nanoparticles and ultrafine particles is that, during the 1970s and 80s, when the first thorough fundamental studies with "nanoparticles" were underway in the USA (by Granqvist and Buhrman) and Japan, (within an ERATO Project) they were called "ultrafine particles" (UFP). However, during the 1990s before the National Nanotechnology Initiative was launched in the USA, the new name, "nanoparticle," had become more common . Nanoparticles can exhibit size-related properties significantly different from those of either fine particles or bulk materials.

Nanoclusters have at least one dimension between 1 and 10 nanometers and a narrow size distribution. Nanopowders are agglomerates of ultrafine particles, nanoparticles, or nanoclusters. Nanometer-sized single crystals, or single-domain ultrafine particles, are often referred to as nanocrystals.

Background

Although nanoparticles are associated with modern science, they have a long history. Nanoparticles were used by artisans as far back as the ninth century in Mesopotamia for creating a glittering effect on the surface of pots.

In modern times, pottery from the Middle Ages and Renaissance often retains a distinct gold- or copper-colored metallic glitter. This luster is caused by a metallic film that was applied to the transparent surface of a glazing. The luster can still be visible if the film has resisted atmospheric oxidation and other weathering.

The luster originates within the film itself, which contains silver and copper nanoparticles dispersed homogeneously in the glassy matrix of the ceramic glaze. These nanoparticles are created by the artisans by adding copper and silver salts and oxides together with vinegar, ochre, and clay on the surface of previously-glazed pottery. The object is then placed into a kiln and heated to about 600 °C in a reducing atmosphere.

In the heat the glaze softens, causing the copper and silver ions to migrate into the outer layers of the glaze. There the reducing atmosphere reduced the ions back to metals, which then came together forming the nanoparticles that give the color and optical effects.

Luster technique showed that ancient craftsmen had a sophisticated empirical knowledge of materials. The technique originated in the Muslim world. As Muslims were not allowed to use gold in artistic representations, they sought a way to create a similar effect without using real gold. The solution they found was using luster.

Michael Faraday provided the first description, in scientific terms, of the optical properties of nanometer-scale metals in his classic 1857 paper. In a subsequent paper, the author (Turner) points out that: "It is well known that when thin leaves of gold or silver are mounted upon glass and heated to a temperature that is well below a red heat (~500 °C), a remarkable change of properties takes place, whereby the continuity of the metallic film is destroyed. The result is that white light is now freely transmitted, reflection is correspondingly diminished, while the electrical resistivity is enormously increased."

Uniformity

The chemical processing and synthesis of high-performance technological components for the private, industrial, and military sectors requires the use of high-purity ceramics, polymers, glass-ceramics, and composite materials. In condensed bodies formed from fine powders, the irregular particle sizes and shapes in a typical powder often lead to non-uniform packing morphologies that result in packing density variations in the powder compact.

Uncontrolled agglomeration of powders due to attractive van der Waals forces can also give rise to microstructural heterogeneity. Differential stresses that develop as a result of non-uniform drying shrinkage are directly related to the rate at which the solvent can be removed, and thus highly dependent upon the distribution of porosity. Such stresses have been associated with a plastic-to-brittle transition in consolidated bodies, and can yield to crack propagation in the unfired body if not relieved.

In addition, any fluctuations in packing density in the compact as it is prepared for the kiln are often amplified during the sintering process, yielding inhomogeneous densification. Some pores and other structural defects associated with density variations have been shown to play a detrimental role in the sintering process by growing and thus limiting end-point densities. Differential stresses arising from inhomogeneous densification have also been shown to result in the propagation of internal cracks, thus becoming the strength-controlling flaws.

Inert gas evaporation and inert gas deposition are free many of these defects due to the distillation (cf. purification) nature of the process and having enough time to form single crystal particles, however even their non-aggreated deposits have lognormal size distribution, which is typical with nanoparticles. The reason why modern gas evaporation techniques can produce a relatively narrow size distribution is that aggregation can be avoided. However, even in this case, random residence times in the growth zone, due to the combination of drift and diffusion, result in a size distribution appearing lognormal.

It would, therefore, appear desirable to process a material in such a way that it is physically uniform with regard to the distribution of components and porosity, rather than using particle size distributions that will maximize the green density. The containment of a uniformly dispersed assembly of strongly interacting particles in suspension requires total control over interparticle forces. Monodisperse nanoparticles and colloids provide this potential.

Monodisperse powders of colloidal silica, for example, may therefore be stabilized sufficiently to ensure a high degree of order in the colloidal crystal or polycrystalline colloidal solid that results from aggregation. The degree of order appears to be limited by the time and space allowed for longer-range correlations to be established. Such defective polycrystalline colloidal structures would appear to be the basic elements of submicrometer colloidal materials science and, therefore, provide the first step in developing a more rigorous understanding of the mechanisms involved in microstructural evolution in high performance materials and components.

Properties

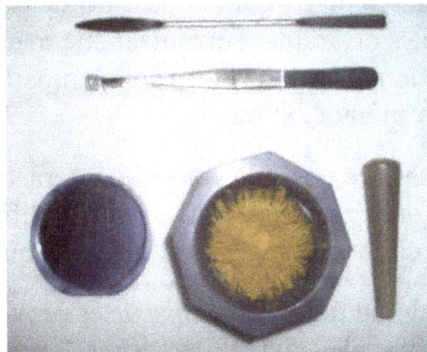

Silicon nanopowder

Nanoparticles are of great scientific interest as they are, in effect, a bridge between bulk materials and atomic or molecular structures. A bulk material should have constant physical properties regardless of its size, but at the nano-scale size-dependent properties are often observed. Thus, the properties of materials change as their size approaches the nanoscale and as the percentage of the surface in relation to the percentage of the volume of a material becomes significant. For bulk materials larger than one micrometer (or micron), the percentage of the surface is insignificant in relation to the volume in the bulk of the material. *The interesting and sometimes unexpected properties of nanoparticles are therefore largely due to the large surface area of the material, which dominates the contributions made by the small bulk of the material.*

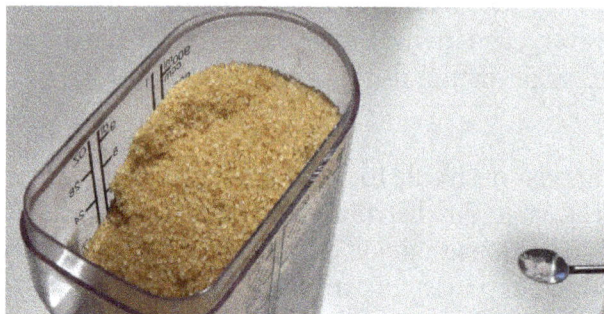

1 kg of particles of 1 mm³ has the same surface area as 1 mg of particles of 1 nm³

Nanoparticles often possess unexpected optical properties as they are small enough to confine their electrons and produce quantum effects. For example, gold nanoparticles appear deep-red to black in solution. Nanoparticles of yellow gold and grey silicon are red in color. Gold nanoparticles melt at much lower temperatures (~300 °C for 2.5 nm size) than the gold slabs (1064 °C);. Absorption of solar radiation is much higher in materials composed of nanoparticles than it is in thin films of continuous sheets of material. In both solar PV and solar thermal applications, controlling the size, shape, and material of the particles, it is possible to control solar absorption.

Other size-dependent property changes include quantum confinement in semiconductor particles, surface plasmon resonance in some metal particles and superparamagnetism in magnetic materials. What would appear ironic is that the changes in physical properties are not always desirable. Ferromagnetic materials smaller than 10 nm can switch their magnetisation direction using room temperature thermal energy, thus making them unsuitable for memory storage.

Suspensions of nanoparticles are possible since the interaction of the particle surface with the solvent is strong enough to overcome density differences, which otherwise usually result in a material either sinking or floating in a liquid.

The high surface area to volume ratio of nanoparticles provides a tremendous driving force for diffusion, especially at elevated temperatures. Sintering can take place at lower temperatures, over shorter time scales than for larger particles. In theory, this does not affect the density of the final product, though flow difficulties and the tendency of nanoparticles to agglomerate complicates matters. Moreover, nanoparticles have been found to impart some extra properties to various day to day products. For example, the presence of titanium dioxide nanoparticles imparts what we call the self-cleaning effect, and, the size being nano-range, the particles cannot be observed. Zinc oxide particles have been found to have superior UV blocking properties compared to its bulk substitute. This is one of the reasons why it is often used in the preparation of sunscreen lotions, is completely photostable and toxic.

Clay nanoparticles when incorporated into polymer matrices increase reinforcement, leading to stronger plastics, verifiable by a higher glass transition temperature and other mechanical property tests. These nanoparticles are hard, and impart their properties to the polymer (plastic). Nanoparticles have also been attached to textile fibers in order to create smart and functional clothing.

Metal, dielectric, and semiconductor nanoparticles have been formed, as well as hybrid structures (e.g., core–shell nanoparticles). Nanoparticles made of semiconducting material may also be labeled quantum dots if they are small enough (typically sub 10 nm) that quantization of electronic energy levels occurs. Such nanoscale particles are used in biomedical applications as drug carriers or imaging agents.

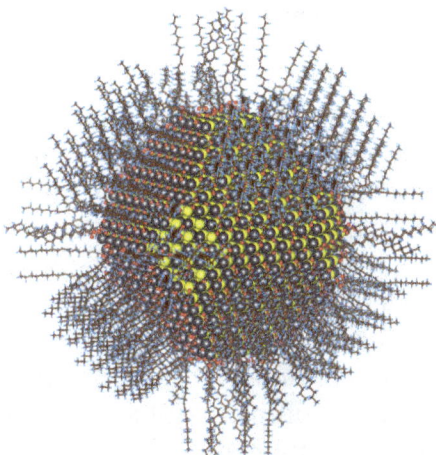

Semiconductor nanoparticle (quantum dot) of lead sulfide with complete passivation by oleic acid, oleyl amine and hydroxyl ligands (size ~5nm)

Semi-solid and soft nanoparticles have been manufactured. A prototype nanoparticle of semi-solid nature is the liposome. Various types of liposome nanoparticles are currently used clinically as delivery systems for anticancer drugs and vaccines.

Nanoparticles with one half hydrophilic and the other half hydrophobic are termed Janus particles and are particularly effective for stabilizing emulsions. They can self-assemble at water/oil interfaces and act as solid surfactants.

Synthesis

There are several methods for creating nanoparticles, including attrition, pyrolysis and hydrothermal synthesis. In attrition, macro- or micro-scale particles are ground in a ball mill, a planetary ball mill, or other size-reducing mechanism. The resulting particles are air classified to recover nanoparticles. In pyrolysis, a vaporous precursor (liquid or gas) is forced through an orifice at high pressure and burned. The resulting solid (a version of soot) is air classified to recover oxide particles from by-product gases. Traditional pyrolysis often results in aggregates and agglomerates rather than single primary particles. Ultrasonic nozzle spray pyrolysis (USP) on the other hand aids in preventing agglomerates from forming.

A thermal plasma can deliver the energy to vaporize small micrometer-size particles. The thermal plasma temperatures are in the order of 10,000 K, so that solid powder easily evaporates. Nanoparticles are formed upon cooling while exiting the plasma region. The main types of the thermal plasma torches used to produce nanoparticles are dc plasma jet, dc arc plasma, and radio frequency (RF) induction plasmas. In the arc plasma reactors, the energy necessary for evaporation and reaction is provided by an electric arc formed between the anode and the cathode. For example, silica sand can be vaporized with an arc plasma at atmospheric pressure, or thin aluminum wires can be vaporized by exploding wire method. The resulting mixture of plasma gas and silica vapour can be rapidly cooled by quenching with oxygen, thus ensuring the quality of the fumed silica produced.

In RF induction plasma torches, energy coupling to the plasma is accomplished through the electromagnetic field generated by the induction coil. The plasma gas does not come in contact with electrodes, thus eliminating possible sources of contamination and allowing the operation of such plasma torches with a wide range of gases including inert, reducing, oxidizing, and other corrosive atmospheres. The working frequency is typically between 200 kHz and 40 MHz. Laboratory units run at power levels in the order of 30–50 kW, whereas the large-scale industrial units have been tested at power levels up to 1 MW. As the residence time of the injected feed droplets in the plasma is very short, it is important that the droplet sizes are small enough in order to obtain complete evaporation. The RF plasma method has been used to synthesize different nanoparticle materials, for example synthesis of various ceramic nanoparticles such as oxides, carbours/carbides, and nitrides of Ti and Si.

Inert-gas condensation is frequently used to make nanoparticles from metals with low melting points. The metal is vaporized in a vacuum chamber and then supercooled with an inert gas stream. The supercooled metal vapor condenses into nanometer-size particles, which can be entrained in the inert gas stream and deposited on a substrate or studied in situ.

Nanoparticles can also be formed using radiation chemistry. Radiolysis from gamma rays can create strongly active free radicals in solution. This relatively simple technique uses a minimum number of chemicals. These including water, a soluble metallic salt, a radical scavenger (often a secondary alcohol), and a surfactant (organic capping agent). High gamma doses on the order of 10^4 Gray are required. In this process, reducing radicals will drop metallic ions down to the zero-valence state. A scavenger chemical will preferentially interact with oxidizing radicals to prevent the re-oxidation of the metal. Once in the zero-valence state, metal atoms begin to coalesce into particles. A chemical surfactant surrounds the particle during formation and regulates its growth. In sufficient concentrations, the surfactant molecules stay attached to the particle. This prevents it from dissociating or forming clusters with other particles. Formation of nanoparticles using the radiolysis method allows for tailoring of particle size and shape by adjusting precursor concentrations and gamma dose.

Sol-gel

The sol-gel process is a wet-chemical technique (also known as chemical solution deposition) widely used recently in the fields of materials science and ceramic engineering. Such methods are used primarily for the fabrication of materials (typically a metal oxide) starting from a chemical solution (*sol*, short for solution), which acts as the precursor for an integrated network (or *gel*) of either discrete particles or network polymers.

Typical precursors are metal alkoxides and metal chlorides, which undergo hydrolysis and polycondensation reactions to form either a network "elastic solid" or a colloidal suspension (or dispersion) – a system composed of discrete (often amorphous) submicrometer particles dispersed to various degrees in a host fluid. Formation of a metal oxide involves connecting the metal centers with oxo (M-O-M) or hydroxo (M-OH-M) bridges, therefore generating metal-oxo or metal-hydroxo polymers in solution. Thus, the sol evolves toward the formation of a gel-like diphasic system containing both a liquid phase and solid phase whose morphologies range from discrete particles to continuous polymer networks.

In the case of the colloid, the volume fraction of particles (or particle density) may be so low that a significant amount of fluid may need to be removed initially for the gel-like properties to be recognized. This can be accomplished in a number of ways. The most simple method is to allow time for sedimentation to occur, and then pour off the remaining liquid. Centrifugation can also be used to accelerate the process of phase separation.

Removal of the remaining liquid (solvent) phase requires a drying process, which is typically causes shrinkage and densification. The rate at which the solvent can be removed is ultimately determined by the distribution of porosity in the gel. The ultimate microstructure of the final component will clearly be strongly influenced by changes implemented during this phase of processing. Afterward, a thermal treatment, or firing process, is often necessary in order to favor further polycondensation and enhance mechanical properties and structural stability via final sintering, densification, and grain growth. One of the distinct advantages of using this methodology as opposed to the more traditional processing techniques is that densification is often achieved at a much lower temperature.

The precursor sol can be either deposited on a substrate to form a film (e.g., by dip-coating or spin-coating), cast into a suitable container with the desired shape (e.g., to obtain a monolithic ceramics, glasses, fibers, membranes, aerogels), or used to synthesize powders (e.g., microspheres, nanospheres). The sol-gel approach is a cheap and low-temperature technique that allows for the fine control of the product's chemical composition. Even small quantities of dopants, such as organic dyes and rare earth metals, can be introduced in the sol and end up uniformly dispersed in the final product. It can be used in ceramics processing and manufacturing as an investment casting material, or as a means of producing very thin films of metal oxides for various purposes. Sol-gel derived materials have diverse applications in optics, electronics, energy, space, (bio)sensors, medicine (e.g., controlled drug release) and separation (e.g., chromatography) technology.

Colloids

The term colloid is used primarily to describe a range of mixtures which have solid particles dispersed in a liquid medium. The term applies only if the particles are larger than atomic dimensions but small enough to exhibit Brownian motion. If the particles are large enough, their dynamic behavior in any given period of time in suspension would be governed by forces of gravity and sedimentation. If the particles are small enough, their irregular motion in suspension can be attributed to the collective bombardment of a myriad of thermally agitated molecules in the liquid suspending medium, as described originally by Albert Einstein in his dissertation. Einstein showed evidence that water was made up of discrete molecules by characterizing the observed erratic particle behavior using the theory of Brownian motion, with sedimentation being a possible result. The critical size range (or particle diameter) typically ranges from nanometers (10^{-9} m) to micrometers (10^{-6} m).

Morphology

Scientists have taken to naming their particles after the real-world shapes that they might represent. Nanospheres, nanochains, nanoreefs, nanoboxes and more have appeared in the literature. These morphologies sometimes arise spontaneously as an effect of a templating or directing agent present in the synthesis such as miscellar emulsions or anodized alumina pores, or from the innate crystallographic growth patterns of the materials themselves. Some of these morphologies may serve a purpose, such as long carbon nanotubes used to bridge an electrical junction, or just a scientific curiosity like the stars shown at right.

Nanostars of vanadium(IV) oxide

Amorphous particles usually adopt a spherical shape (due to their microstructural isotropy), whereas the shape of anisotropic microcrystalline whiskers corresponds to their particular crystal habit. At the small end of the size range, nanoparticles are often referred to as clusters. Spheres, rods, fibers, and cups are just a few of the shapes that have been grown. The study of fine particles is called micromeritics.

Characterization

The majority of nanoparticle characterization techniques are light-based, but a non-optical nanoparticle characterization technique called Tunable Resistive Pulse Sensing (TRPS) has been developed that enables the simultaneous measurement of size, concentration and surface charge for a wide variety of nanoparticles. This technique, which applies the Coulter Principle, allows for particle-by-particle quantification of these three nanoparticle characteristics with high resolution.

Functionalization

The surface coating of nanoparticles determines many of their properties, notably stability, solubility, and targeting. A coating that is multivalent or polymeric confers high stability. Functionalized nanomaterial-based catalysts can be used for catalysis of many known organic reactions.

Surface Coating for Biological Applications

For biological applications, the surface coating should be polar to give high aqueous solubility and prevent nanoparticle aggregation. In serum or on the cell surface, highly charged coatings promote non-specific binding, whereas polyethylene glycol linked to terminal hydroxyl or methoxy groups repel non-specific interactions. Nanoparticles can be linked to biological molecules that can act as address tags, to direct the nanoparticles to specific sites within the body, specific organelles within the cell, or to follow specifically the movement of individual protein or RNA molecules in living cells. Common address tags are monoclonal antibodies, aptamers, streptavidin or peptides. These targeting agents should ideally be covalently linked to the nanoparticle and should be present in a controlled number per nanoparticle. Multivalent nanoparticles, bearing multiple targeting groups, can cluster receptors, which can activate cellular signaling pathways, and give stronger anchoring. Monovalent nanoparticles, bearing a single binding site, avoid clustering and so are preferable for tracking the behavior of individual proteins.

Red blood cell coatings can help nanoparticles evade the immune system.

Safety

Nanoparticles present possible dangers, both medically and environmentally. Most of these are due to the high surface to volume ratio, which can make the particles very reactive or catalytic. They are also able to pass through cell membranes in organisms, and their interactions with biological systems are relatively unknown. However, it is unlikely the particles would enter the cell nucleus, Golgi complex, endoplasmic reticulum or other internal cellular components due to the particle size and intercellular agglomeration. A recent study looking at the effects of ZnO nanoparticles on human immune cells has found varying levels of susceptibility to cytotoxicity. There are concerns that pharmaceutical companies, seeking regulatory approval for nano-reformulations of existing medicines, are relying on safety data produced during clinical studies of the earlier, pre-reformulation version of the medicine. This could result in regulatory bodies, such as the FDA, missing new side effects that are specific to the nano-reformulation.

Whether cosmetics and sunscreens containing nanomaterials pose health risks remains largely unknown at this stage. However considerable research has demonstrated that zinc nanoparticles are not absorbed into the bloodstream in vivo.

Concern has also been raised over the health effects of respirable nanoparticles from certain combustion processes. As of 2013 the U.S. Environmental Protection Agency was investigating the safety of the following nanoparticles:

- Carbon Nanotubes: Carbon materials have a wide range of uses, ranging from composites for use in vehicles and sports equipment to integrated circuits for electronic components. The interactions between nanomaterials such as carbon nanotubes and natural organic matter strongly influence both their aggregation and deposition, which strongly affects their transport, transformation, and exposure in aquatic environments. In past research, carbon nanotubes exhibited some toxicological impacts that will be evaluated in various environmental settings in current EPA chemical safety research. EPA research will provide data, models, test methods, and best practices to discover the acute health effects of carbon nanotubes and identify methods to predict them.

- Cerium oxide: Nanoscale cerium oxide is used in electronics, biomedical supplies, energy, and fuel additives. Many applications of engineered cerium oxide nanoparticles naturally disperse themselves into the environment, which increases the risk of exposure. There is ongoing exposure to new diesel emissions using fuel additives containing CeO_2 nanoparticles, and the environmental and public health impacts of this new technology are unknown. EPA's chemical safety research is assessing the environmental, ecological, and health implications of nanotechnology-enabled diesel fuel additives.

- Titanium dioxide: Nano titanium dioxide is currently used in many products. Depending on the type of particle, it may be found in sunscreens, cosmetics, and paints and coatings. It is also being investigated for use in removing contaminants from drinking water.

- Nano Silver: Nano silver is being incorporated into textiles, clothing, food packaging, and other materials to eliminate bacteria. EPA and the U.S. Consumer Product Safety Commission are studying certain products to see whether they transfer nano-size silver particles in real-world scenarios. EPA is researching this topic to better understand how much nano-silver children come in contact with in their environments.

- Iron: While nano-scale iron is being investigated for many uses, including "smart fluids" for uses such as optics polishing and as a better-absorbed iron nutrient supplement, one of its more prominent current uses is to remove contamination from groundwater. This use, supported by EPA research, is being piloted at a number of sites across the country.

Laser Applications

The use of nanoparticles in laser dye-doped poly(methyl methacrylate) (PMMA) laser gain media was demonstrated in 2003 and it has been shown to improve conversion efficiencies and to decrease laser beam divergence. Researchers attribute the reduction in beam divergence to improved dn/dT characteristics of the organic-inorganic dye-doped nanocomposite. The optimum composition reported by these researchers is 30% w/w of SiO_2 (~ 12 nm) in dye-doped PMMA.

Silver Nanoparticle

Silver nanoparticles are nanoparticles of silver of between 1 nm and 100 nm in size. While frequently described as being 'silver' some are composed of a large percentage of silver oxide due to their large ratio of surface-to-bulk silver atoms. Numerous shapes of nanoparticles can be constructed depending on the application at hand. Commonly used are spherical silver nanoparticles but diamond, octagonal and thin sheets are also popular.

Their extremely large surface area permits the coordination of a vast number of ligands. The properties of silver nanoparticles applicable to human treatments are under investigation in laboratory and animal studies, assessing potential efficacy, toxicity, and costs.

Synthesis

Methods

Silver nanoparticles can be synthesized by several methods. They can be divided into three broad categories: wet chemistry, ion implantation, or biogenic synthesis.

Wet Chemistry

Several wet chemical methods have been developed to form silver nanoparticles. Each of these methods use the same basic mechanism: a silver ion complex is reduced to colloidal silver in the presence of a reducing agent, which oxidizes and gives back the missing electron, so that the silver ion is reduced back to its metallic state. The newly formed silver nanoparticles can then also be stabilized with a stabilizing/capping agent, which stabilizes the energy of the surface of the newly formed nanoparticle to prevent them from aggregate to larger particles. These methods include the use of reducing sugars, citrate reduction, reduction via sodium borohydride, the silver mirror reaction, the polyol process, seed-mediated growth, and light-mediated growth. Each of these methods, or combination of methods, will offer differing degrees of control over the size distribution as well as distributions of geometric arrangements of the nanoparticle. A new, very promising wet-chemical technique was found by Elsupikhe et al. (2015). They have developed a green ultrasonically-assisted synthesis. Under ultrasound treatment, silver nanoparticles (AgNP) are synthesized with κ-carrageenan as a natural stabilizer. The reaction is performed at ambient temperature and produces silver nanoparticles with fcc crystal structure without impurities. The concentration of κ-carrageenan is used to influence particle size distribution of the AgNPs.

A one-pot synthesis of crystalline silver nanoparticles was also reported by using hydrazine hydrate as the reducing agent.

Reduction with Reducing Sugars

Reducing sugars includes glucose, fructose, maltose, maltodextrin, etc. but not sucrose, and is the simpliest method to reduce silver ions back to silver nanoparticles. And if the pH is over 7 the glucose molucule in the reducing sugar will "open up" its ring structure and attach to the surface of the nanoparticles, and thereby act as a weak stabilizing agent as well.

Citrate Reduction

An early, and very common, method for synthesizing silver nanoparticles is citrate reduction. Citrate reduction involves the reduction of silver nitrate, $AgNO_3$, to colloidal silver using trisodium citrate, $Na_3C_6H_5O_7$ at elevated temperature (~100 °C). In this synthesis, the citrate ion acts as both the reducing and capping agent. This method is useful due to its relative ease and short reaction time. However, the silver particles formed may exhibit broad size distributions and/or form several different particle geometries simultaneously.

Reduction Via Sodium Borohydride

A similar method is the reduction of silver nitrate using sodium borohydride, NaBH4. Sodium borohydride is a stronger reducing agent than citrate, but it is not thought to participate in surface stabilization. Thus, a separate capping molecule must be used in addition to the sodium borohydride. Examples of stabilizers used include citrate, poly(vinyl pyrrolidone) (PVP), bovine serum albumin, and cetyltrimethylammonium bromide (CTAB).

Polyol Process

The polyol process is a particularly useful method because it yields a high degree of control over both the size and geometry of the resulting nanoparticles. In general, the polyol synthesis begins with the heating of a polyol compound such as ethylene glycol, 1,5-pentanediol, or 1,2-propylene glycol7. An Ag+ species and a capping agent are added (although the polyol itself is also often the capping agent). The Ag+ species is then reduced by the polyol to colloidal nanoparticles. The polyol process is highly sensitive to reaction conditions such as temperature, chemical environment, and concentration of substrates.(add refs again here) Therefore, by changing these variables, various sizes and geometries can be selected for such as quasi-spheres, pyramids, spheres, and wires. Further study has examined the mechanism for this process as well as resulting geometries under various reaction conditions in greater detail.

Seed-mediated Growth

Seed-mediated growth is the process in which seed nanoparticles, formed from a reduction of Ag+ to colloidal silver by any method, are placed in a growth solution. The growth solution contains an additional silver species that deposits more silver atoms onto the original seeds. Several experimental factors are manipulated to achieve size and geometry control in this method such as concentration of the substrates and the capping agent used. Parameters controlling the size and geometry of the particles is described in greater detail in various studies.

Light-mediated Growth

Light-mediated syntheses have also been explored where light can promote formation of various silver nanoparticle morphologies.

Silver Mirror Reaction

The silver mirror reaction involves the conversion of silver nitrate to Ag(NH3)OH. Ag(NH3)OH is subsequently reduced into colloidal silver using an aldehyde containing molecule such as a sugar. The silver mirror reaction is as follows:

$$2(Ag(NH_3)_2)^+ + RCHO + 2OH^- \rightarrow RCOOH + 2Ag + 4NH_3$$

The size and shape of the nanoparticles produced are difficult to control and often have wide distributions. However, this method is often used to apply thin coatings of silver particles onto surfaces and further study into producing more uniformly sized nanoparticles is being done.

Ion Implantation

Ion implantation has been used to create silver nanoparticles embedded in glass, polyurethane, silicone, polyethylene, and poly(methyl methacrylate). Particles are embedded in the substrate by means of bombardment at high accelerating voltages. At a fixed current density of the ion beam up to a certain value, the size of the embedded silver nanoparticles has been found to be monodisperse within the population, after which only an increase in the ion concentration is observed. A further increase in the ion beam dose has been found to reduce both the nanoparticle size and density in the target substrate, whereas an ion beam operating at a high accelerating voltage with a gradually increasing current density has been found to result in a gradual increase in the nanoparticle size. There are a few competing mechanisms which may result in the decrease in nanoparticle size; destruction of NPs upon collision, sputtering of the sample surface, particle fusion upon heating and dissociation.

The formation of embedded nanoparticles is complex, and all of the controlling parameters and factors have not yet been investigated. Computer simulation is still difficult as it involves processes of diffusion and clustering, however it can be broken down into a few different sub-processes such as implantation, diffusion, and growth. Upon implantation, silver ions will reach different depths within the substrate which approaches a Gaussian distribution with the mean centered at X depth. High temperature conditions during the initial stages of implantation will increase the impurity diffusion in the substrate and as a result limit the impinging ion saturation, which is required for nanoparticle nucleation. Both the implant temperature and ion beam current density are crucial to control in order to obtain a monodisperse nanoparticle size and depth distribution. A low current density may be used to counter the thermal agitation from the ion beam and a buildup of surface charge. After implantation on the surface, the beam currents may be raised as the surface conductivity will increase. The rate at which impurities diffuse drops quickly after the formation of the nanoparticles, which act as a mobile ion trap. This suggests that the beginning of the implantation process is critical for control of the spacing and depth of the resulting nanoparticles, as well as control of the substrate temperature and ion beam density. The presence and nature of these particles can be analyzed using numerous spectroscopy and microscopy instruments. Nanoparticles synthesized in the substrate exhibit surface plasmon resonances as evidenced by characteristic absorption bands; these features undergo spectral shifts depending on the nanoparticle size and surface asperities, however the optical properties also strongly depend on the substrate material of the composite.

Biogenic Synthesis

The biological synthesis of nanoparticles has provided a means for improved techniques compared to the traditional methods that call for the use of harmful reducing agents like sodium borohydride. Many of these methods could improve their environmental footprint by replacing these relatively strong reducing agents. The problems with the chemical production of silver nanoparticles is usually involves high cost and the longevity of the particles is short lived due to aggregation. The harshness of standard chemical methods has sparked the use of using biological organisms to reduce silver ions in solution into colloidal nanoparticles.

In addition, precise control over shape and size is vital during nanoparticle synthesis since the NPs therapeutic properties are intimately dependent on such factors. Hence, the primary focus of research in biogenic synthesis is in developing methods that consistently reproduce NPs with precise properties.

Use of fungi and bacteria

A general representation of the synthesis and applications of biogenically synthesized silver nanoparticles using plant extract.

Bacterial and fungal synthesis of nanoparticles is practical because bacteria and fungi are easy to handle and can be modified genetically with ease. This provides a means to develop biomolecules that can synthesize AgNPs of varying shapes and sizes in high yield, which is at the forefront of current challenges in nanoparticle synthesis. Fungal strains such as Verticillium and bacterial strains such as K. pneumoniae can be used in the synthesis of silver nanoparticles. When the fungus/bacteria is added to solution, protein biomass is released into the solution. Electron donating residues such as tryptophan and tyrosine reduce silver ions in solution contributed by silver nitrate. These methods have been found to effectively create stable monodisperse nanoparticles without the use of harmful reducing agents.

A method has been found of reducing silver ions by the introduction of the fungus *Fusarium oxysporum*. The nanoparticles formed in this method have a size range between 5 and 15 nm and consist of silver hydrosol. The reduction of the silver nanoparticles is thought to come from an enzymatic process and silver nanoparticles produced are extremely stable due to interactions with proteins that are excreted by the fungi.

Bacterium found in silver mines, Pseudomonas stutzeri AG259, were able to construct silver particles in the shapes of triangles and hexagons. The size of these nanoparticles had a large range in size and some of them reached sizes larger than the usual nanoscale with a size of 200 nm. The silver nanoparticles were found in the organic matrix of the bacteria.

Lactic acid producing bacteria have been used to produce silver nanoparticles. The bacteria *Lactobacillus* spp., *Pediococcus pentosaceus*, *Enteroccus faeciumI*, and *Lactococcus garvieae* have been found to be able to reduce silver ions into silver nanoparticles. The production of the nanoparticles takes place in the cell from the interactions between the silver ions and the organic compounds of the cell. It was found that the bacterium *Lactobacillus fermentum* created the smallest silver nanoparticles with an average size of 11.2 nm. It was also found that this bacterium produced the nanoparticles with the smallest size distribution and the nanoparticles were found mostly on the outside of the cells. It was also found that there was an increase in the pH increased the rate of which the nanoparticles were produced and the amount of particles produced.

Use of Plants

The reduction of silver ions into silver nanoparticles has also been achieved using geranium leaves. It has been found that adding geranium leaf extract to silver nitrate solutions causes there silver ions to be quickly reduced and that the nanoparticles produced are particularly stable. The silver nanoparticles produced in solution had a size range between 16 and 40 nm.

In another study different plant leaf extracts were used to reduce silver ions. It was found that out of Camellia sinensis (green tea), pine, persimmon, ginko, magnolia, and platanus that the magnolia leaf extract was the best at creating silver nanoparticles. This method created particles with a disperse size range of 15 to 500 nm, but it was also found that the particle size could be controlled by varying the reaction temperature. The speed at which the ions were reduced by the magnolia leaf extract was comparable to those of using chemicals to reduce.

The use of plants, microbes, and fungi in the production of silver nanoparticles is leading the way to more environmentally sound production of silver nanoparticles.

A green method is available for synthesizing silver nanoparticles using Amaranthus gangeticus Linn leaf extract.

Products and Functionalization

Synthetic protocols for silver nanoparticle production can be modified to produce silver nanoparticles with non-spherical geometries and also to functionalize nanoparticles with different materials, such as silica. Creating silver nanoparticles of different shapes and surface coatings allows for greater control over their size-specific properties.

Anisotropic Structures

Silver nanoparticles can be synthesized in a variety of non-spherical (anisotropic) shapes. Because silver, like other noble metals, exhibits a size and shape dependent optical effect known as localized surface plasmon resonance (LSPR) at the nanoscale, the ability to synthesize Ag nanoparticles in different shapes vastly increases the ability to tune their optical behavior. For example, the wavelength at which LSPR occurs for a nanoparticle of one morphology (e.g. a sphere) will be different if that sphere is changed into a different shape. This shape dependence allows a silver nanoparticle to experience optical enhancement at a range of different wavelengths, even by keeping the size relatively constant, just by changing its shape. The applications of this shape-exploited expansion of optical behavior range from developing more sensitive biosensors to increasing the longevity of textiles.

Triangular Nanoprisms

Triangular shaped nanoparticles are a canonical type of anisotropic morphology studied for both gold and silver.

Though many different techniques for silver nanoprism synthesis exist, several methods employ a seed-mediated approach, which involves first synthesizing small (3-5 nm diameter) silver nanoparticles that offer a template for shape-directed growth into triangular nanostructures.

The silver seeds are synthesized by mixing silver nitrate and sodium citrate in aqueous solution and then rapidly adding sodium borohydride. Additional silver nitrate is added to the seed solution at low temperature, and the prisms are grown by slowly reducing the excess silver nitrate using ascorbic acid.

With the seed-mediated approach to silver nanoprism synthesis, selectivity of one shape over another can in part be controlled by the capping ligand. Using essentially the same procedure above but changing citrate to poly (vinyl pyrrolidone) (PVP) yields cube and rod-shaped nanostructures instead of triangular nanoprisms.

In addition to the seed mediated technique, silver nanoprisms can also be synthesized using a photo-mediated approach, in which preexisting spherical silver nanoparticles are transformed into triangular nanoprisms simply by exposing the reaction mixture to high intensities of light.

Nanocubes

Silver nanocubes can be synthesized using ethylene glycol as a reducing agent and PVP as a capping agent, in a polyol synthesis reaction (vide supra). A typical synthesis using these reagents involves adding fresh silver nitrate and PVP to a solution of ethylene glycol heated at 140 °C.

This procedure can actually be modified to produce another anisotropic silver nanostructure, nanowires, by just allowing the silver nitrate solution to age before using it in the synthesis. By allowing the silver nitrate solution to age, the initial nanostructure formed during the synthesis is slightly different than that obtained with fresh silver nitrate, which influences the growth process, and therefore, the morphology of the final product.

Coating with Silica

In this method, polyvinylpyrrolidone (PVP) is dissolved in water by sonication and mixed with silver colloid particles. Active stirring ensures the PVP has adsorbed to the nanoparticle surface. Centrifuging separates the PVP coated nanoparticles which are then transferred to a solution of ethanol to be centrifuged further and placed in a solution of ammonia, ethanol and Si(OEt$_4$) (TES). Stirring for twelve hours results in the silica shell being formed consisting of a surrounding layer of silicon oxide with an ether linkage available to add functionality. Varying the amount of TES allows for different thicknesses of shells formed. This technique is popular due to the ability to add a variety of functionality to the exposed silica surface.

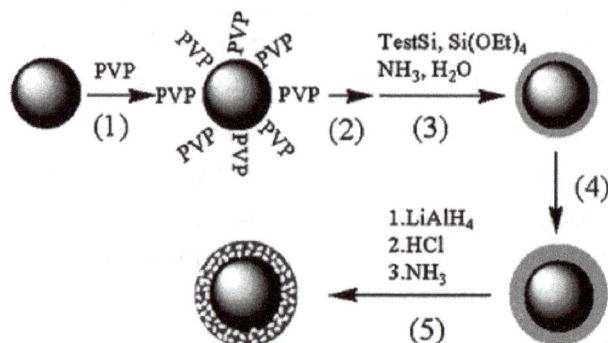

General procedure for coating colloid particles in silica. First PVP is absorbed onto the colloidal surface. These particles are put into a solution of ammonia in ethanol. the particle then begins to grow by addition of Si(OET4).

Use

Catalysis

Using silver nanoparticles for catalysis has been gaining attention in recent years. Although the most common applications are for medicinal or antibacterial purposes, silver nanoparticles have been demonstrated to show catalytic redox properties for dyes, benzene, carbon monoxide, and likely other compounds.

NOTE: This paragraph is a general description of nanoparticle properties for catalysis; it is not exclusive to silver nanoparticles. The size of a nanoparticle greatly determines the properties that it exhibits due to various quantum effects. Additionally, the chemical environment of the nanoparticle plays a large role on the catalytic properties. With this in mind, it is important to note that heterogeneous catalysis takes place by adsorption of the reactant species to the catalytic substrate. When polymers, complex ligands, or surfactants are used to prevent coalescence of the nanoparticles, the catalytic ability is frequently hindered due to reduced adsorption ability. However, these compounds can also be used in such a way that the chemical environment enhances the catalytic ability.

Supported on Silica Spheres – Reduction of Dyes

Silver nanoparticles have been synthesized on a support of inert silica spheres. The support plays virtually no role in the catalytic ability and serves as a method of preventing coalescence of the silver nanoparticles in colloidal solution. Thus, the silver nanoparticles were stabilized and it was possible to demonstrate the ability of them to serve as an electron relay for the reduction of dyes by sodium borohydride. Without the silver nanoparticle catalyst, virtually no reaction occurs between sodium borohydride and the various dyes: methylene blue, eosin, and rose Bengal.

Mesoporous Aerogel – Selective Oxidation of Benzene

Silver nanoparticles supported on aerogel are advantageous due to the higher number of active sites. The highest selectivity for oxidation of benzene to phenol was observed at low weight percent of silver in the aerogel matrix (1% Ag). This better selectivity is believed to be a result of the higher monodispersity within the aerogel matrix of the 1% Ag sample. Each weight percent solution formed different sized particles with a different width of size range.

Silver Alloy – Synergistic Oxidation of Carbon Monoxide

Au-Ag alloy nanoparticles have been shown to have a synergistic effect on the oxidation of carbon monoxide (CO). On its own, each pure-metal nanoparticle shows very poor catalytic activity for CO oxidation; together, the catalytic properties are greatly enhanced. It is proposed that the gold acts as a strong binding agent for the oxygen atom and the silver serves as a strong oxidizing catalyst, although the exact mechanism is still not completely understood. When synthesized in an Au/Ag ratio from 3:1 to 10:1, the alloyed nanoparticles showed complete conversion when 1% CO was fed in air at ambient temperature. Interestingly, the size of the alloyed particles did not play a big role in the catalytic ability. It is well known that gold nanoparticles only show catalytic properties for CO when they are ~3 nm in size, but alloyed particles up to 30 nm demonstrated excellent catalytic activity – catalytic activity better than that of gold nanoparticles on active support such as TiO_2, Fe_2O_3, etc.

Light-enhanced

Plasmonic effects have been studied quite extensively. Until recently, there have not been studies investigating the oxidative catalytic enhancement of a nanostructure via excitation of its surface plasmon resonance. The defining feature for enhancing the oxidative catalytic ability has been identified as the ability to convert a beam of light into the form of energetic electrons that can be transferred to adsorbed molecules. The implication of such a feature is that photochemical reactions can be driven by low-intensity continuous light can be coupled with thermal energy.

The coupling of low-intensity continuous light and thermal energy has been performed with silver nanocubes. The important feature of silver nanostructures that are enabling for photocatalysis is their nature to create resonant surface plasmons from light in the visible range.

The addition of light enhancement enabled the particles to perform to the same degree as particles that were heated up to 40K greater. This is a profound finding when noting that a reduction in temperature of 25K can increase the catalyst lifetime by nearly ten-fold, when comparing the photothermal and thermal process.

Biological Research

Researchers have explored the use of silver nanoparticles as carriers for delivering various payloads such as small drug molecules or large biomolecules to specific targets. Once the AgNP has had sufficient time to reach its target, release of the payload could potentially be triggered by an internal or external stimulus. The targeting and accumulation of nanoparticles may provide high payload concentrations at specific target sites and could minimize side effects.

Chemotherapy

The introduction of nanotechnology into medicine is expected to advance diagnostic cancer imaging and the standards for therapeutic drug design. Nanotechnology may uncover insight about the structure, function and organizational level of the biosystem at the nanoscale.

Silver nanoparticles can undergo coating techniques that offer a uniform functionalized surface to which substrates can be added. When the nanoparticle is coated, for example, in silica the surface exists as silicic acid. Substrates can thus be added through stable ether and ester linkages that are not degraded immediately by natural metabolic enzymes. Recent chemotherapeutic applications have designed anti cancer drugs with a photo cleavable linker, such as an ortho-nitrobenzyl bridge, attaching it to the substrate on the nanoparticle surface. The low toxicity nanoparticle complex can remain viable under metabolic attack for the time necessary to be distributed throughout the bodies systems. If a cancerous tumor is being targeted for treatment, ultraviolet light can be introduced over the tumor region. The electromagnetic energy of the light causes the photo responsive linker to break between the drug and the nanoparticle substrate. The drug is now cleaved and released in an unaltered active form to act on the cancerous tumor cells. Advantages anticipated for this method is that the drug is transported without highly toxic compounds, the drug is released without harmful radiation or relying on a specific chemical reaction to occur and the drug can be selectively released at a target tissue.

A second approach is to attach a chemotherapeutic drug directly to the functionalized surface of the silver nanoparticle combined with a nucelophilic species to undergo a displacement reaction. For example, once the nanoparticle drug complex enters or is in the vicinity of the target tissue or cells, a glutathione monoester can be administered to the site. The nucleophilic ester oxygen will attach to the functionalized surface of the nanoparticle through a new ester linkage while the drug is released to its surroundings.

The drug is now active and can exert its biological function on the cells immediate to its surroundings limiting non-desirable interactions with other tissues.

Multiple Drug Resistance

A major cause for the ineffectiveness of current chemotherapy treatments is multiple drug resistance which can arise from several mechanisms.

Nanoparticles can provide a means to overcome MDR. In general, when using a targeting agent to deliver nanocarriers to cancer cells, it is imperative that the agent binds with high selectivity to molecules that are uniquely expressed on the cell surface. Hence NPs can be designed with proteins that specifically detect drug resistant cells with over-expressed transporter proteins on their surface. A pitfall of the commonly used nano-drug delivery systems is that free drugs that are released from the nanocarriers into the cytosol get exposed to the MDR transporters once again, and are exported. To solve this, 8 nm nano crystalline silver particles were modified by the addition of trans-activating transcriptional activator (TAT), derived from the HIV-1 virus, which acts as a cell penetrating peptide (CPP). Generally, AgNP effectiveness is limited due to the lack of efficient cellular uptake; however, CPP-modification has become one of the most efficient methods for improving intracellular delivery of nanoparticles. Once ingested, the export of the AgNP is prevented based on a size exclusion. The concept is simple: the nanoparticles are too large to be effluxed by the MDR transporters, because the efflux function is strictly subjected to the size of its substrates, which is generally limited to a range of 300-2000 Da. Thereby the nanoparticulates remain insusceptible to the efflux, providing a means to accumulate in high concentrations.

Antimicrobial

Introduction of silver into bacterial cells induces a high degree of structural and morphological changes, which can lead to cell death. As the silver nano particles come in contact with the bacteria, they adhere to the cell wall and cell membrane. Once bound, some of the silver passes through to the inside, and interacts with phosphate-containing compounds like DNA and RNA, while another portion adheres to the sulphur-containing proteins on the membrane. The silver-sulphur interactions at the membrane cause the cell wall to undergo structural changes, like the formation of pits and pores. Through these pores, cellular components are released into the extracellular fluid, simply due to the osmotic difference. Within the cell, the integration of silver creates a low molecular weight region where the DNA then condenses. Having DNA in a condensed state inhibits the cell's replication proteins contact with the DNA. Thus the introduction of silver nanoparticles inhibits replication and is sufficient to cause the death of the cell. Further increasing their effect, when silver comes in contact with fluids, it tends to ionize which increases the nanoparticles bactericidal activity. This has been correlated to the suppression of enzymes and inhibited expression of proteins that relate to the cell's ability to produce ATP.

Although it varies for every type of cell proposed, as their cell membrane composition varies greatly, It has been seen that in general, silver nano particles with an average size of 10 nm or less show electronic effects that greatly increase their bactericidal activity. This could also be partly due to the fact that as particle size decreases, reactivity increases due to the surface area to volume ratio increasing.

It has been noted that the introduction of silver nano particles has shown to have synergistic activity with common antibiotics already used today, such as; penicillin G, ampicillin, erythromycin, clindamycin, and vancomycin against E. coli and S. aureus. In medical equipment, it has been shown that silver nano particles drastically lower the bacterial count on devices used. However, the problem arises when the procedure is over and a new one must be done. In the process of washing the instruments a large portion of the silver nano particles become less effective due to the loss of silver ions. They are more commonly used in skin grafts for burn victims as the silver nano particles embedded with the graft provide better antimicrobial activity and result in significantly less scarring of the victim. They also show promising application as water treatment method to form clean potable water.

Other

Samsung has created and marketed a material called Silver Nano, that includes silver nanoparticles on the surfaces of household appliances.

Safety

Although silver nanoparticles are widely used in a variety of commercial products, there has only recently been a major effort to study their effects on human health. There have been several studies that describe the *in vitro* toxicity of silver nanoparticles to a variety of different organs, including the lung, liver, skin, brain, and reproductive organs. The mechanism of the toxicity of silver nanoparticles to human cells appears to be derived from oxidative stress and inflammation that is caused by the generation of reactive oxygen species (ROS) stimulated by either the Ag NPs, Ag ions, or both. For example, Park *et al.* showed that exposure of a mouse peritoneal macrophage cell line (RAW267.7) to silver nanoparticles decreased the cell viability in a concentration- and time-dependent manner. They further showed that the intracellular reduced glutathione (GSH), which is a ROS scavenger, decreased to 81.4% of the control group of silver nanoparticles at 1.6 ppm.

Modes of Toxicity

Since silver nanoparticles undergo dissolution releasing silver ions, which is well-documented to have toxic effects, there have been several studies that have been conducted to determine whether the toxicity of silver nanoparticles is derived from the release of silver ions or from the nanoparticle itself. Several studies suggest that the toxicity of silver nanoparticles is attributed to their release of silver ions in cells as both silver nanoparticles and silver ions have been reported to have similar cytotoxicity. For example, In some cases it is reported that silver nanoparticles facilitate the release of toxic free silver ions in cells via a "Trojan-horse type mechanism," where the particle enters cells and is then ionized within the cell. However, there have been reports that suggest that a combination of silver nanoparticles and ions is responsible for the toxic effect of silver nanoparticles. Navarro *et al.* using cysteine ligands as a tool to measure the concentration of free silver in solution, determined that although initially silver ions were 18 times more likely to inhibit the photosynthesis of an algae, Chlamydomanas reinhardtii, but after 2 hours of incubation it was revealed that the algae containing silver nanoparticles were more toxic than just silver ions alone. Furthermore, there are studies that suggest that silver nanoparticles induce toxicity independent of free silver ions. For example, Asharani *et al.* compared phenotypic defects observed in zebrafish treated with silver nanoparticles and silver ions and determined that the phenotypic defects observed with silver nanoparticle treatment was not observed with silver ion-treated embryos, suggesting that the toxicity of silver nanoparticles are independent of silver ions.

Protein channels and nuclear membrane pores can often be in the size range of 9 nm to 10 nm in diameter. Small silver nanoparticles constructed of this size have the ability to not only pass through the membrane to interact with internal structures but also to be become lodged within the membrane. Silver nanoparticle depositions in the membrane can impact regulation of solutes, exchange of proteins and cell recognition. Exposure to silver nanoparticles has been associated with "inflammatory, oxidative, genotoxic, and cytotoxic consequences"; the silver particulates primarily accumulate in the liver. but have also been shown to be toxic in other organs including the brain. Nano-silver applied to tissue-cultured human cells leads to the formation of free radicals, raising concerns of potential health risks.

- Allergic reaction: There have been several studies conducted that show a precedence for allerginicity of silver nanoparticles.

- Argyria and staining: Ingested silver or silver compounds, including colloidal silver, can cause a condition called argyria, a discoloration of the skin and organs.In 2006, there was a case study of a 17-year-old man, who sustained burns to 30% of his body, and experienced a temporary bluish-grey hue after several days of treatment with Acticoat, a brand of wound dressing containing silver nanoparticles. Argyria is the deposition of silver in deep tissues, a condition that cannot happen on a temporary basis, raising the question of whether the cause of the man's discoloration was argyria or even a result of the silver treatment. Silver dressings are known to cause a "transient discoloration" that dissipates in 2–14 days, but not a permanent discoloration.

- Silzone heart valve: St. Jude Medical released a mechanical heart valve with a silver coated sewing cuff (coated using ion beam-assisted deposition) in 1997.

The valve was designed to reduce the instances of endocarditis. The valve was approved for sale in Canada, Europe, the United States, and most other markets around the world. In a post-commercialization study, researchers showed that the valve prevented tissue ingrowth, created paravalvular leakage, valve loosening, and in the worst cases explantation. After 3 years on the market and 36,000 implants, St. Jude discontinued and voluntarily recalled the valve.

Platinum Nanoparticles

Platinum nanoparticles are usually in the form of a suspension or colloid of submicrometre-size particles of platinum in a fluid, usually water. A colloid is technically defined as a stable dispersion of particles in a fluid medium (liquid or gas).

Spherical Platinum nanoparticles can be made with sizes between about 2 and 100 nanometres (nm), depending on reaction conditions,. Platinum nanoparticles are suspended in the colloidal solution of brownish-red or black color. Nanoparticles come in wide variety of shapes including spheres, rods, cubes, and tetrahedra.

Platinum nanoparticles are the subject of substantial research, with potential applications in a wide variety of areas. These include catalysis, medicine, and the synthesis of novel materials with unique properties.

Synthesis

Platinum nanoparticles are typically synthesized either by the reduction of platinum ion precursors in solution with a stabilizing or capping agent to form colloidal nanoparticles, or by the impregnation and reduction of platinum ion precursors in a micro-porous support such as alumina.

Some common examples of platinum precursors include potassium hexachloroplatinate (K_2PtCl_6) or platinous chloride (K_2PtCl_6) Different combinations of precursors, such as ruthenium chloride ($RuCl_3$) and chloroplatinic acid (H_2PtCl_6), have been used to synthesize mixed-metal nanoparticles Some common examples of reducing agents include hydrogen gas (H_2), sodium borohydride ($NaBH_4$) and ethylene glycol ($C_2H_6O_2$), although other alcohols and plant-derived compounds have also been used.

As the platinum metal precursor is reduced to neutral platinum metal (Pt^o), the reaction mixture becomes supersaturated with platinum metal and the Pt^o begins to precipitate in the form of nanoscale particles. A capping agent or stabilizing agent such as sodium polyacrylic acid or sodium citrate is often used to stabilize the nanoparticle surfaces, and prevents the aggregation and coalescence of the nanoparticles.

The size of nanoparticles synthesized colloidally may be controlled by changing the platinum precursor, the ratio of capping agent to precursor, and/or the reaction temperature. The size of the nanoparticles can also be controlled with small deviation by using a stepwise seed-mediated growth procedure as outlined by Bigall et al. (2008). The size of nanoparticles synthesized onto a substrate such as alumina depends on various parameters such as the pore size of the support.

Platinum nanoparticles can also be synthesized by decomposing $Pt_2(dba)_3$ (dba = dibenzylideneacetone) under a CO or H_2 atmosphere, in the presence of a capping agent. The size and shape distributions of the resulting nanoparticles depend on the solvent, the reaction atmosphere, the types of capping agents and their relative concentrations, the specific platinum ion precursor, as well at the temperature of the system and reaction time.

Shape and Size Control

Ramirez et al. reported the influence of ligand and solvent effects on the size and shape of platinum nanoparticles. Platinum nanoparticle seeds were prepared by the decomposition of $Pt_2(dba)_3$ in tetrahydrofuran (THF) under carbon monoxide (CO). These conditions produced Pt nanoparticles with weakly bound THF and CO ligands and an approximate diameter on 1.2 nm. Hexadecylamine (HDA) was added to the purified reaction mixture and allowed to displace the THF and CO ligands over the course of approximately seven days, producing monodispersed spherical crystalline Pt nanoparticles with an average diameter of 2.1 nm. After the seven-day period, an elongation of the Pt nanoparticles occurred. When the same procedure was followed using a stronger capping agent such as triphenyl phosphine or octanethiol, the nanoparticles remained spherical, suggesting that the HDA ligand affects particle shape.

When $Pt_2(dba)_3$ was decomposed in THF under hydrogen gas in the presence HDA, the reaction took much longer, and formed nanowires with diameters between 1.5 and 2 nm. Decomposition of $Pt_2(dba)_3$ under hydrogen gas in toluene yielded the formation of nanowires with 2-3 nm diameter independent of HDA concentration. The length of these nanowires was found to be inversely proportional to the concentration of HDA present in solution. When these nanowire syntheses were repeated using reduced concentrations of $Pt_2(dba)_3$, there was little effect on the size, length or distribution of the nanowires formed.

Platinum nanoparticles of controlled shape and size have also been accessed through varying the ratio of polymer capping agent concentration to precursor concentration. Reductive colloidal syntheses as such have yielded tetrahedral, cubic, irregular-prismatic, icosahedral, and cubo-octahedral nanoparticles, whose dispersity is also dependent on the concentration ratio of capping agent to precursor, and which may be applicable to catalysis. The precise mechanism of shape-controlled colloidal synthesis is not yet known; however, it is known that the relative growth rate of crystal facets within the growing nanostructure determines its final shape. Polyol syntheses of platinum nanoparticles, in which chloroplatinic acid is reduced to $PtCl_4^{2-}$ and Pt^0 by ethylene glycol, have also been a means to shape-controlled fabrication. Addition of varying amounts of sodium nitrate to these reactions was shown to yield tetrahedra and octahedra at high concentration ratios of sodium nitrate to chloroplatinic acid. Spectroscopic studies suggest that nitrate is reduced to nitrite by $PtCl_4^{2-}$ early in this reaction, and that the nitrite may then coordinate both Pt(II) and Pt(IV), greatly slowing the polyol reduction and altering the growth rates of distinct crystal facets within the nanoparticles, ultimately yielding morphological differentiation.

Green Synthesis

An ecologically-friendly synthesis of platinum nanoparticles from chloroplatinic acid was achieved through the use of a leaf extract of *Diospyros kaki* as a reducing agent. Nanoparticles synthesized as such were spherical with an average diameter ranging from 2-12 nm depending on reaction temperature and concentration of leaf extract used. Spectroscopic analysis suggests that this reaction is not enzyme-mediated and proceeds instead through plant-derived reductive small molecules. Another eco-friendly synthesis from chloroplatinic acid was reported using leaf extract from *Ocimum sanctum* and tulsi as reducing agents. Spectroscopic analysis suggested that ascorbic acid, gallic acid, various terpenes, and certain amino acids were active in the reduction. Particles synthesized as such were shown through scanning electron microscopy to consist in aggregates with irregular shape. It has been shown that tea extracts with high polyphenol content may be used both as reducing agents and capping agents for platinum nanoparticle synthesis.

Properties and Applications

The chemical and physical properties of platinum nanoparticles (NP) make them applicable for a wide variety of research applications. Extensive experimentation has been done to create new species of platinum NPs, and study their properties. Platinum NP applications include electronics, optics, catalysts, and enzyme immobilization.

Catalytic Properties of Platinum Nanoparticles

Platinum NPs are used as catalysts for proton exchange membrane fuel cell (PEMFC), for industrial synthesis of nitric acid, reduction of exhaust gases from vehicles and as catalytic nucleating agents for synthesis of magnetic NPs. The catalytic reactivity of the NP is dependent on the shape, size and morphology of the particle

One type of platinum NPs that have been researched on are colloidal platinum NPs. Monometallic and bimetallic colloids have been used as catalysts in a wide range of organic chemistry, including, oxidation of carbon monoxide in aqueous solutions, hydrogenation of alkenes in organic or biphasic solutions and hydrosilylation of olefins in organic solutions. Collodial platinum NPs protected by Poly(N-isopropylacrylamide) were synthesised and their catalytic properties measured. It was determined that they were more active in solution and inactive when phase separated due to its solubility being inversely proportional to temperature.

Optical Properties of Platinum Nanoparticles

Platinum NPs exhibit fascinating optical properties. Being a free electron metal NP like Ag and Au, its linear optical response is mainly controlled by the surface plasmon resonance. Surface plasmon resonance occurs when the electrons in the metal surface are subject to an electromagnetic field that exerts a force on the electrons and cause them to displace from their original positions. The nuclei then exert a restoring force that results in oscillation of the electrons, which increase in strength when frequency of oscillations is in resonance with the incident electromagnetic wave.

Surface plasmon resonance (SPR) of platinum NPs occur in the ultraviolet region (215 nm). The resonance peak is very broad because the imaginary dielectric constant of platinum, $Im(\varepsilon_{Pt})$, is large. Experiments were done and the spectra obtained are similar for most platinum particles regardless of size. However, there is an exception. Platinum NPs synthesized via citrate reduction do not have a surface plasmon resonance peak around 215 nm.

Modifying Conductivity of Other Materials

Platinum NPs can be used to dope zinc oxide (ZnO) materials to improve their conductivity. ZnO has several characteristics that allow it to be used in several novel devices such as development of light-emitting assemblies and solar cells. However, because ZnO is of slightly lower conductivity than metal and indium tin oxide (ITO), it can be doped and hybridized with metal NPs like platinum to improve its conductivity. A method to do so would be to synthesize ZnO NPs using methanol reduction and incorporate at 0.25 at.% platinum NPs. This boosts the electrical properties of ZnO films while preserving its transmittance for application in transparent conducting oxides.

Biological Effects

One of the biological effects that researchers are currently investigating is how platinum nanoparticles affect lifespan. Research by Yusei Miyamoto at University of Tokyo, Japan found that 2-3 nm platinum nanoparticles have been found to increase the lifespan of the roundworm Caenorhabditis elegans by 20-25%. This is due to platinum nanoparticles having antioxidant properties. Antioxidant properties help to combat reactive oxygen species (ROS). ROS include, but are not limited superoxide anions, hydroxyl radicals, and hydrogen peroxide. The amount of ROS produced directly correlates with the metabolic rate and life span of an organism. An increase in oxidative stress has been observed in correspondence with age.

In the roundworm, C. elegans, it was noted that the intracellular level of lipofuscin increased with oxidative stress. Lipofuscin is a fluorescent molecule, and the fluorescence emitted was measured to show the amount of ROS in the roundworm. Platinum nanoparticle solutions were introduced to C. elegans in varying concentrations. It was observed that at a concentration of 0.1 mM the nano-Pt solution was too low to hinder aging or decrease the accumulation of lipofuscin. At 0.5 mM nano-Pt solution there was a significant decrease in the fluorescence observed, and the lifespan of C. elegans was extended by 22.3±2.8%. However, at a 1 mM concentration of nano-Pt solution the lifespan of C. elegans drastically decreased. This did not correspond with what the authors had initially hypothesized. They later theorized that the toxicity of the platinum nanoparticles could have attributed to the shortened lifespan of C. elegans.

Toxicology

In one study, mice injected with platinum nanoparticles of less than 1 nm size developed symptoms of liver damage (elevated ALT and AST levels) while mice injected with 15 nm platinum nanoparticles did not. In another study, mice that received a single injection of platinum nanoparticles of less than 1 nm size developed necrosis of tubular epithelial cells and urinary casts in the kidney, while mice repeatedly injected with 8 nm platinum nanoparticles did not. In yet another study, human cells exposed to platinum nanoparticles ~5–8 nm in size experienced DNA damage.

Nanoparticles are a new class of materials that are used commercially in cosmetic, pharmaceutical, and electronic industries. Nanoparticles are becoming increasingly more common for drug use. This is due to their ability to provide specific drug targeting and delivery. Nanoparticles have the benefits of being able to pass through cellular membranes, as well as to bind and stabilize proteins. It's also been noted that generally, nanoparticles can decrease the overall toxicity of the incorporated drug. Initially, it wasn't realized that the carrier system in drug delivery (nanoparticle) could be toxic in itself. This requires further study of the toxicology of nanoparticles.

It has become vital to research the toxicological impact of nanoparticles. This is primarily due to nanoparticle use in the pharmaceutical industry. Nanoparticles can have a high surface-volume ratio, which increases the reactivity and the toxicity of the surface. Small nanoparticles (<8 nm) have the ability to pass through cellular membranes, adversely affecting the human respiratory system, central nervous system, and the gastrointestinal tract, amongst other things. The inhalation of nanoparticles is considered the most antagonistic trait, because it can cause inflammation, and disease in the lungs. Nanotoxicology has been developed to study the adverse effects of nanoparticles. Currently, toxicological impacts haven't been extensively studied in nanoparticles. Platinum nanoparticles in themselves have not been evaluated for the advantages or disadvantages they could bring to the table. Research needs to be further conducted to analyze all of the impacts that platinum nanoparticles contribute to a system.

Magnetic Nanoparticles

Magnetic nanoparticles are a class of nanoparticle that can be manipulated using magnetic fields. Such particles commonly consist of two components, a magnetic material, often iron, nickel and cobalt, and a chemical component that has functionality. While nanoparticles are smaller than 1 micrometer in diameter (typically 5–500 nanometers), the larger microbeads are 0.5–500 micrometer in diameter. Magnetic nanoparticle clusters that are composed of a number of individual magnetic nanoparticles are known as magnetic nanobeads with a diameter of 50–200 nanometers. Magnetic nanoparticle clusters are a basis for their further magnetic assembly into magnetic nanochains. The magnetic nanoparticles have been the focus of much research recently because they possess attractive properties which could see potential use in catalysis including nanomaterial-based catalysts, biomedicine and tissue specific targeting, magnetically tunable colloidal photonic crystals, microfluidics, magnetic resonance imaging, magnetic particle imaging, data storage, environmental remediation, nanofluids, and optical filters, defect sensor and cation sensors.

Properties

The physical and chemical properties of magnetic nanoparticles largely depend on the synthesis method and chemical structure. In most cases, the particles range from 1 to 100 nm in size and may display superparamagnetism.

Types of Magnetic Nanoparticles

Oxides: Ferrites

Ferrite nanoparticles or iron oxide nanoparticles (iron oxides in crystal structure of maghemite or magnetite) are the most explored magnetic nanoparticles up to date. Once the ferrite particles become smaller than 128 nm they become superparamagnetic which prevents self agglomeration since they exhibit their magnetic behavior only when an external magnetic field is applied. The magnetic moment of ferrite nanoparticles can be greatly increased by controlled clustering of a number of individual superparamagnetic nanoparticles into superparamagnetic nanoparticle clusters, namely magnetic nanobeads. With the external magnetic field switched off, the remanence falls back to zero. Just like non-magnetic oxide nanoparticles, the surface of ferrite nanoparticles is often modified by surfactants, silica, silicones or phosphoric acid derivatives to increase their stability in solution.

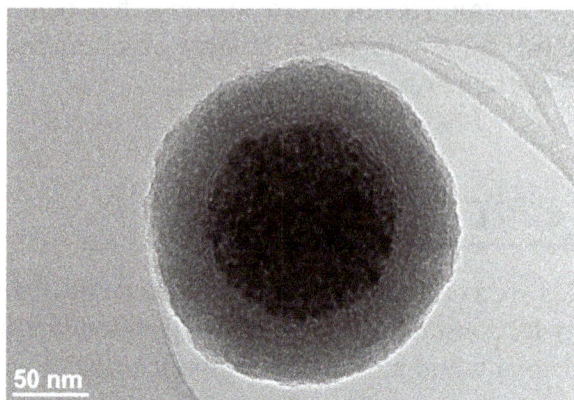

TEM image of a maghemite magnetic nanoparticle cluster with silica shell.

Ferrites with A Shell

The surface of a maghemite or magnetite magnetic nanoparticle is relatively inert and does not usually allow strong covalent bonds with functionalization molecules. However, the reactivity of the magnetic nanoparticles can be improved by coating a layer of silica onto their surface. The silica shell can be easily modified with various surface functional groups via covalent bonds between organo-silane molecules and silica shell. In addition, some fluorescent dye molecules can be covalently bonded to the functionalized silica shell.

Ferrite nanoparticle clusters with narrow size distribution consisting of superparamagnetic oxide nanoparticles (~ 80 maghemite superparamagnetic nanoparticles per bead) coated with a silica shell have several advantages over metallic nanoparticles:

- Higher chemical stability (crucial for biomedical applications)

- Narrow size distribution (crucial for biomedical applications)

- Higher colloidal stability since they do not magnetically agglomerate

- Magnetic moment can be tuned with the nanoparticle cluster size

- Retained superparamagnetic properties (independent of the nanoparticle cluster size)

- Silica surface enables straightforward covalent functionalization

Metallic

Metallic nanoparticles may be beneficial for some technical applications due to their higher magnetic moment whereas oxides (maghemite, magnetite) would be beneficial for biomedical applications. This also implies that for the same moment, metallic nanoparticles can be made smaller than their oxide counterparts. On the other hand, metallic nanoparticles have the great disadvantage of being pyrophoric and reactive to oxidizing agents to various degrees. This makes their handling difficult and enables unwanted side reactions which makes them less appropriate for biomedical applications. Colloid formation for metallic particles is also much more challenging.

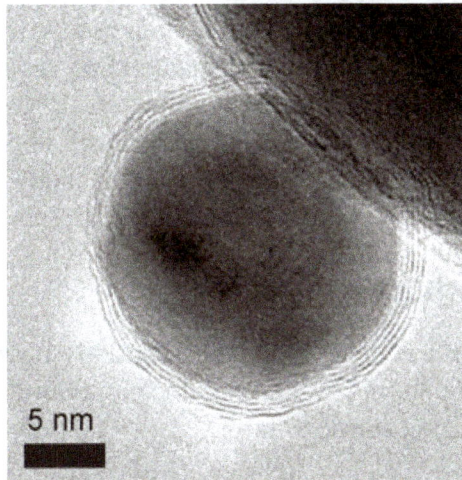

Cobalt nanoparticle with graphene shell (note: The individual graphene layers are visible)

Metallic with A Shell

The metallic core of magnetic nanoparticles may be passivated by gentle oxidation, surfactants, polymers and precious metals. In an oxygen environment, Co nanoparticles form an anti-ferromagnetic CoO layer on the surface of the Co nanoparticle. Recently, work has explored the synthesis and exchange bias effect in these Co core CoO shell nanoparticles with a gold outer shell. Nanoparticles with a magnetic core consisting either of elementary Iron or Cobalt with a nonreactive shell made of graphene have been synthesized recently. The advantages compared to ferrite or elemental nanoparticles are:

- Higher magnetization

- Higher stability in acidic and basic solution as well as organic solvents

- Chemistry on the graphene surface via methods already known for carbon nanotubes

Synthesis

Several methods exist for preparing magnetic nanoparticle.

Co-precipitation

Co-precipitation is a facile and convenient way to synthesize iron oxides (either Fe_3O_4 or γ-Fe_2O_3) from aqueous Fe^{2+}/Fe^{3+} salt solutions by the addition of a base under inert atmosphere at room temperature or at elevated temperature. The size, shape, and composition of the magnetic nanoparticles very much depends on the type of salts used (e.g.chlorides, sulfates, nitrates), the Fe^{2+}/Fe^{3+} ratio, the reaction temperature, the pH value and ionic strength of the media, and the mixing rate with the base solution used to provoke the precipitation. The co-precipitation approach has been used extensively to produce ferrite nanoparticles of controlled sizes and magnetic properties. A variety of experimental arrangements have been reported to facilitate continuous and large–scale co–precipitation of magnetic particles by rapid mixing. Recently, the growth rate of the magnetic nanoparticles was measured in real-time during the precipitation of magnetite nanoparticles by an integrated AC magnetic susceptometer within the mixing zone of the reactants.

Thermal Decomposition

Magnetic nanocrystals with smaller size can essentially be synthesized through the thermal decomposition of alkaline organometallic compounds in high-boiling organic solvents containing stabilizing surfactants.

Microemulsion

Using the microemulsion technique, metallic cobalt, cobalt/platinum alloys, and gold-coated cobalt/platinum nanoparticles have been synthesized in reverse micelles of cetyltrimethlyammonium bromide, using 1-butanol as the cosurfactant and octane as the oil phase.,

Flame Spray Synthesis

Using flame spray pyrolysis and varying the reaction conditions, oxides, metal or carbon coated nanoparticles are produced at a rate of > 30 g/h .

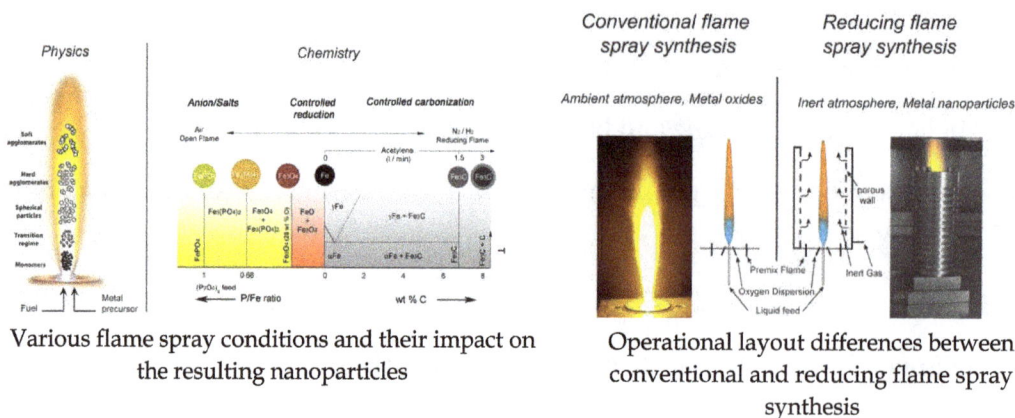

Various flame spray conditions and their impact on the resulting nanoparticles

Operational layout differences between conventional and reducing flame spray synthesis

Potential Applications

A wide variety of potential applications have been envisaged. Since magnetic nanoparticles are expensive to produce, there is interest in their recycling or for highly specialized applications. They are only used in scientific research. An industrial use has yet to be established.

The potential and versatility of magnetic chemistry arises from the fast and easy separation of the magnetic nanoparticles, eliminating tedious and costly separation processes usually applied in chemistry. Furthermore the magnetic nanoparticles can be guided via a magnetic field to the desired location which could, for example, enable pinpoint precision in fighting cancer.

Medical Diagnostics and Treatments

Magnetic nanoparticles have been examined for use in an experimental cancer treatment called magnetic hyperthermia in which the fact that nanoparticles heat when they are placed in an alternative magnetic field is used.

Affinity ligands such as epidermal growth factor (EGF), folic acid, aptamers, lectins etc. can be attached to the magnetic nanoparticle surface with the use of various chemistries. This enables targeting of magnetic nanoparticles to specific tissues or cells. This strategy is used in cancer research to target and treat tumors in combination with magnetic hyperthermia or nanoparticle-delivered cancer drugs.

Another potential treatment of cancer includes attaching magnetic nanoparticles to free-floating cancer cells, allowing them to be captured and carried out of the body. The treatment has been tested in the laboratory on mice and will be looked at in survival studies.

Magnetic nanoparticles can be used for the detection of cancer. Blood can be inserted onto a microfluidic chip with magnetic nanoparticles in it. These magnetic nanoparticles are trapped inside due to an externally applied magnetic field as the blood is free to flow through. The magnetic nanoparticles are coated with antibodies targeting cancer cells or proteins. The magnetic nanoparticles can be recovered and the attached cancer-associated molecules can be assayed to test for their existence.

Magnetic nanoparticles can be conjugated with carbohydrates and used for detection of bacteria. Iron oxide particles have been used for the detection of Gram negative bacteria like Escherichia coli and for detection of Gram positive bacteria like Streptococcus suis

Magnetic Immunoassay

Magnetic immunoassay (MIA) is a novel type of diagnostic immunoassay utilizing magnetic nanobeads as labels in lieu of conventional, enzymes, radioisotopes or fluorescent moieties. This assay involves the specific binding of an antibody to its antigen, where a magnetic label is conjugated to one element of the pair. The presence of magnetic nanobeads is then detected by a magnetic reader (magnetometer) which measures the magnetic field change induced by the beads. The signal measured by the magnetometer is proportional to the analyte (virus, toxin, bacteria, cardiac marker,etc.) quantity in the initial sample.

Waste Water Treatment

Thanks to the easy separation by applying a magnetic field and the very large surface to volume ratio, magnetic nanoparticles have a potential for treatment of contaminated water. In this method, attachment of EDTA-like chelators to carbon coated metal nanomagnets results in a magnetic reagent for the rapid removal of heavy metals from solutions or contaminated water by three orders of magnitude to concentrations as low as micrograms per Litre. Magnetic nanobeads or nanoparticle clusters composed of FDA-approved oxide superparamagnetic nanoparticles (e.g. maghemite, magnetite) hold much potential for waste water treatment since they express excellent biocompatibility which concerning the environmental impacts of the material is an advantage compared to metallic nanoparticles.

Supported Enzymes and Peptides

Enzymes, proteins, and other biologically and chemically active substances have been immobilized on magnetic nanoparticles. They are of interest as possible supports for solid phase synthesis.

This technology is potentially relevant to cellular labelling/cell separation, detoxification of biological fluids, tissue repair, drug delivery, magnetic resonance imaging, hyperthermia and magnetofection.

Catalyst Support

Magnetic nanoparticles are of potential use as a catalyst or catalyst supports. In chemistry, a catalyst support is the material, usually a solid with a high surface area, to which a catalyst is affixed. The reactivity of heterogeneous catalysts occurs at the surface atoms. Consequently, great effort is made to maximize the surface area of a catalyst by distributing it over the support. The support may be inert or participate in the catalytic reactions. Typical supports include various kinds of carbon, alumina, and silica. Immobilizing the catalytic center on top of nanoparticles with a large surface to volume ratio addresses this problem. In the case of magnetic nanoparticles it adds the property of facile a separation. An early example involved a rhodium catalysis attached to magnetic nanoparticles .

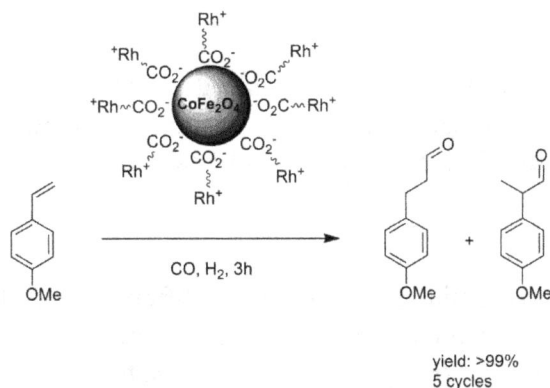

yield: >99%
5 cycles

In another example, the stable radical TEMPO was attached to the graphene-coated cobalt nanoparticles via a diazonium reaction. The resulting catalyst was then used for the chemoselective oxidation of primary and secondary alcohols.

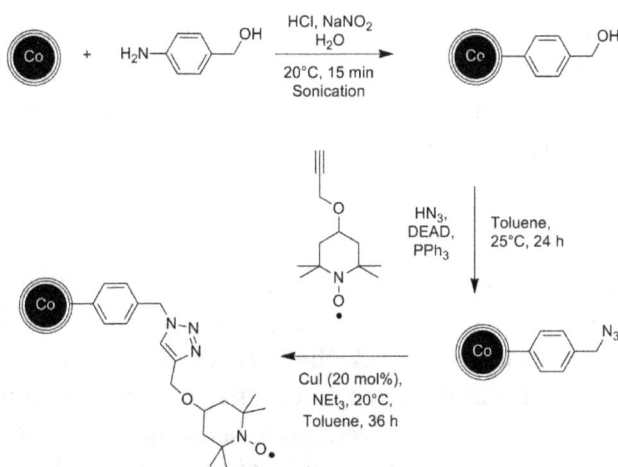

The catalytic reaction can be conducted in a continuous flow reactor instead of a batch reactor with no remains of the catalyst in the end product. Graphene coated cobalt nanoparticles have been used for that experiment since they exhibit a higher magnetization than Ferrite nanoparticles, which is essential for a fast and clean separation via external magnetic field.

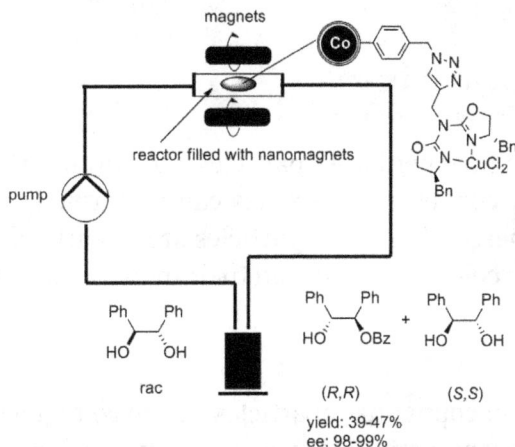

yield: 39-47%
ee: 98-99%

Biomedical Imaging

There are many applications for iron-oxide based nanoparticles in concert with magnetic resonance imaging. Magnetic CoPt nanoparticles are being used as an MRI contrast agent for transplanted neural stem cell detection.

Cancer Therapy

In magnetic fluid hyperthermia, nanoparticles of different types like Iron oxide, magnetite, maghemite or even gold are injected in tumor and then subjected under a high frequency magnetic field. These nanoparticles produce heat that can increase temperature up to 42 degree which can be useful for destroying cancerous cells.

Information Storage

A promising candidate for high-density storage is the face-centered tetragonal phase FePt alloy. Grain sizes can be as small as 3 nanometers. If it's possible to modify the MNPs at this small scale, the information density that can be achieved with this media could easily surpass 1 Terabyte per square inch.

Genetic Engineering

Magnetic nanoparticles can be used for a variety of genetics applications. One application is the rapid isolation of mRNA. In one application, the magnetic bead is attached to a poly T tail. When mixed with mRNA, the poly A tail of the mRNA will attach to the bead's poly T tail and the isolation takes place simply by placing a magnet on the side of the tube and pouring out the liquid. Magnetic beads have also been used in plasmid assembly. Rapid genetic circuit construction has been achieved by the sequential addition of genes onto a growing genetic chain, using nanobeads as an anchor. This method has been shown to be much faster than previous methods, taking less than an hour to create functional multi-gene constructs in vitro.

Copper Nanoparticles

Copper nanoparticles are copper based particles 1 to 100 nm in size. Like many other forms of nanoparticles, copper nanoparticles can be formed by natural processes or through chemical synthesis. These nanoparticles are of particular interest due to their historical application as coloring agents and their modern-day biomedical ones.

Historical Uses

One of the earliest uses of copper nanoparticles was to color glass and ceramics during the ninth century in Mesopotamia. This was done by creating a glaze with copper and silver salts and applying it to clay pottery. When the pottery was baked at high temperatures in reducing conditions, the metal ions migrated to the outer part of the glaze and were reduced to metals. The end result was a double layer of metal nanoparticles with a small amount of glaze in between them. When the finished pottery was exposed to light, the light would penetrate and reflect off the first layer. The light penetrating the first layer would reflect off the second layer of nanoparticles and cause interference effects with light reflecting off the first layer, creating a luster effect that results from both constructive and destructive interference.

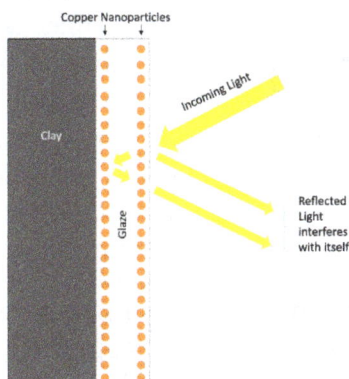

Synthesis

Various methods have been described to chemically synthesize copper nanoparticles. An older method involves the reduction of copper hydrazine carboxylate in an aqueous solution using reflux or by heating through ultrasound under an inert argon atmosphere. This results in a combination of copper oxide and pure copper nanoparticle clusters, depending on the method used. A more modern synthesis utilizes copper chloride in a room temperature reaction with sodium citrate or myristic acid in an aqueous solution containing sodium formaldehyde sulfoxylate (SFS) to obtain a pure copper nanoparticle powder. While these syntheses generate fairly consistent copper nanoparticles, the possibility of controlling the sizes and shapes of copper nanoparticles has also been reported. The reduction of copper(II) acetylacetonate in organic solvent with oleyl amine and oleic acid causes the formation of rod and cube-shaped nanoparticles while variations in reaction temperature affect the size of the synthesized particles.

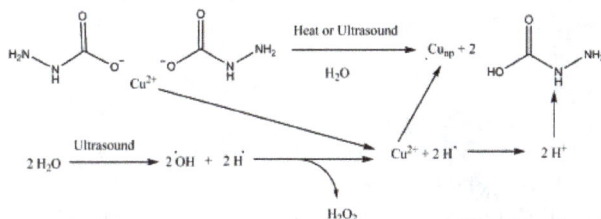

One method of synthesizing copper nanoparticles involves the copper (II) hydrazine carboxylate salt which undergoes a radical reaction with radical hydrogen produced by ultrasounds to form nanoparticles, hydrogen peroxide, and hydrazine carboxylic acid.

Another method of synthesis involves using copper (II) hydrazine carboxylate salt with ultrasound or heat in water to generate a radical reaction, as shown in the figure to the right. Copper nanoparticles can also be synthesized using green chemistry to reduce the environmental impact of the reaction. Copper chloride can be reduced using only L-ascorbic acid in a heated aqueous solution to produce stable copper nanoparticles.

Characteristics

Copper nanoparticles display unique characteristics including catalytic and antifungal/ antibacterial activities that are not observed in commercial copper. First of all, copper nanoparticles demonstrate a very strong catalytic activity, a property that can be attributed to their large catalytic surface area. With the small size and great porosity, the nanoparticles are able to achieve a higher reaction yield and a shorter reaction time when utilized as reagents in organic and organometallic synthesis. In fact, copper nanoparticles that are used in a condensation reaction of iodobenzene attained about 88% conversion to biphenyl, while the commercial copper exhibited only a conversion of 43%.

Copper nanoparticles that are extremely small and have a high surface to volume ratio can also serve as antifungal/antibacterial agents. The antimicrobial activity is induced by their close interaction with microbial membranes and their metal ions released in solutions. As the nanoparticles oxidize slowly in solutions, cupric ions are released from them and they can create toxic hydroxyl free radicals when the lipid membrane is nearby. Then, the free radicals disassemble lipids in cell membranes through oxidation to degenerate the membranes. As a result, the intracellular substances seep out of cells through the destructed membranes; the cells are no longer able to sustain fundamental biochemical processes. In the end, all these alterations inside of the cell caused by the free radicals lead to cell death.

Applications

Copper nanoparticles with great catalytic activities can be applied to biosensors and electrochemical sensors. Redox reactions utilized in those sensors are generally irreversible and also require high overpotentials (more energy) to run. In fact, the nanoparticles have the ability to make the redox reactions reversible and to lower the overpotentials when applied to the sensors.

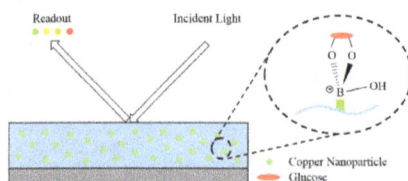

Figure 3: A polyacrylamide hydrogel with copper nanoparticles inside is able to determine glucose levels in a sample added to the gel. As phenylboronic acid groups on the hydrogel polymers bind the glucose molecules, the gel becomes swollen. As a result, the copper nanoparticles move apart, changing how incident light is diffracted by the gel. As the glucose levels decrease, the color of gel changes from red to orange to yellow to green.

One of the examples is a glucose sensor. With the use of copper nanoparticles, the sensor does not require any enzyme and therefore has no need to deal with enzyme degradation and denaturation. As described in Figure 3, depending on the level of glucose, the nanoparticles in the sensor diffract the incident light at a different angle. Consequently, the resulting diffracted light gives a different color based on the level of glucose. In fact, the nanoparticles enable the sensor to be more stable at high temperatures and varying pH, and more resistant to toxic chemicals. Moreover, using nanoparticles, native amino acids can be detected. A copper nanoparticle-plated screen-printed carbon electrode functions as a stable and effective sensing system for all 20 amino acid detection.

Colloidal Gold

Colloidal gold is a sol or colloidal suspension of submicrometre-size nanoparticles of gold in a fluid, usually water. The liquid is usually either an intense red colour (for particles less than 100 nm) or blue/purple (for larger particles). Due to the unique optical, electronic, and molecular-recognition properties of gold nanoparticles, they are the subject of substantial research, with applications in a wide variety of areas, including electron microscopy, electronics, nanotechnology, and materials science.

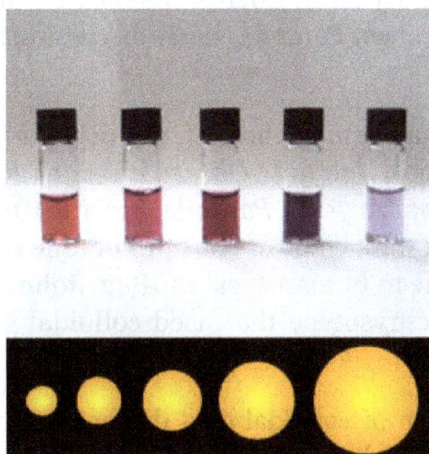

Suspensions of gold nanoparticles of various sizes. The size difference causes the difference in colors.

The properties of colloidal gold nanoparticles, and thus their applications, depend strongly upon their size and shape. For example, rodlike particles have both transverse and longitudinal absorption peak, and anisotropy of the shape affects their self-assembly.

History

Known, or at least used (perhaps proceeding by accident without much understanding of the process) since ancient times, the synthesis of colloidal gold was crucial to the 4th-century Lycurgus Cup, which changes color depending on the location of light source. Later it was used as a method of staining glass.

This cranberry glass bowl was made by adding a gold salt (probably gold chloride) to molten glass.

During the Middle Ages, soluble gold, a solution containing gold salt, had a reputation for its curative property for various diseases. In 1618, Francis Anthony, a philosopher and member of the medical profession, published a book called *Panacea Aurea, sive tractatus duo de ipsius Auro Potabili* (Latin: gold potion, or two treatments of potable gold). The book introduces information on the formation of colloidal gold and its medical uses. About half a century later, English botanist Nicholas Culpepper published book in 1656, *Treatise of Aurum Potabile*, solely discussing the medical uses of colloidal gold.

In 1676, Johann Kunckel, a German chemist, published a book on the manufacture of stained glass. In his book *Valuable Observations or Remarks About the Fixed and Volatile Salts-Auro and Argento Potabile, Spiritu Mundi and the Like*, Kunckel assumed that the slight pink color of Aurum Potabile came from small particles of metallic gold, not visible to human eyes. In 1842, John Herschel invented a photographic process called chrysotype that used colloidal gold to record images on paper.

Modern scientific evaluation of colloidal gold did not begin until Michael Faraday's work in the 1850s. In 1856, in a basement laboratory of Royal Institution, Faraday accidentally created a ruby red solution while mounting pieces of gold leaf onto microscope slides. Since he was already interested in the properties of light and matter, Faraday further investigated the optical properties of the colloidal gold. He prepared the first pure sample of colloidal gold, which he called 'activated gold', in 1857. He used phosphorus to reduce a solution of gold chloride. The colloidal gold Faraday made 150 years ago is still optically active. For a long time, the composition of the 'ruby' gold was unclear. Several chemists suspected it to be a gold tin compound, due to its preparation. Faraday recognized that the color was actually due to the miniature size of the gold particles. He noted the light scattering properties of suspended gold microparticles, which is now called Faraday-Tyndall effect.

In 1898, Richard Adolf Zsigmondy prepared the first colloidal gold in diluted solution. Apart from Zsigmondy, Theodor Svedberg, who invented ultracentrifugation, and Gustav Mie, who provided the theory for scattering and absorption by spherical particles, were also interested in the synthesis and properties of colloidal gold.

With advances in various analytical technologies in the 20th century, studies on gold nanoparticles has accelerated. Advanced microscopy methods, such as atomic force microscopy and electron microscopy, have contributed the most to nanoparticle research. Due to their comparably easy synthesis and high stability, various gold particles have been studied for their practical uses. Different types of gold nanoparticle are already used in many industries, such as medicine and electronics. For example, several FDA-approved nanoparticles are currently used in drug delivery.

Synthesis

Generally, gold nanoparticles are produced in a liquid ("liquid chemical methods") by reduction of chloroauric acid ($H[AuCl_4]$). After dissolving $H[AuCl_4]$, the solution is rapidly stirred while a reducing agent is added. This causes Au^{3+} ions to be reduced to Au^+ ions. Then a disproportionation reaction occurs whereby 3 Au^+ ions give rise to Au^{3+} and 2 Au^0 atoms. The Au^0 atoms act as center of nucleation around which further Au^+ ions gets reduced. To prevent the particles from aggregating, some sort of stabilizing agent that sticks to the nanoparticle surface is usually added. In the Turkevich method of Au NP synthesis, citrate initially acts as the reducing agent and finally as the capping agent which stabilizes the Au NP through electrostatic interactions between the lone pair of electrons on the oxygen and the metal surface. Also, gold colloids can be synthesised without stabilizers by laser ablation in liquids.

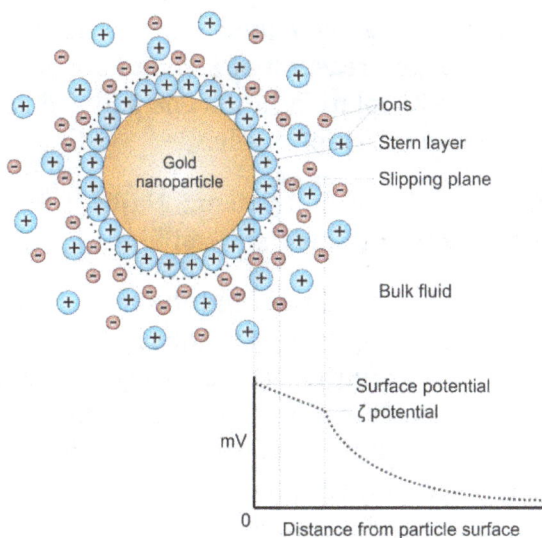

Potential difference as a function of distance from particle surface.

They can be functionalized with various organic ligands to create organic-inorganic hybrids with advanced functionality.

Turkevich Method

The method pioneered by J. Turkevich et al. in 1951 and refined by G. Frens in the 1970s, is the simplest one available. In general, it is used to produce modestly monodisperse spherical gold nanoparticles suspended in water of around 10–20 nm in diameter. Larger particles can be produced, but this comes at the cost of monodispersity and shape. It involves the reaction of small amounts of hot chloroauric acid with small amounts of sodium citrate solution. The colloidal gold will form because the citrate ions act as both a reducing agent and a capping agent.

Recently, the evolution of the spherical gold nanoparticles in the Turkevich reaction has been elucidated. It is interesting to note that extensive networks of gold nanowires are formed as a transient intermediate. These gold nanowires are responsible for the dark appearance of the reaction solution before it turns ruby-red.

To produce larger particles, less sodium citrate should be added (possibly down to 0.05%, after which there simply would not be enough to reduce all the gold). The reduction in the amount of sodium citrate will reduce the amount of the citrate ions available for stabilizing the particles, and this will cause the small particles to aggregate into bigger ones (until the total surface area of all particles becomes small enough to be covered by the existing citrate ions).

Brust Method

This method was discovered by Brust and Schiffrin in the early 1990s, and can be used to produce gold nanoparticles in organic liquids that are normally not miscible with water (like toluene). It involves the reaction of a chlorauric acid solution with tetra-octylammonium bromide (TOAB) solution in toluene and sodium borohydride as an anti-coagulant and a reducing agent, respectively.

Here, the gold nanoparticles will be around 5–6 nm. $NaBH_4$ is the reducing agent, and TOAB is both the phase transfer catalyst and the stabilizing agent.

It is important to note that TOAB does not bind to the gold nanoparticles particularly strongly, so the solution will aggregate gradually over the course of approximately two weeks. To prevent this, one can add a stronger binding agent, like a thiol (in particular, alkanethiols), which will bind to gold, producing a near-permanent solution. Alkanethiol protected gold nanoparticles can be precipitated and then redissolved. Some of the phase transfer agent may remain bound to the purified nanoparticles, this may affect physical properties such as solubility. In order to remove as much of this agent as possible, the nanoparticles must be further purified by soxhlet extraction.

Perrault Method

This approach, discovered by Perrault and Chan in 2009, uses hydroquinone to reduce $HAuCl_4$ in an aqueous solution that contains 15 nm gold nanoparticle seeds. This seed-based method of synthesis is similar to that used in photographic film development, in which silver grains within the film grow through addition of reduced silver onto their surface. Likewise, gold nanoparticles can act in conjunction with hydroquinone to catalyze reduction of ionic gold onto their surface. The presence of a stabilizer such as citrate results in controlled deposition of gold atoms onto the particles, and growth. Typically, the nanoparticle seeds are produced using the citrate method. The hydroquinone method complements that of Frens, as it extends the range of monodispersed spherical particle sizes that can be produced. Whereas the Frens method is ideal for particles of 12–20 nm, the hydroquinone method can produce particles of at least 30–300 nm.

Martin Method

This simple method, discovered by Martin and Eah in 2010, generates nearly monodisperse "naked" gold nanoparticles in water. Precisely controlling the reduction stoichiometry by simply adjusting the ratio of $NaBH_4$-NaOH ions to $HAuCl_4$-HCl ions within the "sweet zone," along with heating, enables reproducible diameter tuning between 3–6 nm. The aqueous particles are colloidally stable due to their high charge from the excess ions in solution. These particles can be coated with various hydrophilic functionalities, or mixed with hydrophobic molecules for applications in non-polar solvents. Interestingly, in non-polar solvents the nanoparticles curiously remain highly charged, and self-assemble on liquid droplets to form 2D monolayer films of monodisperse nanoparticles.

Nanotech Applications

Bacillus licheniformis can be used in synthesis of gold nanocubes. Researchers have synthesized gold nanoparticles with sizes between 10 and 100 nanometres. Gold nanoparticles are usually synthesized at high temperatures, in organic solvents and using toxic reagents. The bacteria produce them in much milder conditions.

Navarro Et Al. Method

As pointed out before, the precise size control with a low polydispersity of spherical gold nanoparticles remains difficult for particles larger than 30 nm. In order to provide maximum control on the NP structure along with synthetic straightforwardness, Navarro and co-workers used a 'one-pot' protocol, relying on a modified Turkevitch-Frens procedure. The method consist in the addition of sodium acetylacetonate (Na(acac) allowing the complexation and reduction of Au^{3+} to Au^{1+}) immediately followed by the addition of sodium citrate (reduction of AuI to Au0). For this experiment, sodium acetylacetonate purity is really important. The concentration of sodium acetylacetonate directly impacts the nuclei numbers inducing a size evolution of the gold core up to 90 nm with a narrow size distribution. Practically, the tetrachloroaurate and sodium citrate concentration are fixed to 0.30 mM and 0.255 mM (ratio sodium citrate to gold: 0.85). In these conditions, and without sodium acetylacetonate, the obtained nanoparticles precipitate immediately. The concentration of sodium acetylacetonate can be tuned from 0.33 mM to 1.0 mM. The addition of the Na(acac) and sodium citrate to the yellow boiling gold ions solution results in a colorless one indicating the complexation and reduction of AuIII to AuI. Within a few minutes, the solution turns light blue-purple, featuring the formation of the gold nuclei. The small crystalline structures will then diffuse over the solution allowing the growth of the final spherical particles.

Sonolysis

Another method for the experimental generation of gold particles is by sonolysis. The first method of this type was invented by Baigent and Müller. This work pioneered the use of ultrasound to provide the energy for the processes involved and allowed the creation of gold particles with a diameter of under 10 nm. In another method using ultrasound, the reaction of an aqueous solution of $HAuCl_4$ with glucose, the reducing agents are hydroxyl radicals and sugar pyrolysis radicals (forming at the interfacial region between the collapsing cavities and the bulk water) and the morphology obtained is that of nanoribbons with width 30–50 nm and length of several micrometers. These ribbons are very flexible and can bend with angles larger than 90°. When glucose is replaced by cyclodextrin (a glucose oligomer), only spherical gold particles are obtained, suggesting that glucose is essential in directing the morphology toward a ribbon.

Block Copolymer-mediated Method

An economical, environmentally benign and fast synthesis methodology for gold nanoparticles using block copolymer has been developed by Sakai et al. In this synthesis methodology, block copolymer plays the dual role of a reducing agent as well as a stabilizing agent. The formation of gold nanoparticles comprises three main steps: reduction of gold salt ion by block copolymers in the solution and formation of gold clusters, adsorption of block copolymers on gold clusters and further reduction of gold salt ions on the surfaces of these gold clusters for the growth of gold particles in steps, and finally its stabilization by block copolymers. But this method usually has a limited-yield (nanoparticle concentration), which does not increase with the increase in the gold salt concentration. Recently, Ray et al. demonstrated that the presence of an additional reductant (trisodium citrate) in 1:1 molar ratio with gold salt enhances the yield by manyfold at ambient conditions and room temperature.

"Green Chemistry" Based Methods

Phytochemicals found in various plant sources have been utilized as a means of developing a more economical and environmentally friendly synthetic pathway in the formation of gold nanoparticles. In accordance to the principles of "green chemistry," these methods employ the use of nontoxic chemicals, marginal energy consumption, renewable materials, and environmentally benign solvents to minimize the use, disposal, and health repercussions of hazardous chemicals. Additionally, these methods provide a more efficient synthetic pathway through a one-step process without the use of supplementary surfactants or polymers, capping agents, or templates to restrict agglomeration of the gold nanoparticles. This method is effective in producing well-defined gold nanoparticles since the phytochemicals perform a dual role as both a reducing agent of gold and as a stabilizer in the formation of a sturdy coating on the nanoparticles. One "green" method that has been employed in the formation of gold nanoparticles utilizes the phytochemicals and polyphenols in Darjeeling black tea leaves, with water acting as a benign solvent at room temperature. The phytochemicals in the black tea reduce $HAuCl_4$ to its salt and stabilize the aggregation of the gold atoms as the nanoparticle is formed. In addition to the "green" benefits of using black tea, the size of the nanoparticle is influenced by the concentration of the tea, and the absorbance and size of the gold nanoparticles formed can be easily determined using UV-Vis spectrometry abiding that an increase in λmax correlates to an increase in the size of the nanoparticle. Other "green" methods that have been studied and employed include the use of Elettaria cardamomum (cardamom) and cinnamon in the synthesis of gold nanoparticles, as well as Syzygium aromaticum (cloves) in the formation of copper nanoparticles and table sugar in the formation of silver nanoparticles.

Electron Microscopy

Colloidal gold and various derivatives have long been among the most widely used labels for antigens in biological electron microscopy. Colloidal gold particles can be attached to many traditional biological probes such as antibodies, lectins, superantigens, glycans, nucleic acids, and receptors. Particles of different sizes are easily distinguishable in electron micrographs, allowing simultaneous multiple-labelling experiments.

In addition to biological probes, gold nanoparticles can be transferred to various mineral substrates, such as mica, single crystal silicon, and atomically flat gold(III), to be observed under atomic force microscopy (AFM).

Medical Research

Drug Delivery System

Gold nanoparticles can be used to optimize the biodistribution of drugs to diseased organs, tissues or cells, in order to improve and target drug delivery. It is important to realize that the nanoparticle-mediated drug delivery is feasible only if the drug distribution is otherwise inadequate. These cases include drug targeting of difficult, unstable molecules (proteins, siRNA, DNA), delivery to the difficult sites (brain, retina, tumors, intracellular organelles) and drugs with serious side effects (e.g. anti-cancer agents).

The performance of the nanoparticles depends on the size and surface functionalities in the particles. Also, the drug release and particle disintegration can vary depending on the system (e.g. biodegradable polymers sensitive to pH). An optimal nanodrug delivery system ensures that the active drug is available at the site of action for the correct time and duration, and their concentration should be above the minimal effective concentration (MEC) and below the minimal toxic concentration (MTC).

Gold nanoparticles are being investigated as carriers for drugs such as Paclitaxel. The administration of hydrophobic drugs require molecular encapsulation and it is found that nanosized particles are particularly efficient in evading the reticuloendothelial system.

Gold nanoparticles are also used to circumvent multidrug resistance (MDR) mechanisms. Mechanisms of MDR include decreased uptake of drugs, reduced intracellular drug concentration by activation of the efflux transporters, modifications in cellular pathways by altering cell cycle checkpoints, increased metabolism of drugs, induced emergency response genes to impair apoptotic pathways and altered DNA repair mechanisms.

Tumor Detection

In cancer research, colloidal gold can be used to target tumors and provide detection using SERS (Surface Enhanced Raman Spectroscopy) *in vivo*. These gold nanoparticles are surrounded with Raman reporters, which provide light emission that is over 200 times brighter than quantum dots. It was found that the Raman reporters were stabilized when the nanoparticles were encapsulated with a thiol-modified polyethylene glycol coat. This allows for compatibility and circulation *in vivo*. To specifically target tumor cells, the pegylated gold particles are conjugated with an antibody (or an antibody fragment such as scFv), against, e.g. Epidermal growth factor receptor, which is sometimes overexpressed in cells of certain cancer types. Using SERS, these pegylated gold nanoparticles can then detect the location of the tumor.

Gold nanoparticles accumulate in tumors, due to the leakiness of tumor vasculature, and can be used as contrast agents for enhanced imaging in a time-resolved optical tomography system using short-pulse lasers for skin cancer detection in mouse model. It is found that intravenously administrated spherical gold nanoparticles broadened the temporal profile of reflected optical signals and enhanced the contrast between surrounding normal tissue and tumors.

Therefore, gold nanoparticles have the potential to join numerous therapeutic functions into a single platform, by targeting specific tumor cells, tissues and organs. Actually, Conde et al. reported the evaluation of the inflammatory response and therapeutic siRNA silencing via RGD-nanoparticles in a lung cancer mouse model. This study reported the use of siRNA/RGD gold nanoparticles capable of targeting tumor cells in two lung cancer xenograft mouse models, resulting in successful and significant *c-Myc* oncogene downregulation followed by tumor growth inhibition and prolonged survival of the animals. This delivery system can achieve translocation of siRNA duplexes directly into the tumour cell cytoplasm and accomplish successful silencing of an oncogene expression. Actually, RGD/siRNA-AuNPs can target preferentially and be taken up by tumor cells via integrin αvβ3-receptor-mediated endocytosis with no cytotoxicity, showing that can accumulate in tumor tissues overexpressing αvβ3 integrins and selectively delivered *c-Myc* siRNA to suppress tumor growth and angiogenesis.

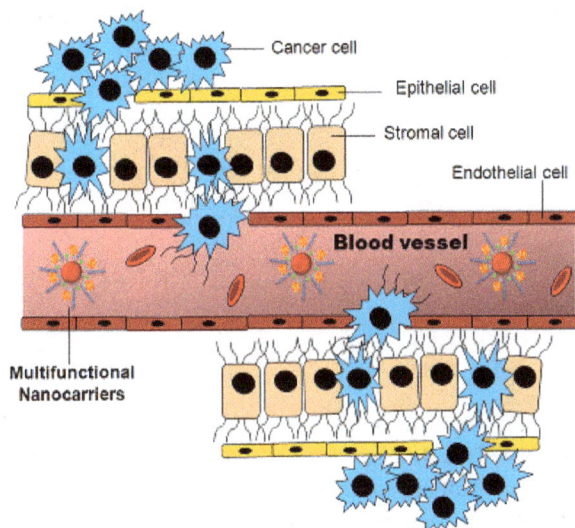

Tumor targeting via multifunctional nanocarriers. Cancer cells reduce adhesion to neighboring cells and migrate into the vasculature-rich stroma. Once at the vasculature, cells can freely enter the bloodstream. Once the tumor is directly connected to the main blood circulation system, multifunctional nanocarriers can interact directly with cancer cells and effectively target tumors.

Gene Therapy

Gene therapy is receiving increasing attention and, in particular, small-interference RNA (siRNA) shows importance in novel molecular approaches in the knockdown of specific gene expression in cancerous cells. The major obstacle to clinical application is the uncertainty about how to deliver therapeutic siRNAs with maximal therapeutic impact. Gold nanoparticles have shown potential as intracellular delivery vehicles for siRNA oligonucleotides with maximal therapeutic impact.

Recently, Conde et al. provided evidence of *in vitro* and *in vivo* RNAi triggering via the synthesis of a library of novel multifunctional gold nanoparticles, using a hierarchical approach including three biological systems of increasing complexity: *in vitro* cultured human cells, *in vivo* freshwater polyp (*Hydra vulgaris*), and *in vivo* mice models. The authors developed effective conjugation strategies to combine, in a highly controlled way, specific biomolecules to the surface of gold nanoparticles such as: (a) biofunctional spacers: Poly(ethylene glycol) (PEG) spacers used to increase solubility and biocompatibility; (b) cell penetrating peptides such as TAT and RGD peptides: A novel class of membrane translocating agents named cell penetrating peptides (CPPs) that exploit more than one mechanism of endocytosis to overcome the lipophilic barrier of the cellular membranes and deliver large molecules and even small particles inside the cell for their biological actions; and (c) siRNA complementary to a master regulator gene, the protooncogene c-myc, were bond covalently (thiol-siRNA) and ionically (naked/unmodified siRNA) to gold nanoparticles.

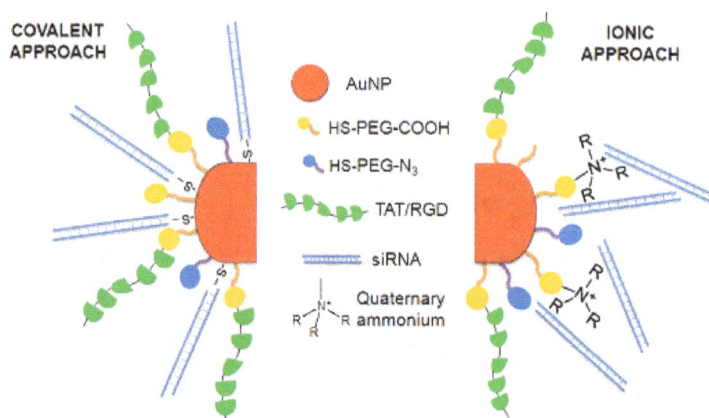

Multifunctional siRNA-gold nanoparticles with several biomolecules: PEG, cell penetration and cell adhesion peptides and siRNA. Two different approaches were employed to conjugate the siRNA to the gold nanoparticle: (1) Covalent approach: use of thiolated siRNA for gold-thiol binding to the nanoparticle; (2) Ionic approach: interaction of the negatively charged siRNA to the modified surface of the AuNP through ionic interactions.

Gold nanoparticles have also shown potential as intracellular delivery vehicles for antisense oligonucleotides (ssDNA,dsDNA) by providing protection against intracellular nucleases and ease of functionalization for selective targeting. Recently, Conde et al. developed a new theranostic system capable of intersecting all RNA pathways: from gene specific downregulation to silencing the silencers, i.e. siRNA and miRNA pathways. The authors reported the development gold nanoparticles functionalized with a fluorophore labeled hairpin-DNA, i.e. gold nanobeacons, capable of efficiently silencing single gene expression, exogenous siRNA and endogenous miRNAs while yielding a quantifiable fluorescence signal directly proportional to the level of silencing. This method describes a gold nanoparticle-based nanobeacon as an innovative theranostic approach for detection and inhibition of sequence-specific DNA and RNA for *in vitro* and *ex vivo* applications. Under hairpin configuration, proximity to gold nanoparticles leads to fluorescence quenching; hybridization to a complementary target restores fluorescence emission due to the gold nanobeacons' conformational reorganization that causes the fluorophore and the gold nanoparticle to part from each other. This concept can easily be extended and adapted to assist the in vitro evaluation of silencing potential of a given sequence to be later used for *ex vivo* gene silencing and RNAi approaches, with the ability to monitor real-time gene delivery action.

Photothermal Agents

Gold nanorods are being investigated as photothermal agents for in-vivo applications. Gold nanorods are rod-shaped gold nanoparticles whose aspect ratios tune the surface plasmon resonance (SPR) band from the visible to near-infrared wavelength. The total extinction of light at the SPR is made up of both absorption and scattering. For the smaller axial diameter nanorods (~10 nm), absorption dominates, whereas for the larger axial diameter nanorods (>35 nm) scattering can dominate. As a consequence, for in-vivo applications, small diameter gold nanorods are being used as photothermal converters of near-infrared light due to their high absorption cross-sections. Since near-infrared light transmits readily through human skin and tissue, these nanorods can be used as ablation components for cancer, and other targets. When coated with polymers, gold nanorods have been known to circulate in-vivo for greater than 15 hours half-life. Apart from rod-like gold nanoparticles, also spherical colloidal gold nanoparticles are recently used as markers in combination with photothermal single particle microscopy.

Radiotherapy Dose Enhancer

Following work by Hainfield et al. there has been considerable interest in the use of gold and other heavy-atom containing nanoparticles to enhance the dose delivered to tumors. Since the gold nanoparticles are taken up by the tumors more than the nearby healthy tissue, the dose is selectively enhanced. The biological effectiveness of this type of therapy seems to be due to the local deposition of the radiation dose near the nanoparticles. This mechanism is the same as occurs in heavy ion therapy.

Detection of Toxic Gas

Researchers have developed simple inexpensive methods for on-site detection of hydrogen sulfide H_2S present in air based on the antiaggregation of gold nanoparticles (AuNPs). Dissolving H_2S into a weak alkaline buff solution leads to the formation of $HS-$, which can stabilize AuNPs and ensure they maintain their red color allowing for visual detection of toxic levels of H_2S.

Gold Nanoparticle Based Biosensor

Gold nanoparticles are incorporated into biosensors to enhance its stability, sensitivity, and selectivity. Nanoparticle properties such as small size, high surface-to-volume ratio, and high surface energy allow immobilization of large range of biomolecules. Gold nanoparticle, in particular, could also act as "electron wire" to transport electrons and its amplification effect on electromagnetic light allows it to function as signal amplifiers. Main types of gold nanoparticle based biosensors are optical and electrochemical biosensor.

Optical Biosensor

Gold nanoparticles improve the sensitivity of optical sensor by response to the change in local refractive index. The angle of the incidence light for Surface Plasmon resonance, an interaction between light wave and conducting electrons in metal, changes when other substances are bounded to the metal surface. Because gold is very sensitive to its surroundings' dielectric constant, binding of an analyte would significantly shift gold nanoparticle's SPR and therefore allow more sensitive detection. Gold nanoparticle could also amplify the SPR signal. When the Plasmon wave pass through the gold nanoparticle, the charge density in the wave and the electron I the gold interacted and resulted in higher energy response, so called electron coupling. Since the analyte and bio-receptor now bind to the gold, it increases the apparent mass of the analyte and therefore amplified the signal. These properties had been used to build DNA sensor with 1000-fold sensitive than without the Au NP. Humidity senor was also built by altering the atom interspacing between molecules with humidity change, the interspacing change would also result in a change of the Au NP's LSPR.

Electrochemical Biosensor

Electrochemical sensor covert biological information into electrical signals that could be detected. The conductivity and biocompatibility of Au NP allow it to act as "electron wire". It transfers electron between the electrode and the active site of the enzyme. It could be accomplished in two ways: attach the Au NP to either the enzyme or the electrode. GNP-glucose oxidase monolayer electrode was constructed use these two methods. The Au NP allowed more freedom in the enzyme's orientation and therefore more sensitive and stable detection. Au NP also acts as immobilization platform for the enzyme. Most biomolecules denatures or lose its activity when interacted with the electrode. The biocompatibility and high surface energy of Au allow it to bind to a large amount of protein without altering its activity and results in a more sensitive sensor. Moreover, Au NP also catalyzes biological reactions. Gold nanoparticle under 2 nm has shown catalytic activity to the oxidation of styrene.

References

- Onoda, G.Y. Jr.; Hench, L.L., eds. (1979). Ceramic Processing Before Firing. New York: Wiley & Sons. ISBN 0-471-65410-8.

- Engheta, Nader; Richard W. Ziolkowski (June 2006). Metamaterials: Physics and Engineering Explorations. Wiley & Sons. pp. xv, 3–30, 37, 143–150, 215–234, 240–256. ISBN 978-0-471-76102-0.

- Zouhdi, Saïd; Ari Sihvola; Alexey P. Vinogradov (December 2008). Metamaterials and Plasmonics: Fundamentals, Modelling, Applications. New York: Springer-Verlag. pp. 3–10, Chap. 3, 106. ISBN 978-1-4020-9406-4.

- Eleftheriades, George V.; Keith G. Balmain (2005). Negative-refraction metamaterials: fundamental principles and applications. Wiley, John & Sons. p. 340. ISBN 978-0-471-60146-3.

- Engheta, Nader; Richard W. Ziolkowski (June 2006). Metamaterials: physics and engineering explorations (added this reference on 2009-12-14.). Wiley & Sons. pp. 211–221. ISBN 978-0-471-76102-0.

- Capolino, Filippo (October 2009). "32". Theory and Phenomena of Metamaterials. Taylor & Francis. pp. 32–1. ISBN 978-1-4200-5425-5.

- MacNaught, Alan D.; Wilkinson, Andrew R., eds. (1997). Compendium of Chemical Terminology: IUPAC Recommendations (2nd ed.). Blackwell Science. ISBN 0865426848.

- Reiss, Gunter; Hutten, Andreas (2010). "Magnetic Nanoparticles". In Sattler, Klaus D. Handbook of Nanophysics: Nanoparticles and Quantum Dots. CRC Press. pp. 2–1. ISBN 9781420075458.

- Brinker, C.J. & Scherer, G.W. (1990). Sol-Gel Science: The Physics and Chemistry of Sol-Gel Processing. Academic Press. ISBN 0-12-134970-5.

- Corriu, Robert & Anh, Nguyên Trong (2009). Molecular Chemistry of Sol-Gel Derived Nanomaterials. John Wiley and Sons. ISBN 0-470-72117-0.

Carbon Nanotube: An Overview

Carbon nanotubes are allotropes of carbon. They have unusual properties and they are very important for nanotechnology and other fields of material science. This chapter helps the reader in understanding the basic concepts of carbon nanotubes.

Carbon Nanotube

Carbon nanotubes (CNTs) are allotropes of carbon with a cylindrical nanostructure. These cylindrical carbon molecules have unusual properties, which are valuable for nanotechnology, electronics, optics and other fields of materials science and technology. Owing to the material's exceptional strength and stiffness, nanotubes have been constructed with length-to-diameter ratio of up to 132,000,000:1, significantly larger than for any other material.

In addition, owing to their extraordinary thermal conductivity, mechanical, and electrical properties, carbon nanotubes find applications as additives to various structural materials. For instance, nanotubes form a tiny portion of the material(s) in some (primarily carbon fiber) baseball bats, golf clubs, car parts or damascus steel.

Nanotubes are members of the fullerene structural family. Their name is derived from their long, hollow structure with the walls formed by one-atom-thick sheets of carbon, called graphene. These sheets are rolled at specific and discrete ("chiral") angles, and the combination of the rolling angle and radius decides the nanotube properties; for example, whether the individual nanotube shell is a metal or semiconductor. Nanotubes are categorized as single-walled nanotubes (SWNTs) and multi-walled nanotubes (MWNTs). Individual nanotubes naturally align themselves into "ropes" held together by van der Waals forces, more specifically, pi-stacking.

Applied quantum chemistry, specifically, orbital hybridization best describes chemical bonding in nanotubes. The chemical bonding of nanotubes is composed entirely of sp^2 bonds, similar to those of graphite. These bonds, which are stronger than the sp^3 bonds found in alkanes and diamond, provide nanotubes with their unique strength.

Types of Carbon Nanotubes and Related Structures

Terminology

There is no consensus on some terms describing carbon nanotubes in scientific literature: both "-wall" and "-walled" are being used in combination with "single", "double", "triple" or "multi", and the letter C is often omitted in the abbreviation; for example, multi-walled carbon nanotube (MWNT).

Single-walled

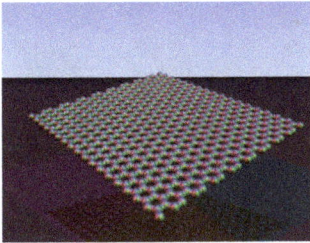

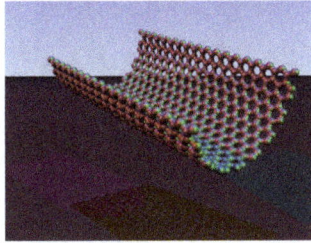

| Graphene nanoribbon | The translation vector is bent, while the chiral vector stays straight | Armchair (*n,n*) i.e.: m=n |

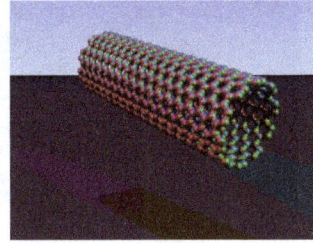

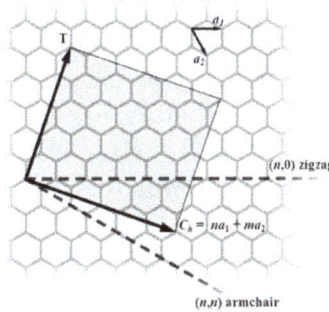

The (*n,m*) nanotube naming scheme can be thought of as a vector (C$_h$) in an infinite graphene sheet that describes how to "roll up" the graphene sheet to make the nanotube. T denotes the tube axis, and a$_1$ and a$_2$ are the unit vectors of graphene in real space.

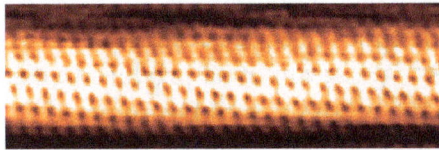

A scanning tunneling microscopy image of single-walled carbon nanotube

Most single-walled nanotubes (SWNTs) have a diameter of close to 1 nanometer, and can be many millions of times longer. The structure of a SWNT can be conceptualized by wrapping a one-atom-thick layer of graphite called graphene into a seamless cylinder. The way the graphene sheet is wrapped is represented by a pair of indices (*n,m*). The integers *n* and *m* denote the number of unit vectors along two directions in the honeycomb crystal lattice of graphene. If *m* = 0, the nanotubes are called zigzag nanotubes, and if *n* = *m*, the nanotubes are called armchair nanotubes. Otherwise, they are called chiral. The diameter of an ideal nanotube can be calculated from its (n,m) indices as follows

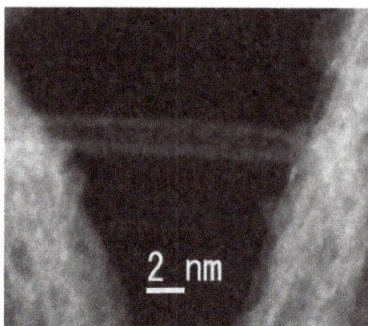

A transmission electron microscopy image of a single-walled carbon nanotube

$$d = \frac{a}{\pi}\sqrt{(n^2 + nm + m^2)} = 78.3\sqrt{((n+m)^2 - nm)}\text{pm},$$

where a = 0.246 nm.

SWNTs are an important variety of carbon nanotube because most of their properties change significantly with the (n,m) values, and this dependence is non-monotonic. In particular, their band gap can vary from zero to about 2 eV and their electrical conductivity can show metallic or semiconducting behavior. Single-walled nanotubes are likely candidates for miniaturizing electronics. The most basic building block of these systems is the electric wire, and SWNTs with diameters of an order of a nanometer can be excellent conductors. One useful application of SWNTs is in the development of the first intermolecular field-effect transistors (FET). The first intermolecular logic gate using SWCNT FETs was made in 2001. A logic gate requires both a p-FET and an n-FET. Because SWNTs are p-FETs when exposed to oxygen and n-FETs otherwise, it is possible to expose half of an SWNT to oxygen and protect the other half from it. The resulting SWNT acts as a *not* logic gate with both p and n-type FETs in the same molecule.

Prices for single-walled nanotubes declined from around $1500 per gram as of 2000 to retail prices of around $50 per gram of as-produced 40–60% by weight SWNTs as of March 2010. As of 2016 the retail price of as-produced 75% by weight SWNTs were $2 per gram, cheap enough for widespread use. SWNTs are forecast to make a large impact in electronics applications by 2020 according to the *The Global Market for Carbon Nanotubes* report.

Multi-walled

Multi-walled nanotubes (MWNTs) consist of multiple rolled layers (concentric tubes) of graphene. There are two models that can be used to describe the structures of multi-walled nanotubes. In the *Russian Doll* model, sheets of graphite are arranged in concentric cylinders, e.g., a (0,8) single-walled nanotube (SWNT) within a larger (0,17) single-walled nanotube. In the *Parchment* model, a single sheet of graphite is rolled in around itself, resembling a scroll of parchment or a rolled newspaper. The interlayer distance in multi-walled nanotubes is close to the distance between graphene layers in graphite, approximately 3.4 Å. The Russian Doll structure is observed more commonly. Its individual shells can be described as SWNTs, which can be metallic or semiconducting. Because of statistical probability and restrictions on the relative diameters of the individual tubes, one of the shells, and thus the whole MWNT, is usually a zero-gap metal.

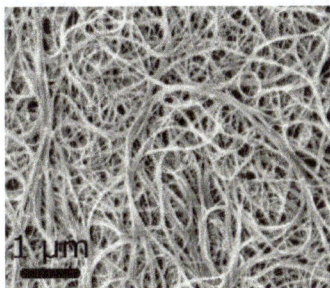

A scanning electron microscopy image of carbon nanotubes bundles

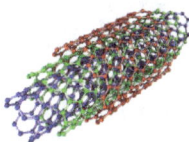

Triple-walled armchair carbon nanotube

Double-walled carbon nanotubes (DWNTs) form a special class of nanotubes because their morphology and properties are similar to those of SWNTs but they are more resistant to chemicals. This is especially important when it is necessary to graft chemical functions to the surface of the nanotubes (functionalization) to add properties to the CNT. Covalent functionalization of SWNTs will break some C=C double bonds, leaving "holes" in the structure on the nanotube, and thus modifying both its mechanical and electrical properties. In the case of DWNTs, only the outer wall is modified. DWNT synthesis on the gram-scale was first proposed in 2003 by the CCVD technique, from the selective reduction of oxide solutions in methane and hydrogen.

The telescopic motion ability of inner shells and their unique mechanical properties will permit the use of multi-walled nanotubes as main movable arms in coming nanomechanical devices. Retraction force that occurs to telescopic motion caused by the Lennard-Jones interaction between shells and its value is about 1.5 nN.

Torus

In theory, a nanotorus is a carbon nanotube bent into a torus (doughnut shape). Nanotori are predicted to have many unique properties, such as magnetic moments 1000 times larger than previously expected for certain specific radii. Properties such as magnetic moment, thermal stability, etc. vary widely depending on radius of the torus and radius of the tube.

Nanobud

Carbon nanobuds are a newly created material combining two previously discovered allotropes of carbon: carbon nanotubes and fullerenes. In this new material, fullerene-like "buds" are covalently bonded to the outer sidewalls of the underlying carbon nanotube. This hybrid material has useful properties of both fullerenes and carbon nanotubes. In particular, they have been found to be exceptionally good field emitters. In composite materials, the attached fullerene molecules may function as molecular anchors preventing slipping of the nanotubes, thus improving the composite's mechanical properties.

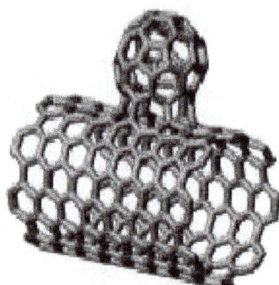

A stable nanobud structure

Three-dimensional Carbon Nanotube Architectures

Recently, several studies have highlighted the prospect of using carbon nanotubes as building blocks to fabricate three-dimensional macroscopic (>100 nm in all three dimensions) all-carbon devices. Lalwani et al. have reported a novel radical initiated thermal crosslinking method to fabricate macroscopic, free-standing, porous, all-carbon scaffolds using single- and multi-walled carbon nanotubes as building blocks. These scaffolds possess macro-, micro-, and nano- structured pores and the porosity can be tailored for specific applications. These 3D all-carbon scaffolds/architectures may be used for the fabrication of the next generation of energy storage, supercapacitors, field emission transistors, high-performance catalysis, photovoltaics, and biomedical devices and implants.

3D carbon scaffolds

Graphenated Carbon Nanotubes (g-CNTs)

SEM series of graphenated CNTs with varying foliate density

Graphenated CNTs are a relatively new hybrid that combines graphitic foliates grown along the sidewalls of multiwalled or bamboo style CNTs. Yu *et al.* reported on "chemically bonded graphene leaves" growing along the sidewalls of CNTs. Stoner *et al.* described these structures as "graphenated CNTs" and reported in their use for enhanced supercapacitor performance. Hsu *et al.* further reported on similar structures formed on carbon fiber paper, also for use in supercapacitor applications. Pham *et al.* also reported a similar structure, namely "graphene-carbon nanotube hybrids", grown directly onto carbon fiber paper to form an integrated, binder free, high surface area conductive catalyst support for Proton Exchange Membrane Fuel Cells electrode applications with enhanced performance and durability. The foliate density can vary as a function of deposition conditions (e.g. temperature and time) with their structure ranging from few layers of graphene (< 10) to thicker, more graphite-like.

The fundamental advantage of an integrated graphene-CNT structure is the high surface area three-dimensional framework of the CNTs coupled with the high edge density of graphene. Graphene edges provide significantly higher charge density and reactivity than the basal plane, but they are difficult to arrange in a three-dimensional, high volume-density geometry. CNTs are readily aligned in a high density geometry (i.e., a vertically aligned forest) but lack high charge density surfaces—the sidewalls of the CNTs are similar to the basal plane of graphene and exhibit low charge density except where edge defects exist. Depositing a high density of graphene foliates along the length of aligned CNTs can significantly increase the total charge capacity per unit of nominal area as compared to other carbon nanostructures.

Nitrogen-doped Carbon Nanotubes

Nitrogen doped carbon nanotubes (N-CNTs) can be produced through five main methods, chemical vapor deposition, high-temperature and high-pressure reactions, gas-solid reaction of amorphous carbon with NH_3 at high temperature, solid reaction, and solvothermal synthesis.

N-CNTs can also be prepared by a CVD method of pyrolyzing melamine under Ar at elevated temperatures of 800–980 °C. However synthesis by CVD of melamine results in the formation of bamboo-structured CNTs. XPS spectra of grown N-CNTs reveal nitrogen in five main components, pyridinic nitrogen, pyrrolic nitrogen, quaternary nitrogen, and nitrogen oxides. Furthermore, synthesis temperature affects the type of nitrogen configuration.

Nitrogen doping plays a pivotal role in lithium storage, as it creates defects in the CNT walls allowing for Li ions to diffuse into interwall space. It also increases capacity by providing more favorable bind of N-doped sites. N-CNTs are also much more reactive to metal oxide nanoparticle deposition which can further enhance storage capacity, especially in anode materials for Li-ion batteries. However boron-doped nanotubes have been shown to make batteries with triple capacity.

Peapod

A carbon peapod is a novel hybrid carbon material which traps fullerene inside a carbon nanotube. It can possess interesting magnetic properties with heating and irradiation. It can also be applied as an oscillator during theoretical investigations and predictions.

Cup-stacked Carbon Nanotubes

Cup-stacked carbon nanotubes (CSCNTs) differ from other quasi-1D carbon structures, which normally behave as quasi-metallic conductors of electrons. CSCNTs exhibit semiconducting behaviors due to the stacking microstructure of graphene layers.

Extreme Carbon Nanotubes

The observation of the *longest* carbon nanotubes grown so far are over 1/2 m (550 mm long) was reported in 2013. These nanotubes were grown on Si substrates using an improved chemical vapor deposition (CVD) method and represent electrically uniform arrays of single-walled carbon nanotubes.

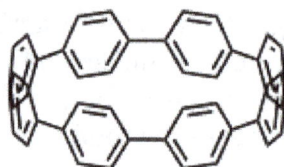

Cycloparaphenylene

The *shortest* carbon nanotube is the organic compound cycloparaphenylene, which was synthesized in 2008.

The *thinnest* carbon nanotube is the armchair (2,2) CNT with a diameter of 0.3 nm. This nanotube was grown inside a multi-walled carbon nanotube. Assigning of carbon nanotube type was done by a combination of high-resolution transmission electron microscopy (HRTEM), Raman spectroscopy and density functional theory (DFT) calculations.

The *thinnest freestanding* single-walled carbon nanotube is about 0.43 nm in diameter. Researchers suggested that it can be either (5,1) or (4,2) SWCNT, but the exact type of carbon nanotube remains questionable. (3,3), (4,3) and (5,1) carbon nanotubes (all about 0.4 nm in diameter) were unambiguously identified using aberration-corrected high-resolution transmission electron microscopy inside double-walled CNTs.

The *highest density* of CNTs was achieved in 2013, grown on a conductive titanium-coated copper surface that was coated with co-catalysts cobalt and molybdenum at lower than typical temperatures of 450 °C. The tubes averaged a height of 380 nm and a mass density of 1.6 g cm^{-3}. The material showed ohmic conductivity (lowest resistance -22 kΩ).

Properties

Strength

Carbon nanotubes are the strongest and stiffest materials yet discovered in terms of tensile strength and elastic modulus respectively. This strength results from the covalent sp^2 bonds formed between the individual carbon atoms. In 2000, a multi-walled carbon nanotube was tested to have a tensile strength of 63 gigapascals (9,100,000 psi). (For illustration, this translates into the ability to endure tension of a weight equivalent to 6,422 kilograms-force (62,980 N; 14,160 lbf) on a cable with cross-section of 1 square millimetre (0.0016 sq in).) Further studies, such as one conducted in 2008, revealed that individual CNT shells have strengths of up to ~100 gigapascals (15,000,000 psi), which is in agreement with quantum/atomistic models. Since carbon nanotubes have a low density for a solid of 1.3 to 1.4 g/cm^3, its specific strength of up to 48,000 kN·m·kg^{-1} is the best of known materials, compared to high-carbon steel's 154 kN·m·kg^{-1}.

Under excessive tensile strain, the tubes will undergo plastic deformation, which means the deformation is permanent. This deformation begins at strains of approximately 5% and can increase the maximum strain the tubes undergo before fracture by releasing strain energy.

Although the strength of individual CNT shells is extremely high, weak shear interactions between adjacent shells and tubes lead to significant reduction in the effective strength of multi-walled carbon nanotubes and carbon nanotube bundles down to only a few GPa. This limitation has been recently addressed by applying high-energy electron irradiation, which crosslinks inner shells and tubes, and effectively increases the strength of these materials to ~60 GPa for multi-walled carbon nanotubes and ~17 GPa for double-walled carbon nanotube bundles.

CNTs are not nearly as strong under compression. Because of their hollow structure and high aspect ratio, they tend to undergo buckling when placed under compressive, torsional, or bending stress.

Comparison of mechanical properties			
Material	**Young's modulus (TPa)**	**Tensile strength (GPa)**	**Elongation at break (%)**
SWNT[E]	~1 (from 1 to 5)	13–53	16
Armchair SWNT[T]	0.94	126.2	23.1
Zigzag SWNT[T]	0.94	94.5	15.6–17.5
Chiral SWNT	0.92		
MWNT[E]	0.2–0.8–0.95	11–63–150	
Stainless steel[E]	0.186–0.214	0.38–1.55	15–50
Kevlar–29&149[E]	0.06–0.18	3.6–3.8	~2

[E]Experimental observation; [T]Theoretical prediction

The above discussion referred to axial properties of the nanotube, whereas simple geometrical considerations suggest that carbon nanotubes should be much softer in the radial direction than along the tube axis. Indeed, TEM observation of radial elasticity suggested that even the van der Waals forces can deform two adjacent nanotubes. Nanoindentation experiments, performed by several groups on multiwalled carbon nanotubes and tapping/contact mode atomic force microscope measurements performed on single-walled carbon nanotubes, indicated a Young's modulus of the order of several GPa, confirming that CNTs are indeed rather soft in the radial direction.

Hardness

Standard single-walled carbon nanotubes can withstand a pressure up to 25 GPa without [plastic/permanent] deformation. They then undergo a transformation to superhard phase nanotubes. Maximum pressures measured using current experimental techniques are around 55 GPa. However, these new superhard phase nanotubes collapse at an even higher, albeit unknown, pressure.

The bulk modulus of superhard phase nanotubes is 462 to 546 GPa, even higher than that of diamond (420 GPa for single diamond crystal).

Wettability

The surface wettability of CNT is of importance for its applications in various settings. Although the intrinsic contact angle of graphite is around 90°, the contact angles of most as-synthesized CNT arrays are over 160°, exhibiting a superhydrophobic property. By applying a voltage as low as 1.3V, the extreme water repellant surface can be switched to a superhydrophilic one.

Kinetic Properties

Multi-walled nanotubes are multiple concentric nanotubes precisely nested within one another. These exhibit a striking telescoping property whereby an inner nanotube core may slide, almost without friction, within its outer nanotube shell, thus creating an atomically perfect linear or rotational bearing. This is one of the first true examples of molecular nanotechnology, the precise positioning of atoms to create useful machines. Already, this property has been utilized to create the world's smallest rotational motor. Future applications such as a gigahertz mechanical oscillator are also envisioned.

Electrical Properties

Unlike graphene, which is a two-dimensional semimetal, carbon nanotubes are either metallic or semiconducting along the tubular axis. For a given (n,m) nanotube, if $n = m$, the nanotube is metallic; if $n - m$ is a multiple of 3, then the nanotube is semiconducting with a very small band gap, otherwise the nanotube is a moderate semiconductor. Thus all armchair ($n = m$) nanotubes are metallic, and nanotubes (6,4), (9,1), etc. are semiconducting. Carbon nanotubes are not semimetallic because the degenerate point (that point where the π [bonding] band meets the π^* [anti-bonding] band, at which the energy goes to zero) is slightly shifted away from the K point in the Brillouin zone due to the curvature of the tube surface, casing hybridization between the σ^* and π^* anti-bonding bands, modifying the band dispersion.

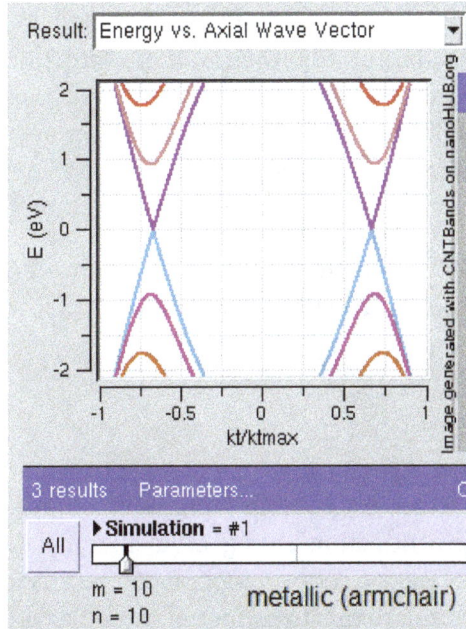

Band structures computed using tight binding approximation for (6,0) CNT (zigzag, metallic), (10,2) CNT (semiconducting) and (10,10) CNT (armchair, metallic).

The rule regarding metallic versus semiconductor behavior has exceptions, because curvature effects in small diameter tubes can strongly influence electrical properties. Thus, a (5,0) SWCNT that should be semiconducting in fact is metallic according to the calculations. Likewise, zigzag and chiral SWCNTs with small diameters that should be metallic have a finite gap (armchair nanotubes remain metallic). In theory, metallic nanotubes can carry an electric current density of 4×10^9 A/cm^2, which is more than 1,000 times greater than those of metals such as copper, where for copper interconnects current densities are limited by electromigration. Carbon nanotubes are thus being explored as conductivity enhancing components in composite materials and many groups are attempting to commercialize highly conducting electrical wire assembled from individual carbon nanotubes. There are significant challenges to be overcome, however, such as the much more resistive nanotube-to-nanotube junctions and impurities, all of which lower the electrical conductivity of the macroscopic nanotube wires by orders of magnitude, as compared to the conductivity of the individual nanotubes.

Because of its nanoscale cross-section, electrons propagate only along the tube's axis.

As a result, carbon nanotubes are frequently referred to as one-dimensional conductors. The maximum electrical conductance of a single-walled carbon nanotube is $2G_0$, where $G_0 = 2e^2/h$ is the conductance of a single ballistic quantum channel.

Due to the role of the π-electron system in determining the electronic properties of graphene, doping in carbon nanotubes differs from that of bulk crystalline semiconductors from the same group of the periodic table (e.g. silicon). Graphitic substitution of carbon atoms in the nanotube wall by boron or nitrogen dopants leads to p-type and n-type behavior, respectively, as would be expected in silicon. However, some non-substitutional (intercalated or adsorbed) dopants introduced into a carbon nanotube, such as alkali metals as well as electron-rich metallocenes, result in n-type conduction because they donate electrons to the π-electron system of the nanotube. By contrast, π-electron acceptors such as $FeCl_3$ or electron-deficient metallocenes function as p-type dopants since they draw π-electrons away from the top of the valence band.

Intrinsic superconductivity has been reported, although other experiments found no evidence of this, leaving the claim a subject of debate.

Optical Properties

Thermal Properties

All nanotubes are expected to be very good thermal conductors along the tube, exhibiting a property known as "ballistic conduction", but good insulators lateral to the tube axis. Measurements show that a SWNT has a room-temperature thermal conductivity along its axis of about 3500 $W \cdot m^{-1} \cdot K^{-1}$; compare this to copper, a metal well known for its good thermal conductivity, which transmits 385 $W \cdot m^{-1} \cdot K^{-1}$. A SWNT has a room-temperature thermal conductivity across its axis (in the radial direction) of about 1.52 $W \cdot m^{-1} \cdot K^{-1}$, which is about as thermally conductive as soil. The temperature stability of carbon nanotubes is estimated to be up to 2800 °C in vacuum and about 750 °C in air.

Defects

As with any material, the existence of a crystallographic defect affects the material properties. Defects can occur in the form of atomic vacancies. High levels of such defects can lower the tensile strength by up to 85%. An important example is the Stone Wales defect, which creates a pentagon and heptagon pair by rearrangement of the bonds. Because of the very small structure of CNTs, the tensile strength of the tube is dependent on its weakest segment in a similar manner to a chain, where the strength of the weakest link becomes the maximum strength of the chain.

Crystallographic defects also affect the tube's electrical properties. A common result is lowered conductivity through the defective region of the tube. A defect in armchair-type tubes (which can conduct electricity) can cause the surrounding region to become semiconducting, and single monatomic vacancies induce magnetic properties.

Crystallographic defects strongly affect the tube's thermal properties. Such defects lead to phonon scattering, which in turn increases the relaxation rate of the phonons. This reduces the mean free path and reduces the thermal conductivity of nanotube structures. Phonon transport simulations indicate that substitutional defects such as nitrogen or boron will primarily lead to scattering of high-frequency optical phonons. However, larger-scale defects such as Stone Wales defects cause phonon scattering over a wide range of frequencies, leading to a greater reduction in thermal conductivity.

Safety and Health

The toxicity of carbon nanotubes has been an important question in nanotechnology. As of 2007, such research had just begun. The data is still fragmentary and subject to criticism. Preliminary results highlight the difficulties in evaluating the toxicity of this heterogeneous material. Parameters such as structure, size distribution, surface area, surface chemistry, surface charge, and agglomeration state as well as purity of the samples, have considerable impact on the reactivity of carbon nanotubes. However, available data clearly show that, under some conditions, nanotubes can cross membrane barriers, which suggests that, if raw materials reach the organs, they can induce harmful effects such as inflammatory and fibrotic reactions.

Synthesis

Techniques have been developed to produce nanotubes in sizable quantities, including arc discharge, laser ablation, high-pressure carbon monoxide disproportionation, and chemical vapor deposition (CVD). Most of these processes take place in a vacuum or with process gases. CVD growth of CNTs can occur in vacuum or at atmospheric pressure. Large quantities of nanotubes can be synthesized by these methods; advances in catalysis and continuous growth are making CNTs more commercially viable.

Chemical Modification

Carbon nanotubes can be functionalized to attain desired properties that can be used in a wide variety of applications. The two main methods of carbon nanotube functionalization are covalent and non-covalent modifications. Because of their hydrophobic nature, carbon nanotubes tend to agglomerate hindering their dispersion is solvents or viscous polymer melts. The resulting nanotube bundles or aggregates reduce the mechanical performance of the final composite. The surface of the carbon nanotubes can be modified to reduce the hydrophobicity and improve interfacial adhesion to a bulk polymer through chemical attachment.

Applications

Current

Current use and application of nanotubes has mostly been limited to the use of bulk nanotubes, which is a mass of rather unorganized fragments of nanotubes. Bulk nanotube materials may never achieve a tensile strength similar to that of individual tubes, but such composites may, nevertheless, yield strengths sufficient for many applications. Bulk carbon nanotubes have already been used as composite fibers in polymers to improve the mechanical, thermal and electrical properties of the bulk product.

- Easton-Bell Sports, Inc. have been in partnership with Zyvex Performance Materials, using CNT technology in a number of their bicycle components—including flat and riser handlebars, cranks, forks, seatposts, stems and aero bars.

- Zyvex Technologies has also built a 54' maritime vessel, the Piranha Unmanned Surface Vessel, as a technology demonstrator for what is possible using CNT technology. CNTs help improve the structural performance of the vessel, resulting in a lightweight 8,000 lb boat that can carry a payload of 15,000 lb over a range of 2,500 miles.

- Amroy Europe Oy manufactures Hybtonite carbon nanoepoxy resins where carbon nanotubes have been chemically activated to bond to epoxy, resulting in a composite material that is 20% to 30% stronger than other composite materials. It has been used for wind turbines, marine paints and variety of sports gear such as skis, ice hockey sticks, baseball bats, hunting arrows, and surfboards.

Other current applications include:

- tips for atomic force microscope probes

- in tissue engineering, carbon nanotubes can act as scaffolding for bone growth

There is also ongoing research in using carbon nanotubes as a scaffold for diverse microfabrication techniques.

Potential

The strength and flexibility of carbon nanotubes makes them of potential use in controlling other nanoscale structures, which suggests they will have an important role in nanotechnology engineering. The highest tensile strength of an individual multi-walled carbon nanotube has been tested to be 63 GPa. Carbon nanotubes were found in Damascus steel from the 17th century, possibly helping to account for the legendary strength of the swords made of it. Recently, several studies have highlighted the prospect of using carbon nanotubes as building blocks to fabricate three-dimensional macroscopic (>1mm in all three dimensions) all-carbon devices. Lalwani et al. have reported a novel radical initiated thermal crosslinking method to fabricated macroscopic, free-standing, porous, all-carbon scaffolds using single- and multi-walled carbon nanotubes as building blocks. These scaffolds possess macro-, micro-, and nano- structured pores and the porosity can be tailored for specific applications. These 3D all-carbon scaffolds/architectures maybe used for the fabrication of the next generation of energy storage, supercapacitors, field emission transistors, high-performance catalysis, photo-voltaics, and biomedical devices and implants.

Discovery

The true identity of the discoverers of carbon nanotubes is a subject of some contro-versy. For years, scientists assumed that Sumio Iijima of NEC had discovered carbon nanotubes in 1991. He published a paper describing his discovery which initiated a flurry of excitement and could be credited by inspiring the many scientists now studying applications of carbon nanotubes. Though Iijima has been given much of the credit for discovering carbon nanotubes, it turns out that the timeline of carbon nanotubes goes back much further than 1991. In 1952 L. V. Radushkevich and V. M. Lukyanovich published clear images of 50 nanometer diameter tubes made of carbon in the Soviet *Journal of Physical Chemistry*. This discovery was largely unnoticed, as the article was published in Russian, and Western scientists' access to Soviet press was limited during the Cold War. Before they came to be known as carbon nanotubes, in 1976, Morinobu Endo of CNRS observed hollow tubes of rolled up graphite sheets synthesised by a chemical vapour-growth technique. The first specimens observed would later come to be known as single-walled carbon nanotubes (SWNTs). The three scientists have been the first ones to show images of a nanotube with a solitary graphene wall.

Endo, in his early review of vapor-phase-grown carbon fibers (VPCF), also reminded us that he had observed a hollow tube, linearly extended with parallel carbon layer faces near the fiber core. This appears to be the observation of multi-walled carbon nano-tubes at the center of the fiber. The mass-produced MWCNTs today are strongly related to the VPGCF developed by Endo. In fact, they call it the "Endo-process", out of respect for his early work and patents.

In 1979, John Abrahamson presented evidence of carbon nanotubes at the 14th Bi-ennial Conference of Carbon at Pennsylvania State University. The conference paper described carbon nanotubes as carbon fibers that were produced on carbon anodes during arc discharge. A characterization of these fibers was given as well as hypotheses for their growth in a nitrogen atmosphere at low pressures.

In 1981, a group of Soviet scientists published the results of chemical and structural characterization of carbon nanoparticles produced by a thermocatalytical dispropor-tionation of carbon monoxide. Using TEM images and XRD patterns, the authors sug-gested that their "carbon multi-layer tubular crystals" were formed by rolling graphene layers into cylinders. They speculated that by rolling graphene layers into a cylinder, many different arrangements of graphene hexagonal nets are possible. They suggested two possibilities of such arrangements: circular arrangement (armchair nanotube) and a spiral, helical arrangement (chiral tube).

In 1987, Howard G. Tennent of Hyperion Catalysis was issued a U.S. patent for the production of "cylindrical discrete carbon fibrils" with a "constant diameter between about 3.5 and about 70 nanometers..., length 10^2 times the diameter, and an outer region of multiple essentially continuous layers of ordered carbon atoms and a distinct inner core...."

Iijima's discovery of multi-walled carbon nanotubes in the insoluble material of arc-burned graphite rods in 1991 and Mintmire, Dunlap, and White's independent prediction that if single-walled carbon nanotubes could be made, then they would exhibit remarkable conducting properties helped create the initial buzz that is now associated with carbon nanotubes. Nanotube research accelerated greatly following the independent discoveries by Bethune at IBM and Iijima at NEC of *single-walled* carbon nanotubes and methods to specifically produce them by adding transition-metal catalysts to the carbon in an arc discharge. The arc discharge technique was well-known to produce the famed Buckminster fullerene on a preparative scale, and these results appeared to extend the run of accidental discoveries relating to fullerenes. The discovery of nanotubes remains a contentious issue. Many believe that Iijima's report in 1991 is of particular importance because it brought carbon nanotubes into the awareness of the scientific community as a whole.

Optical Properties of Carbon Nanotubes

Within materials science, the optical properties of carbon nanotubes refer specifically to the absorption, photoluminescence (fluorescence), and Raman spectroscopy of carbon nanotubes. Spectroscopic methods offer the possibility of quick and non-destructive characterization of relatively large amounts of carbon nanotubes. There is a strong demand for such characterization from the industrial point of view: numerous parameters of the nanotube synthesis can be changed, intentionally or unintentionally, to alter the nanotube quality. As shown below, optical absorption, photoluminescence and Raman spectroscopies allow quick and reliable characterization of this "nanotube quality" in terms of non-tubular carbon content, structure (chirality) of the produced nanotubes, and structural defects. Those features determine nearly any other properties such as optical, mechanical, and electrical properties.

Carbon nanotubes are unique "one-dimensional systems" which can be envisioned as rolled single sheets of graphite (or more precisely graphene). This rolling can be done at different angles and curvatures resulting in different nanotube properties. The diameter typically varies in the range 0.4–40 nm (i.e. "only" ~100 times), but the length can vary ~10,000 times, reaching 55.5 cm. The nanotube aspect ratio, or the length-to-diameter ratio, can be as high as 132,000,000:1, which is unequalled by any other material. Consequently, all the properties of the carbon nanotubes relative to those of typical semiconductors are extremely anisotropic (directionally dependent) and tunable.

Whereas mechanical, electrical and electrochemical (supercapacitor) properties of the carbon nanotubes are well established and have immediate applications, the practical use of optical properties is yet unclear. The aforementioned tunability of properties is potentially useful in optics and photonics. In particular, light-emitting diodes (LEDs) and photo-detectors based on a single nanotube have been produced in the lab. Their unique feature is not the efficiency, which is yet relatively low, but the narrow selectivity in the wavelength of emission and detection of light and the possibility of its fine tuning through the nanotube structure. In addition, bolometer and optoelectronic memory devices have been realised on ensembles of single-walled carbon nanotubes.

Terminology

This article uses the following abbreviations:

- Carbon nanotube (CNT)

- Single wall carbon nanotube (SWCNT)

- Multiwall carbon nanotube (MWCNT)

However, C is often omitted in scientific literature, so NT, SWNT and MWNT are more commonly used. Also, "wall" is often exchanged with "walled".

Electronic Structure of Carbon Nanotube

A single-wall carbon nanotube can be imagined as graphene sheet rolled at a certain "chiral" angle with respect to a plane perpendicular to the tube's long axis. Consequently, SWCNT can be defined by its diameter and chiral angle. The chiral angle can range from 0 to 30 degrees.

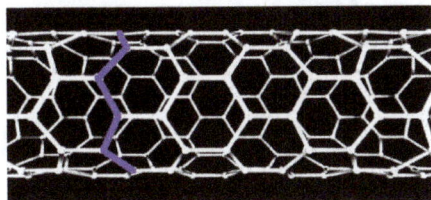

Zigzag nanotube   Armchair nanotube

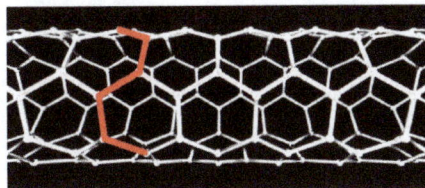

However, more conveniently, a pair of indices (n, m) is used instead. The indices refer to equally long unit vectors at 60° angles to each other across a single 6-member carbon ring. Taking the origin as carbon number 1, the a_1 unit vector may be considered the line drawn from carbon 1 to carbon 3, and the a_2 unit vector is then the line drawn from carbon 1 to carbon 5. To visualize a CNT with indices (n, m), draw n a_1 unit vectors across the graphene sheet, then draw m a_2 unit vectors at a 60° angle to the a_1 vectors, then add the vectors together. The line representing the sum of the vectors will define the circumference of the CNT along the plane perpendicular to its long axis, connecting one end to the other. In the diagram at right, C_h is a (4, 2) vector: the sum of 4 unit vectors from the origin directly to the right, then 2 unit vectors at a 60° angle down and to the right.

Tubes having $n = m$ (chiral angle = 30°) are called "armchair" and those with $m = 0$ (chiral angle = 0°) «zigzag». Those indices uniquely determine whether CNT is a metal, semimetal or semiconductor, as well as its band gap: when $|m - n| = 3k$ (k is integer), the tube is metallic; but if $|m - n| = 3k \pm 1$, the tube is semiconducting. The nanotube diameter d is related to m and n as

$$d = \frac{a}{\pi}\sqrt{(n^2 + nm + m^2)}.$$

In this equation, a = 0.246 nm is the magnitude of either unit vector a_1 or a_2.

The situation in multi-wall CNTs is complicated as their properties are determined by contribution of all individual shells; those shells have different structures, and, because of the synthesis, are usually more defective than SWCNTs. Therefore, optical properties of MWCNTs will not be considered here.

Van Hove Singularities

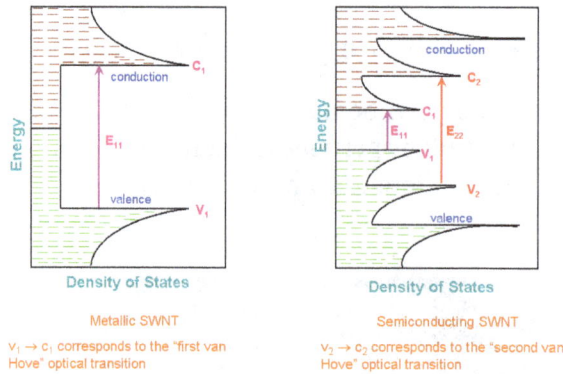

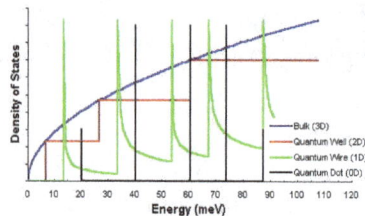

A bulk 3D material (blue) has continuous DOS, but a 1D wire (green) has Van Hove singularities.

Optical properties of carbon nanotubes derive from electronic transitions within one-dimensional density of states (DOS). A typical feature of one-dimensional crystals is that their DOS is not a continuous function of energy, but it descends gradually and then increases in a discontinuous spike. In contrast, three-dimensional materials have continuous DOS. The sharp peaks found in one-dimensional materials are called Van Hove singularities.

Van Hove singularities result in the following remarkable optical properties of carbon nanotubes:

- Optical transitions occur between the $v_1 - c_1$, $v_2 - c_2$, etc., states of semiconducting or metallic nanotubes and are traditionally labeled as S_{11}, S_{22}, M_{11}, etc., or, if the "conductivity" of the tube is unknown or unimportant, as E_{11}, E_{22}, etc. Crossover transitions $c_1 - v_2$, $c_2 - v_1$, etc., are dipole-forbidden and thus are extremely weak, but they were possibly observed using cross-polarized optical geometry.

- The energies between the Van Hove singularities depend on the nanotube structure. Thus by varying this structure, one can tune the optoelectronic properties of carbon nanotube. Such fine tuning has been experimentally demonstrated using UV illumination of polymer-dispersed CNTs.

- Optical transitions are rather sharp (~10 meV) and strong. Consequently, it is relatively easy to selectively excite nanotubes having certain (n, m) indices, as well as to detect optical signals from individual nanotubes.

Kataura Plot

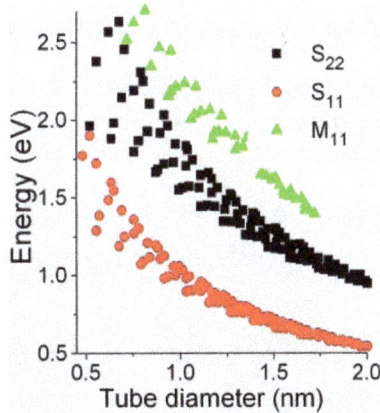

In this Kataura plot, the energy of an electronic transition decreases as the diameter of the nanotube increases.

The band structure of carbon nanotubes having certain (n, m) indexes can be easily calculated. A theoretical graph based on this calculations was designed in 1999 by Hiromichi Kataura to rationalize experimental findings. A Kataura plot relates the nanotube diameter and its bandgap energies for all nanotubes in a diameter range. The oscillating shape of every branch of the Kataura plot reflects the intrinsic strong dependence of the SWCNT properties on the (n, m) index rather than on its diameter. For example, (10, 1) and (8, 3) tubes have almost the same diameter, but very different properties: the former is a metal, but the latter is a semiconductor.

Optical Absorption

Optical absorption in carbon nanotubes differs from absorption in conventional 3D materials by presence of sharp peaks (1D nanotubes) instead of an absorption threshold followed by an absorption increase (most 3D solids). Absorption in nanotubes originates from electronic transitions from the v_2 to c_2 (energy E_{22}) or v_1 to c_1 (E_{11}) levels, etc. The transitions are relatively sharp and can be used to identify nanotube types. Note that the sharpness deteriorates with increasing energy, and that many nanotubes have very similar E_{22} or E_{11} energies, and thus significant overlap occurs in absorption spectra. This overlap is avoided in photoluminescence mapping measurements, which instead of a combination of overlapped transitions identifies individual (E_{22}, E_{11}) pairs.

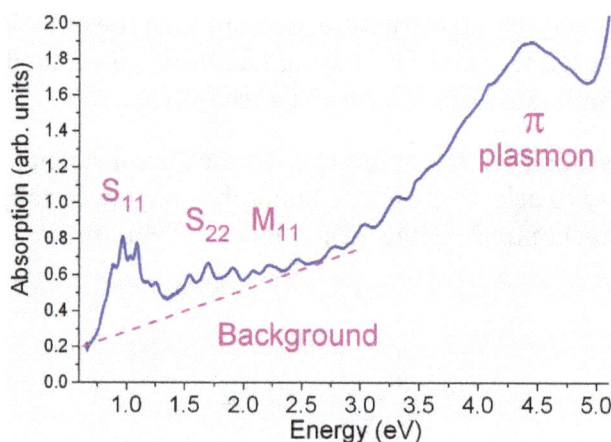

Optical absorption spectrum from dispersed single-wall carbon nanotubes

Interactions between nanotubes, such as bundling, broaden optical lines. While bundling strongly affects photoluminescence, it has much weaker effect on optical absorption and Raman scattering. Consequently, sample preparation for the latter two techniques is relatively simple.

Optical absorption is routinely used to quantify quality of the carbon nanotube powders.

The spectrum is analyzed in terms of intensities of nanotube-related peaks, background and pi-carbon peak; the latter two mostly originate from non-nanotube carbon in contaminated samples. However, it has been recently shown that by aggregating nearly single chirality semiconducting nanotubes into closely packed Van der Waals bundles the absorption background can be attributed to free carrier transition originating from intertube charge transfer.

Carbon Nanotubes as A Black Body

An ideal black body should have emissivity or absorbance of 1.0, which is difficult to attain in practice, especially in a wide spectral range. Vertically aligned "forests" of single-wall carbon nanotubes can have absorbances of 0.98–0.99 from the far-ultraviolet (200 nm) to far-infrared (200 μm) wavelengths.

These SWNT forests (buckypaper) were grown by the super-growth CVD method to about 10 μm height. Two factors could contribute to strong light absorption by these structures: (i) a distribution of CNT chiralities resulted in various bandgaps for individual CNTs. Thus a compound material was formed with broadband absorption. (ii) Light might be trapped in those forests due to multiple reflections.

Reflectance measurements			
	UV-to-near IR	**Near-to-mid IR**	**Mid-to-far IR**
Wavelength, μm	0.2-2	2–20	25–200
Incident angle, °	8	5	10
Reflection	Hemispherical-directional	Hemispherical-directional	Specular
Reference	White reflectance standard	Gold mirror	Aluminum mirror
Average reflectance	0.0160	0.0097	0.0017
Standard deviation	0.0048	0.0041	0.0027

Luminescence

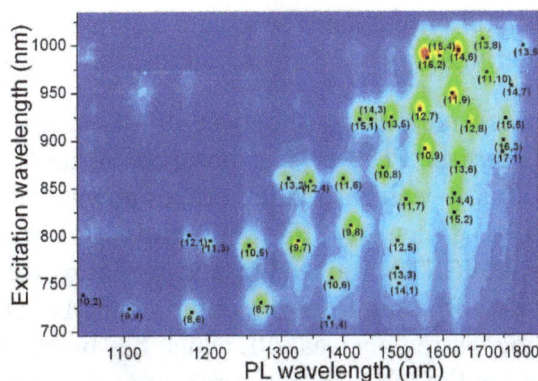

Photoluminescence map from single-wall carbon nanotubes. (n, m) indexes identify certain semiconducting nanotubes. Note that PL measurements do not detect nanotubes with $n = m$ or $m = 0$.

Photoluminescence (Fluorescence)

Semiconducting single-walled carbon nanotubes emit near-infrared light upon photo-excitation, described interchangeably as fluorescence or photoluminescence (PL). The excitation of PL usually occurs as follows: an electron in a nanotube absorbs excitation light via S_{22} transition, creating an electron-hole pair (exciton). Both electron and hole rapidly relax (via phonon-assisted processes) from c_2 to c_1 and from v_2 to v_1 states, respectively. Then they recombine through a $c_1 - v_1$ transition resulting in light emission.

No excitonic luminescence can be produced in metallic tubes. Their electrons can be excited, thus resulting in optical absorption, but the holes are immediately filled by other electrons out of the many available in the metal. Therefore, no excitons are produced.

Salient Properties

- Photoluminescence from SWCNT, as well as optical absorption and Raman scattering, is linearly polarized along the tube axis. This allows monitoring of the SWCNTs orientation without direct microscopic observation.

- PL is quick: relaxation typically occurs within 100 picoseconds.

- PL efficiency was first found to be low (~0.01%), but later studies measured much higher quantum yields. By improving the structural quality and isolation of nanotubes, emission efficiency increased. A quantum yield of 1% was reported in nanotubes sorted by diameter and length through gradient centrifugation, and it was further increased to 20% by optimizing the procedure of isolating individual nanotubes in solution.

- The spectral range of PL is rather wide. Emission wavelength can vary between 0.8 and 2.1 micrometers depending on the nanotube structure.

- Excitons are apparently delocalized over several nanotubes in single chirality bundles as the photoluminescence spectrum displays a splitting consistent with intertube exciton tunneling.

- Interaction between nanotubes or between a nanotube and another material may quench or increase PL. No PL is observed in multi-walled carbon nanotubes. PL from double-wall carbon nanotubes strongly depends on the preparation method: CVD grown DWCNTs show emission both from inner and outer shells. However, DWCNTs produced by encapsulating fullerenes into SWCNTs and annealing show PL only from the outer shells. Isolated SWCNTs lying on the substrate show extremely weak PL which has been detected in few studies only. Detachment of the tubes from the substrate drastically increases PL.

- Position of the (S_{22}, S_{11}) PL peaks depends slightly (within 2%) on the nanotube environment (air, dispersant, etc.). However, the shift depends on the (n, m) index, and thus the whole PL map not only shifts, but also warps upon changing the CNT medium.

Applications

- Photoluminescence is used for characterization purposes to measure the quantities of semiconducting nanotube species in a sample. Nanotubes are isolated (dispersed) using an appropriate chemical agent ("dispersant") to reduce the intertube quenching. Then PL is measured, scanning both the excitation and emission energies and thereby producing a PL map. The ovals in the map define (S_{22}, S_{11}) pairs, which unique identify (n, m) index of a tube. The data of Weisman and Bachilo are conventionally used for the identification.

- Nanotube fluorescence has been investigated for the purposes of imaging and sensing in biomedical applications.

Sensitization

Optical properties, including the PL efficiency, can be modified by encapsulating organic dyes (carotene, lycopene, etc.) inside the tubes. Efficient energy transfer occurs between the encapsulated dye and nanotube — light is efficiently absorbed by the dye and without significant loss is transferred to the SWCNT. Thus potentially, optical properties of a carbon nanotube can be controlled by encapsulating certain molecule inside it. Besides, encapsulation allows isolation and characterization of organic molecules which are unstable under ambient conditions. For example, Raman spectra are extremely difficult to measure from dyes because of their strong PL (efficiency close to 100%). However, encapsulation of dye molecules inside SWCNTs completely quenches dye PL, thus allowing measurement and analysis of their Raman spectra.

Cathodoluminescence

Cathodoluminescence (CL) — light emission excited by electron beam — is a process commonly observed in TV screens. An electron beam can be finely focused and scanned across the studied material. This technique is widely used to study defects in semiconductors and nanostructures with nanometer-scale spatial resolution. It would be beneficial to apply this technique to carbon nanotubes. However, no reliable CL, i.e. sharp peaks assignable to certain (n, m) indices, has been detected from carbon nanotubes yet.

Electroluminescence

If appropriate electrical contacts are attached to a nanotube, electron-hole pairs (excitons) can be generated by injecting electrons and holes from the contacts. Subsequent exciton recombination results in electroluminescence (EL). Electroluminescent devices have been produced from single nanotubes and their macroscopic assemblies. Recombination appears to proceed via triplet-triplet annihilation giving distinct peaks corresponding to E_{11} and E_{22} transitions.

Raman Scattering

Raman spectroscopy has good spatial resolution (~ 0.5 micrometers) and sensitivity (single nanotubes); it requires only minimal sample preparation and is rather informative. Consequently, Raman spectroscopy is probably the most popular technique of carbon nanotube characterization. Raman scattering in SWCNTs is resonant, i.e., only those tubes are probed which have one of the bandgaps equal to the exciting laser energy. Several scattering modes dominate the SWCNT spectrum, as discussed below.

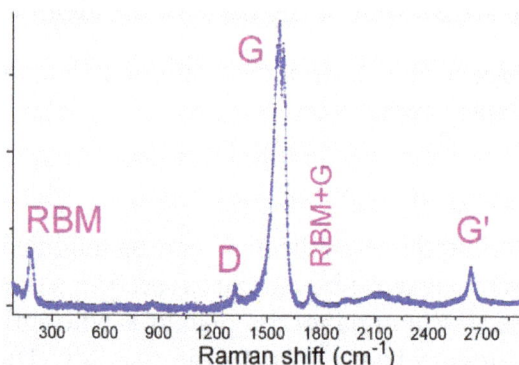

Raman spectrum of single-wall carbon nanotubes

Similar to photoluminescence mapping, the energy of the excitation light can be scanned in Raman measurements, thus producing Raman maps. Those maps also contain oval-shaped features uniquely identifying (n, m) indices. Contrary to PL, Raman mapping detects not only semiconducting but also metallic tubes, and it is less sensitive to nanotube bundling than PL. However, requirement of a tunable laser and a dedicated spectrometer is a strong technical impediment.

Radial Breathing Mode

Radial breathing mode (RBM) corresponds to radial expansion-contraction of the nanotube. Therefore, its frequency ν_{RBM} (in cm^{-1}) depends on the nanotube diameter d as, $\nu_{RBM} = A/d + B$ (where A and B are constants dependent on the environment in which the nanotube is present. For example, B=0 for individual nanotubes.) (in nanometers) and can be estimated as $\nu_{RBM} = 234/d + 10$ for SWNT or $\nu_{RBM} = 248/d$ for DWNT, which is very useful in deducing the CNT diameter from the RBM position. Typical RBM range is 100–350 cm^{-1}. If RBM intensity is particularly strong, its weak second overtone can be observed at double frequency.

Bundling Mode

The bundling mode is a special form of RBM supposedly originating from collective vibration in a bundle of SWCNTs.

G Mode

Another very important mode is the G mode (G from graphite). This mode corresponds to planar vibrations of carbon atoms and is present in most graphite-like materials. G band in SWCNT is shifted to lower frequencies relative to graphite (1580 cm^{-1}) and is split into several peaks. The splitting pattern and intensity depend on the tube structure and excitation energy; they can be used, though with much lower accuracy compared to RBM mode, to estimate the tube diameter and whether the tube is metallic or semiconducting.

D Mode

D mode is present in all graphite-like carbons and originates from structural defects. Therefore, the ratio of the *G/D* modes is conventionally used to quantify the structural quality of carbon nanotubes. High-quality nanotubes have this ratio significantly higher than 100. At a lower functionalisation of the nanotube, the *G/D* ratio remains almost unchanged. This ratio gives an idea of the functionalisation of a nanotube.

G' Mode

The name of this mode is misleading: it is given because in graphite, this mode is usually the second strongest after the G mode. However, it is actually the second overtone of the defect-induced D mode (and thus should logically be named D'). Its intensity is stronger than that of the D mode due to different selection rules. In particular, D mode is forbidden in the ideal nanotube and requires a structural defect, providing a phonon of certain angular momentum, to be induced. In contrast, G' mode involves a "self-annihilating" pair of phonons and thus does not require defects. The spectral position of G' mode depends on diameter, so it can be used roughly to estimate the SWCNT diameter. In particular, G' mode is a doublet in double-wall carbon nanotubes, but the doublet is often unresolved due to line broadening.

Other overtones, such as a combination of RBM+G mode at ~1750 cm^{-1}, are frequently seen in CNT Raman spectra. However, they are less important and are not considered here.

Anti-stokes Scattering

All the above Raman modes can be observed both as Stokes and anti-Stokes scattering. As mentioned above, Raman scattering from CNTs is resonant in nature, i.e. only tubes whose band gap energy is similar to the laser energy are excited. The difference between those two energies, and thus the band gap of individual tubes, can be estimated from the intensity ratio of the Stokes/anti-Stokes lines. This estimate however relies on the temperature factor (Boltzmann factor), which is often miscalculated – a focused laser beam is used in the measurement, which can locally heat the nanotubes without changing the overall temperature of the studied sample.

Rayleigh Scattering

Carbon nanotubes have very large aspect ratio, i.e., their length is much larger than their diameter. Consequently, as expected from the classical electromagnetic theory, elastic light scattering (or Rayleigh scattering) by straight CNTs has anisotropic angular dependence, and from its spectrum, the band gaps of individual nanotubes can be deduced.

Another manifestation of Rayleigh scattering is the "antenna effect", an array of nanotubes standing on a substrate has specific angular and spectral distributions of reflected light, and both those distributions depend on the nanotube length.

Mechanical Properties of Carbon Nanotubes

This article considers the mechanical properties of carbon nanotubes in the radial (transverse) direction. Carbon nanotubes are one of the strongest materials in nature. Carbon nanotubes (CNTs) are long hollow cylinders of graphene. Although graphene sheets have 2D symmetry, carbon nanotubes by geometry have different properties in axial and radial directions. It has been shown that CNTs are very strong in the axial direction. Young's modulus on the order of 270 - 950 GPa and tensile strength of 11 - 63 GPa were obtained.

On the other hand, there was evidence that in the radial direction they are rather soft. The first transmission electron microscope observation of radial elasticity suggested that even the van der Waals forces can deform two adjacent nanotubes. Later, nanoindentations with atomic force microscope were performed by several groups to quantitatively measure radial elasticity of multiwalled carbon nanotubes and tapping/contact mode atomic force microscopy was also performed on single-walled carbon nanotubes. Young's modulus of on the order of several GPa showed that CNTs are in fact very soft in the radial direction.

Radial direction elasticity of CNTs is important especially for carbon nanotube composites where the embedded tubes are subjected to large deformation in the transverse direction under the applied load on the composite structure.

One of the main problems in characterizing the radial elasticity of CNTs is the knowledge about the internal radius of the CNT; carbon nanotubes with identical outer diameter may have different internal diameter (or the number of walls). In 2008, a method using an atomic force microscope was introduced to determine the exact number of layers and hence the internal diameter of the CNT. In this way, mechanical characterization is more accurate.

Synthesis of Carbon Nanotubes

Techniques have been developed to produce carbon nanotubes in sizable quantities, including arc discharge, laser ablation, high-pressure carbon monoxide disproportionation, and chemical vapor deposition (CVD). Most of these processes take place in a vacuum or with process gases. CVD growth of CNTs can occur in vacuum or at atmospheric pressure. Large quantities of nanotubes can be synthesized by these methods; advances in catalysis and continuous growth are making CNTs more commercially viable.

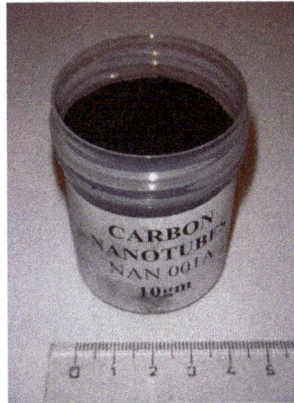

powder of carbon nanotubes

Types

Arc Discharge

Nanotubes were observed in 1991 in the carbon soot of graphite electrodes during an arc discharge, by using a current of 100 amps, that was intended to produce fullerenes. However the first macroscopic production of carbon nanotubes was made in 1992 by two researchers at NEC's Fundamental Research Laboratory. The method used was the same as in 1991. During this process, the carbon contained in the negative electrode sublimates because of the high-discharge temperatures.

The yield for this method is up to 30% by weight and it produces both single- and multi-walled nanotubes with lengths of up to 50 micrometers with few structural defects. Arc-discharge technique uses higher temperatures (above 1,700 °C) for CNT synthesis which typically causes the expansion of CNTs with fewer structural defects in comparison with other methods.

Laser Ablation

In laser ablation, a pulsed laser vaporizes a graphite target in a high-temperature reactor while an inert gas is bled into the chamber. Nanotubes develop on the cooler surfaces of the reactor as the vaporized carbon condenses. A water-cooled surface may be included in the system to collect the nanotubes.

This process was developed by Dr. Richard Smalley and co-workers at Rice University, who at the time of the discovery of carbon nanotubes, were blasting metals with a laser to produce various metal molecules. When they heard of the existence of nanotubes they replaced the metals with graphite to create multi-walled carbon nanotubes. Later that year the team used a composite of graphite and metal catalyst particles (the best yield was from a cobalt and nickel mixture) to synthesize single-walled carbon nanotubes.

The laser ablation method yields around 70% and produces primarily single-walled carbon nanotubes with a controllable diameter determined by the reaction temperature. However, it is more expensive than either arc discharge or chemical vapor deposition.

The effective equation for few cycle optical pulse dynamics was obtained by virtue of the Boltzmann collision-less equation solution for conduction band electrons of semiconductor carbon nanotubes in the case when medium with carbon nanotubes has spatially-modulated refractive index.

Plasma Torch

Single-walled carbon nanotubes can also be synthesized by a thermal plasma method, first invented in 2000 at INRS (Institut national de la recherche scientifique) in Varennes, Canada, by Olivier Smiljanic. In this method, the aim is to reproduce the conditions prevailing in the arc discharge and laser ablation approaches, but a carbon-containing gas is used instead of graphite vapors to supply the necessary carbon. Doing so, the growth of SWNT is more efficient (decomposing the gas can be 10 times less energy-consuming than graphite vaporization). The process is also continuous and low cost. A gaseous mixture of argon, ethylene and ferrocene is introduced into a microwave plasma torch, where it is atomized by the atmospheric pressure plasma, which has the form of an intense 'flame'. The fumes created by the flame contain SWNT, metallic and carbon nanoparticles and amorphous carbon.

Another way to produce single-walled carbon nanotubes with a plasma torch is to use the induction thermal plasma method, implemented in 2005 by groups from the University of Sherbrooke and the National Research Council of Canada. The method is similar to arc discharge in that both use ionized gas to reach the high temperature necessary to vaporize carbon-containing substances and the metal catalysts necessary for the ensuing nanotube growth. The thermal plasma is induced by high-frequency oscillating currents in a coil, and is maintained in flowing inert gas. Typically, a feedstock of carbon black and metal catalyst particles is fed into the plasma, and then cooled down to form single-walled carbon nanotubes. Different single-wall carbon nanotube diameter distributions can be synthesized.

The induction thermal plasma method can produce up to 2 grams of nanotube material per minute, which is higher than the arc discharge or the laser ablation methods.

Chemical Vapor Deposition (CVD)

The catalytic vapor phase deposition of carbon was reported in 1952 and 1959, but it was not until 1993 that carbon nanotubes were formed by this process. In 2007, researchers at the University of Cincinnati (UC) developed a process to grow aligned carbon nanotube arrays of length 18 mm on a FirstNano ET3000 carbon nanotube growth system.

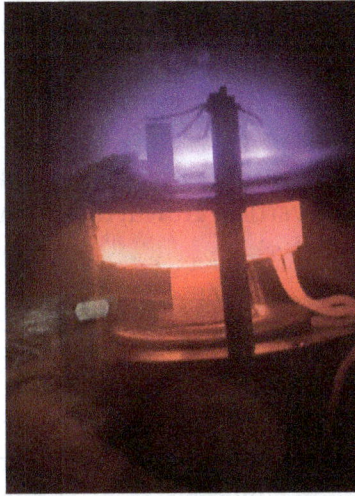

nanotubes being grown by plasma enhanced chemical vapor deposition

During CVD, a substrate is prepared with a layer of metal catalyst particles, most commonly nickel, cobalt, iron, or a combination. The metal nanoparticles can also be produced by other ways, including reduction of oxides or oxides solid solutions. The diameters of the nanotubes that are to be grown are related to the size of the metal particles. This can be controlled by patterned (or masked) deposition of the metal, annealing, or by plasma etching of a metal layer. The substrate is heated to approximately 700 °C. To initiate the growth of nanotubes, two gases are bled into the reactor: a process gas (such as ammonia, nitrogen or hydrogen) and a carbon-containing gas (such as acetylene, ethylene, ethanol or methane). Nanotubes grow at the sites of the metal catalyst; the carbon-containing gas is broken apart at the surface of the catalyst particle, and the carbon is transported to the edges of the particle, where it forms the nanotubes. This mechanism is still being studied. The catalyst particles can stay at the tips of the growing nanotube during growth, or remain at the nanotube base, depending on the adhesion between the catalyst particle and the substrate. Thermal catalytic decomposition of hydrocarbon has become an active area of research and can be a promising route for the bulk production of CNTs. Fluidised bed reactor is the most widely used reactor for CNT preparation. Scale-up of the reactor is the major challenge.

CVD is the most widely used method for the production of carbon nanotubes. For this purpose, the metal nanoparticles are mixed with a catalyst support such as MgO or Al_2O_3 to increase the surface area for higher yield of the catalytic reaction of the carbon feedstock with the metal particles. One issue in this synthesis route is the removal of the catalyst support via an acid treatment, which sometimes could destroy the original structure of the carbon nanotubes. However, alternative catalyst supports that are sol-uble in water have proven effective for nanotube growth.

If a plasma is generated by the application of a strong electric field during growth (plas-ma-enhanced chemical vapor deposition), then the nanotube growth will follow the direction of the electric field. By adjusting the geometry of the reactor it is possible to synthesize vertically aligned carbon nanotubes (i.e., perpendicular to the substrate), a morphology that has been of interest to researchers interested in electron emission from nanotubes. Without the plasma, the resulting nanotubes are often randomly oriented. Under certain reaction conditions, even in the absence of a plasma, closely spaced nanotubes will maintain a vertical growth direction resulting in a dense array of tubes resembling a carpet or forest.

Of the various means for nanotube synthesis, CVD shows the most promise for industri-al-scale deposition, because of its price/unit ratio, and because CVD is capable of grow-ing nanotubes directly on a desired substrate, whereas the nanotubes must be collected in the other growth techniques. The growth sites are controllable by careful deposition of the catalyst. In 2007, a team from Meijo University demonstrated a high-efficiency CVD technique for growing carbon nanotubes from camphor. Researchers at Rice Uni-versity, until recently led by the late Richard Smalley, have concentrated upon finding methods to produce large, pure amounts of particular types of nanotubes. Their ap-proach grows long fibers from many small seeds cut from a single nanotube; all of the resulting fibers were found to be of the same diameter as the original nanotube and are expected to be of the same type as the original nanotube.

Super-growth CVD

Super-growth CVD (water-assisted chemical vapor deposition) was developed by Kenji Hata, Sumio Iijima and co-workers at AIST, Japan. In this process, the activity and lifetime of the catalyst are enhanced by addition of water into the CVD reactor. Dense millimeter-tall nanotube "forests", aligned normal to the substrate, were produced. The forests height could be expressed, as

$$H(t) = \beta \tau_o (1 - e^{-t/\tau_o}).$$

In this equation, β is the initial growth rate and τ_o is the characteristic catalyst lifetime.

Their specific surface exceeds 1,000 m²/g (capped) or 2,200 m²/g (uncapped), surpass-ing the value of 400–1,000 m²/g for HiPco samples. The synthesis efficiency is about 100 times higher than for the laser ablation method. The time required to make SWNT forests of the height of 2.5 mm by this method was 10 minutes in 2004. Those SWNT forests can be easily separated from the catalyst, yielding clean SWNT material (puri-ty >99.98%) without further purification. For comparison, the as-grown HiPco CNTs contain about 5–35% of metal impurities; it is therefore purified through dispersion and centrifugation that damages the nanotubes. Super-growth avoids this problem. Patterned highly organized single-walled nanotube structures were successfully fabricated using the super-growth technique.

The mass density of super-growth CNTs is about 0.037 g/cm³. It is much lower than that of conventional CNT powders (~1.34 g/cm³), probably because the latter contain metals and amorphous carbon.

The super-growth method is basically a variation of CVD. Therefore, it is possible to grow material containing SWNT, DWNTs and MWNTs, and to alter their ratios by tuning the growth conditions. Their ratios change by the thinness of the catalyst. Many MWNTs are included so that the diameter of the tube is wide.

The vertically aligned nanotube forests originate from a "zipping effect" when they are immersed in a solvent and dried. The zipping effect is caused by the surface tension of the solvent and the van der Waals forces between the carbon nanotubes. It aligns the nanotubes into a dense material, which can be formed in various shapes, such as sheets and bars, by applying weak compression during the process. Densification increases the Vickers hardness by about 70 times and density is 0.55 g/cm³. The packed carbon nanotubes are more than 1 mm long and have a carbon purity of 99.9% or higher; they also retain the desirable alignment properties of the nanotubes forest.

Natural, Incidental, and Controlled Flame Environments

Fullerenes and carbon nanotubes are not necessarily products of high-tech laboratories; they are commonly formed in such mundane places as ordinary flames, produced by burning methane, ethylene, and benzene, and they have been found in soot from both indoor and outdoor air. However, these naturally occurring varieties can be highly irregular in size and quality because the environment in which they are produced is often highly uncontrolled. Thus, although they can be used in some applications, they can lack in the high degree of uniformity necessary to satisfy the many needs of both research and industry. Recent efforts have focused on producing more uniform carbon nanotubes in controlled flame environments. Such methods have promise for large-scale, low-cost nanotube synthesis based on theoretical models, though they must compete with rapidly developing large scale CVD production.

Purification

Removal of Catalysts

Nanoscale metal catalysts are important ingredients for fixed- and fluidized-bed CVD synthesis of CNTs. They allow increasing the growth efficiency of CNTs and may give control over their structure and chirality. During synthesis, catalysts can convert carbon precursors into tubular carbon structures but can also form encapsulating carbon overcoats. Together with metal oxide supports they may therefore attach to or become incorporated into the CNT product. The presence of metal impurities can be problematic for many applications. Especially catalyst metals like nickel, cobalt or yttrium may be of toxicological concern. While unencapsulated catalyst metals may be readily removable by acid washing, encapsulated ones require oxidative treatment for opening their carbon shell. The effective removal of catalysts, especially of encapsulated ones, while preserving the CNT structure is a challenge and has been addressed in many studies. A new approach to break carbonaceous catalyst encapsulations is based on rapid thermal annealing.

Application-related Issues

Many electronic applications of carbon nanotubes crucially rely on techniques of selectively producing either semiconducting or metallic CNTs, preferably of a certain chirality. Several methods of separating semiconducting and metallic CNTs are known, but most of them are not yet suitable for large-scale technological processes. The most efficient method relies on density-gradient ultracentrifugation, which separates surfactant-wrapped nanotubes by the minute difference in their density. This density difference often translates into difference in the nanotube diameter and (semi)conducting properties. Another method of separation uses a sequence of freezing, thawing, and compression of SWNTs embedded in agarose gel. This process results in a solution containing 70% metallic SWNTs and leaves a gel containing 95% semiconducting SWNTs. The diluted solutions separated by this method show various colors. The separated carbon nanotubes using this method have been applied to electrodes, e.g. electric double-layer capacitor. Moreover, SWNTs can be separated by the column chromatography method. Yield is 95% in semiconductor type SWNT and 90% in metallic type SWNT.

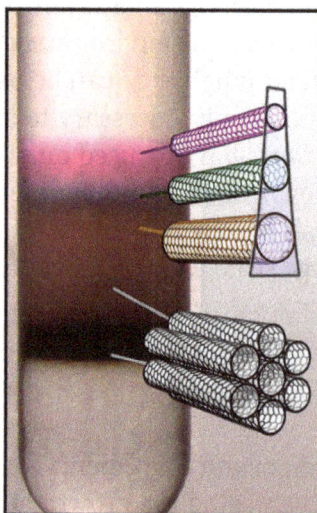

Centrifuge tube with a solution of carbon nanotubes, which were sorted by diameter using density-gradient ultracentrifugation.

In addition to separation of semiconducting and metallic SWNTs, it is possible to sort SWNTs by length, diameter, and chirality. The highest resolution length sorting, with length variation of <10%, has thus far been achieved by size exclusion chromatography (SEC) of DNA-dispersed carbon nanotubes (DNA-SWNT). SWNT diameter separation has been achieved by density-gradient ultracentrifugation (DGU) using surfactant-dispersed SWNTs and by ion-exchange chromatography (IEC) for DNA-SWNT. Purification of individual chiralities has also been demonstrated with IEC of DNA-SWNT: specific short DNA oligomers can be used to isolate individual SWNT chiralities. Thus far, 12 chiralities have been isolated at purities ranging from 70% for (8,3) and (9,5) SWNTs to 90% for (6,5), (7,5) and (10,5)SWNTs. There have been successful efforts to integrate these purified nanotubes into devices, e. g. FETs.

An alternative to separation is development of a selective growth of semiconducting or metallic CNTs. Recently, a new CVD recipe that involves a combination of ethanol and methanol gases and quartz substrates resulting in horizontally aligned arrays of 95–98% semiconducting nanotubes was announced.

Nanotubes are usually grown on nanoparticles of magnetic metal (Fe, Co), which facilitates production of electronic (spintronic) devices. In particular, control of current through a field-effect transistor by magnetic field has been demonstrated in such a single-tube nanostructure.

Potential Applications of Carbon Nanotubes

Carbon nanotubes (CNTs) are cylinders of one or more layers of graphene (lattice). Diameters of single-walled carbon nanotubes (SWNTs) and multi-walled carbon nanotubes (MWNTs) are typically 0.8 to 2 nm and 5 to 20 nm, respectively, although MWNT diameters can exceed 100 nm. CNT lengths range from less than 100 nm to 0.5 m.

Individual CNT walls can be metallic or semiconducting depending on the orientation of the lattice with respect to the tube axis, which is called chirality. MWNT's cross-sectional area offers an elastic modulus approaching 1 TPa and a tensile strength of 100 GPa, over 10-fold higher than any industrial fiber. MWNTs are typically metallic and can carry currents of up to 10^9 A cm^{-2}. SWNTs can display thermal conductivity of 3500 W m^{-1} K^{-1}, exceeding that of diamond.

As of 2013, carbon nanotube production exceeded several thousand tons per year, used for applications in energy storage, automotive parts, boat hulls, sporting goods, water filters, thin-film electronics, coatings, actuators and electromagnetic shields. CNT-related publications more than tripled in the prior decade, while rates of patent issuance also increased. Most output was of unorganized architecture. Organized CNT architectures such as "forests", yarns and regular sheets were produced in much smaller volumes. CNTs have even been proposed as the tether for a purported space elevator.

3D carbon nanotube scaffolds.

Recently, several studies have highlighted the prospect of using carbon nanotubes as building blocks to fabricate three-dimensional macroscopic (>1mm in all three dimensions) all-carbon devices. Lalwani et al. have reported a novel radical initiated thermal cross-linking method to fabricated macroscopic, free-standing, porous, all-carbon scaffolds using single- and multi-walled carbon nanotubes as building blocks. These scaffolds possess macro-, micro-, and nano- structured pores and the porosity can be tailored for specific applications. These 3D all-carbon scaffolds/architectures may be used for the fabrication of the next generation of energy storage, supercapacitors, field emission transistors, high-performance catalysis, photovoltaics, and biomedical devices and implants.

Biological and Biomedical Research

Researchers from Rice University and State University of New York – Stony Brook have shown that the addition of low weight % of carbon nanotubes can lead to significant improvements in the mechanical properties of biodegradable polymeric nanocomposites for applications in tissue engineering including bone, cartilage, muscle and nerve tissue. Dispersion of low weight % of graphene (~0.02 wt.%) results in significant increases in compressive and flexural mechanical properties of polymeric nanocomposites. Researchers at Rice University, Stony Brook University, Radboud University Nijmegen Medical Centre and University of California, Riverside have shown that carbon nanotubes and their polymer nanocomposites are suitable scaffold materials for bone tissue engineering and bone formation.

CNTs exhibit dimensional and chemical compatibility with biomolecules, such as DNA and proteins. CNTs enable fluorescent and photoacoustic imaging, as well as localized heating using near-infrared radiation.

SWNT biosensors exhibit large changes in electrical impedance and optical properties, which is typically modulated by adsorption of a target on the CNT surface. Low detection limits and high selectivity require engineering the CNT surface and field effects, capacitance, Raman spectral shifts and photoluminescence for sensor design. Products under development include printed test strips for estrogen and progesterone detection, microarrays for DNA and protein detection and sensors for NO 2 and cardiac troponin. Similar CNT sensors support food industry, military and environmental applications.

CNTs can be internalized by cells, first by binding their tips to cell membrane receptors. This enables transfection of molecular cargo attached to the CNT walls or encapsulated by CNTs. For example, the cancer drug doxorubicin was loaded at up to 60 wt % on CNTs compared with a maximum of 8 to 10 wt % on liposomes. Cargo release can be triggered by near-infrared radiation. However, limiting the retention of CNTs within the body is critical to prevent undesirable accumulation.

CNT toxicity remains a concern, although CNT biocompatibility may be engineerable. The degree of lung inflammation caused by injection of well-dispersed SWNTs was insignificant compared with asbestos and with particulate matter in air. Medical acceptance of CNTs requires understanding of immune response and appropriate exposure standards for inhalation, injection, ingestion and skin contact. CNT forests immobilized in a polymer did not show elevated inflammatory response in rats relative to controls. CNTs are under consideration as low-impedance neural interface electrodes and for coating of catheters to reduce thrombosis.

CNT enabled x-ray sources for medical imaging are also in development. Relying on the unique properties of the CNTs, researchers have developed field emission cathodes that allow precise x-ray control and close placement of multiple sources. CNT enabled x-ray sources have been demonstrated for pre-clinical, small animal imaging applications, and are currently in clinical trials.

In November 2012 researchers at the American National Institute of Standards and Technology (NIST) proved that single-wall carbon nanotubes may help protect DNA molecules from damage by oxidation.

A highly effective method of delivering carbon nanotubes into cells is Cell squeezing, a high-throughput vector-free microfluidic platform for intracellular delivery developed at the Massachusetts Institute of Technology in the labs of Robert S. Langer.

Carbon nanotubes have furthermore been grown inside microfluidic channels for chemical analysis, based on electrochromatography. Here, the high surface-area-to-volume ratio and high hydrophobicity of CNTs are used in order to greatly decrease the analysis time of small neutral molecules that typically require large bulky equipment for analysis.

Composite Materials

Because of the carbon nanotube's superior mechanical properties, many structures have been proposed ranging from everyday items like clothes and sports gear to combat jackets and space elevators. However, the space elevator will require further efforts in refining carbon nanotube technology, as the practical tensile strength of carbon nanotubes must be greatly improved.

For perspective, outstanding breakthroughs have already been made. Pioneering work led by Ray H. Baughman at the NanoTech Institute has shown that single and multi-walled nanotubes can produce materials with toughness unmatched in the man-made and natural worlds.

Carbon nanotubes being spun to form a yarn, CSIRO

Carbon nanotubes are also a promising material as building blocks in hierarchical composite materials given their exceptional mechanical properties (~1 TPa in modulus, and ~100 GPa in strength). Initial attempts to incorporate CNTs into hierarchical structures (such as yarns, fibres or films) has led to mechanical properties that were significantly lower than these potential limits. Windle *et al.* have used an *in situ* chemical vapor deposition (CVD) spinning method to produce continuous CNT yarns from CVD-grown CNT aerogels. CNT yarns can also be manufactured by drawing out CNT bundles from a CNT forest and subsequently twisting to form the fibre. The Windle group have fabricated CNT yarns with strengths as high as ~9 GPa at small gage lengths of ~1 mm, however, strengths of only about ~1 GPa were reported at the longer gage length of 20 mm. The reason why fibre strengths have been low compared to the strength of individual CNTs is due to a failure to effectively transfer load to the constituent (discontinuous) CNTs within the fibre. One potential route for alleviating this problem is via irradiation (or deposition) induced covalent inter-bundle and inter-CNT cross-linking to effectively 'join up' the CNTs. Espinosa *et al.* developed high performance DWNT-polymer composite yarns by twisting and stretching ribbons of randomly oriented bundles of DWNTs thinly coated with polymeric organic compounds. These DWNT-polymer yarns exhibited an unusually high energy to failure of ~100 $J \cdot g^{-1}$ (comparable to one of the toughest natural materials – spider silk), and strength as high as ~1.4 GPa. Effort is ongoing to produce CNT composites that incorporate tougher matrix materials, such as Kevlar, to further improve on the mechanical properties toward those of individual CNTs.

Because of the high mechanical strength of carbon nanotubes, research is being made into weaving them into clothes to create stab-proof and bulletproof clothing. The nanotubes would effectively stop the bullet from penetrating the body, although the bullet's kinetic energy would likely cause broken bones and internal bleeding.

Mixtures

MWNTs were first used as electrically conductive fillers in metals, at concentrations as high as 83.78 percent by weight (wt %). MWNT-polymer composites reach conductivities as high as 10,000 S m^{-1} at 10 wt % loading. In the automotive industry, CNT plastics are used in electrostatic-assisted painting of mirror housings, as well as fuel lines and filters that dissipate electrostatic charge. Other products include electromagnetic interference (EMI)–shielding packages and silicon wafer carriers.

For load-bearing applications, CNT powders are mixed with polymers or precursor resins to increase stiffness, strength and toughness. These enhancements depend on CNT diameter, aspect ratio, alignment, dispersion and interfacial interaction. Premixed resins and master batches employ CNT loadings from 0.1 to 20 wt %. Nanoscale stick-slip among CNTs and CNT-polymer contacts can increase material damping, enhancing sporting goods, including tennis racquets, baseball bats and bicycle frames.

CNT resins enhance fiber composites, including wind turbine blades and hulls for maritime security boats that are made by enhancing carbon fiber composites with CNT-enhanced resin. CNTs are deployed as additives in the organic precursors of stronger 1-µm diameter carbon fibers. CNTs influence the arrangement of carbon in pyrolyzed fiber.

Toward the challenge of organizing CNTs at larger scales, hierarchical fiber composites are created by growing aligned forests onto glass, silicon carbide (SiC), alumina and carbon fibers, creating so-called "fuzzy" fibers. Fuzzy epoxy CNT-SiC and CNT-alumina fabric showed 69% improved crack-opening (mode I) and/or in-plane shear interlaminar (mode II) toughness. Applications under investigation include lightning-strike protection, deicing, and structural health monitoring for aircraft.

MWNTs can be used as a flame-retardant additive to plastics due to changes in rheology by nanotube loading. Such additives can replace halogenated flame retardants, which face environmental restrictions.

CNT/Concrete blends offer increased tensile strength and reduced crack propagation.

Buckypaper (nanotube aggregate) can significantly improve fire resistance due to efficient heat reflection.

Textiles

The previous studies on the use of CNTs for textile functionalization were focused on fiber spinning for improving physical and mechanical properties. Recently a great deal of attention has been focused on coating CNTs on textile fabrics. Various methods have been employed for modifying fabrics using CNTs. Shim et al. produced intelligent e-textiles for Human Biomonitoring using a polyelectrolyte-based coating with CNTs. Additionally, Panhuis et al. dyed textile material by immersion in either a poly (2-methoxy aniline-5-sulfonic acid) PMAS polymer solution or PMAS-SWNT dispersion with enhanced conductivity and capacitance with a durable behavior. In another study, Hu and coworkers coated single-walled carbon nanotubes with a simple "dipping and drying" process for wearable electronics and energy storage applications. In the recent study, Li and coworkers using elastomeric separator and almost achieved a fully stretchable supercapacitor based on buckled single-walled carbon nanotube macrofilms. The electrospun polyurethane was used and provided sound mechanical stretchability and the whole cell achieve excellent charge-discharge cycling stability. CNTs have an aligned nanotube structure and a negative surface charge. Therefore, they have similar structures to direct dyes, so the exhaustion method is applied for coating and absorbing CNTs on the fiber surface for preparing multifunctional fabric including antibacterial, electric conductive, flame retardant and electromagnetic absorbance properties.

Later, CNT yarns and laminated sheets made by direct chemical vapor deposition (CVD) or forest spinning or drawing methods may compete with carbon fiber for high-end uses, especially in weight-sensitive applications requiring combined electrical and mechanical functionality. Research yarns made from few-walled CNTs have reached a stiffness of 357 GPa and a strength of 8.8 GPa for a gauge length comparable to the millimeter-long CNTs within the yarn. Centimeter-scale gauge lengths offer only 2-GPa gravimetric strengths, matching that of Kevlar.

Because the probability of a critical flaw increases with volume, yarns may never achieve the strength of individual CNTs. However, CNT's high surface area may provide interfacial coupling that mitigates these deficiencies. CNT yarns can be knotted without loss of strength. Coating forest-drawn CNT sheets with functional powder before inserting twist yields weavable, braidable and sewable yarns containing up to 95 wt % powder. Uses include superconducting wires, battery and fuel cell electrodes and self-cleaning textiles.

As yet impractical fibers of aligned SWNTs can be made by coagulation-based spinning of CNT suspensions. Cheaper SWNTs or spun MWNTs are necessary for commercialization. Carbon nanotubes can be dissolved in superacids such as fluorosulfuric acid and drawn into fibers in dry jet-wet spinning.

DWNT-polymer composite yarns have been made by twisting and stretching ribbons of randomly oriented bundles of DWNTs thinly coated with polymeric organic compounds.

Body armor—combat jackets Cambridge University developed the fibres and licensed a company to make them. In comparison, the bullet-resistant fiber Kevlar fails at 27–33 J/g.

Synthetic muscles offer high contraction/extension ratio given an electric current

SWNT are in use as an experimental material for removable, structural bridge panels.

In 2015 researchers incorporated CNTs and graphene into spider silk, increasing its strength and toughness to a new record. They sprayed 15 Pholcidae spiders with water containing the nanotubes or flakes.The resulting silk had a fracture strength up to 5.4 GPa, a Young's modulus up to 47.8 GPa and a toughness modulus up to 2.1 GPa, surpassing both synthetic polymeric high performance fibres (e.g. Kevlar49) and knotted fibers.

Carbon Nanotube Springs

"Forests" of stretched, aligned MWNT springs can achieve an energy density 10 times greater than that of steel springs, offering cycling durability, temperature insensitivity, no spontaneous discharge and arbitrary discharge rate. SWNT forests are expected to be able to store far more than MWNTs.

Alloys

Adding small amounts of CNTs to metals increases tensile strength and modulus with potential in aerospace and automotive structures. Commercial aluminum-MWNT composites have strengths comparable to stainless steel (0.7 to 1 GPa) at one-third the density (2.6 g cm^{-3}), comparable to more expensive aluminium-lithium alloys.

Coatings and Films

CNTs can serve as a multifunctional coating material. For example, paint/MWNT mixtures can reduce biofouling of ship hulls by discouraging attachment of algae and barnacles. They are a possible alternative to environmentally hazardous biocide-containing paints. Mixing CNTs into anticorrosion coatings for metals can enhance coating stiffness and strength and provide a path for cathodic protection.

CNTs provide a less expensive alternative to ITO for a range of consumer devices. Besides cost, CNT's flexible, transparent conductors offer an advantage over brittle ITO coatings for flexible displays. CNT conductors can be deposited from solution and patterned by methods such as screen printing. SWNT films offer 90% transparency and a sheet resistivity of 100 ohm per square. Such films are under development for thin-film heaters, such as for defrosting windows or sidewalks.

Carbon nanotubes forests and foams can also be coated with a variety of different materials to change their functionality and performance. Examples include silicon coated CNTs to create flexible energy-dense batteries, graphene coatings to create highly elastic aerogels and silicon carbide coatings to create a strong structural material for robust high-aspect-ratio 3D-micro architectures.

Optical Power Detectors

A spray-on mixture of carbon nanotubes and ceramic demonstrates unprecedented ability to resist damage while absorbing laser light. Such coatings that absorb as the energy of high-powered lasers without breaking down are essential for optical power detectors that measure the output of such lasers. These are used, for example, in military equipment for defusing unexploded mines. The composite consists of multiwall carbon nanotubes and a ceramic made of silicon, carbon and nitrogen. Including boron boosts the breakdown temperature. The nanotubes and graphene-like carbon transmit heat well, while the oxidation-resistant ceramic boosts damage resistance. Creating the coating involves dispersing the nanotubes in toluene, to which a clear liquid polymer containing boron was added. The mixture was heated to 1,100 °C (2,010 °F). The result is crushed into a fine powder, dispersed again in toluene and sprayed in a thin coat on a copper surface. The coating absorbed 97.5 percent of the light from a far-infrared laser and tolerated 15 kilowatts per square centimeter for 10 seconds. Damage tolerance is about 50 percent higher than for similar coatings, e.g., nanotubes alone and carbon paint.

Radar Absorption

Radars work in the microwave frequency range, which can be absorbed by MWNTs. Applying the MWNTs to the aircraft would cause the radar to be absorbed and therefore seem to have a smaller radar cross-section. One such application could be to paint the nanotubes onto the plane. Recently there has been some work done at the University of Michigan regarding carbon nanotubes usefulness as stealth technology on aircraft. It has been found that in addition to the radar absorbing properties, the nanotubes neither reflect nor scatter visible light, making it essentially invisible at night, much like painting current stealth aircraft black except much more effective. Current limitations in manufacturing, however, mean that current production of nanotube-coated aircraft is not possible. One theory to overcome these current limitations is to cover small particles with the nanotubes and suspend the nanotube-covered particles in a medium such as paint, which can then be applied to a surface, like a stealth aircraft.

Microelectronics

Nanotube-based transistors, also known as carbon nanotube field-effect transistors (CNTFETs), have been made that operate at room temperature and that are capable of digital switching using a single electron. However, one major obstacle to realization of nanotubes has been the lack of technology for mass production. In 2001 IBM researchers demonstrated how metallic nanotubes can be destroyed, leaving semiconducting ones behind for use as transistors. Their process is called "constructive destruction," which includes the automatic destruction of defective nanotubes on the wafer. This process, however, only gives control over the electrical properties on a statistical scale.

SWNTs are attractive for transistors because of their low electron scattering and their bandgap. SWNTs are compatible with field-effect transistor (FET) architectures and high-k dielectrics. Despite progress following the CNT transistor's appearance in 1998, including a tunneling FET with a subthreshold swing of <60 mV per decade (2004), a radio (2007) and an FET with sub-10-nm channel length and a normalized current density of 2.41 mA μm^{-1} at 0.5 V, greater than those obtained for silicon devices.

However, control of diameter, chirality, density and placement remains insufficient for commercial production. Less demanding devices of tens to thousands of SWNTs are more immediately practical. The use of CNT arrays/transistor increases output current and compensates for defects and chirality differences, improving device uniformity and reproducibility. For example, transistors using horizontally aligned CNT arrays achieved mobilities of 80 cm^2 V^{-1} s^{-1}, subthreshold slopes of 140 mV per decade and on/off ratios as high as 10^5. CNT film deposition methods enable conventional semiconductor fabrication of more than 10,000 CNT devices per chip.

Printed CNT thin-film transistors (TFTs) are attractive for driving organic light-emitting diode displays, showing higher mobility than amorphous silicon (~1 cm^2 V^{-1} s^{-1}) and can be deposited by low-temperature, nonvacuum methods. Flexible CNT TFTs with a mobility of 35 cm^2 V^{-1} s^{-1} and an on/off ratio of 6×10^6 were demonstrated. A vertical CNT FET showed sufficient current output to drive OLEDs at low voltage, enabling red-green-blue emission through a transparent CNT network. CNTs are under consideration for radio-frequency identification tags. Selective retention of semiconducting SWNTs during spin-coating and reduced sensitivity to adsorbates were demonstrated.

The International Technology Roadmap for Semiconductors suggests that CNTs could replace Cu in microelectronic interconnects, owing to their low scattering, high current-carrying capacity, and resistance to electromigration. For this, vias comprising tightly packed (>10^{13} cm^{-2}) metallic CNTs with low defect density and low contact resistance are needed. Recently, complementary metal oxide semiconductor (CMOS)–compatible 150-nm-diameter interconnects with a single CNT–contact hole resistance of 2.8 kOhm were demonstrated on full 200-mm-diameter wafers. Also, as a replacement for solder bumps, CNTs can function both as electrical leads and heat dissipaters for use in high-power amplifiers.

Last, a concept for a nonvolatile memory based on individual CNT crossbar electromechanical switches has been adapted for commercialization by patterning tangled CNT thin films as the functional elements. This required development of ultrapure CNT suspensions that can be spin-coated and processed in industrial clean room environments and are therefore compatible with CMOS processing standards.

Transistors

Carbon nanotube field-effect transistors (CNTFETs) can operate at room temperature and are capable of digital switching using a single electron. In 2013, a CNT logic circuit was demonstrated that could perform useful work. Major obstacles to nanotube-based microelectronics include the absence of technology for mass production, circuit density, positioning of individual electrical contacts, sample purity, control over length, chirality and desired alignment, thermal budget and contact resistance.

One of the main challenges was regulating conductivity. Depending on subtle surface features, a nanotube may act as a conductor or as a semiconductor.

Another way to make carbon nanotube transistors has been to use random networks of them. By doing so one averages all of their electrical differences and one can produce devices in large scale at the wafer level. This approach was first patented by Nanomix Inc. (date of original application June 2002). It was first published in the academic literature by the United States Naval Research Laboratory in 2003 through independent research work. This approach also enabled Nanomix to make the first transistor on a flexible and transparent substrate.

Since the electron mean free path in SWCNTs can exceed 1 micrometer, long channel CNTFETs exhibit near-ballistic transport characteristics, resulting in high speeds. CNT devices are projected to operate in the frequency range of hundreds of gigahertz.

Nanotubes can be grown on nanoparticles of magnetic metal (Fe, Co) that facilitates production of electronic (spintronic) devices. In particular control of current through a field-effect transistor by magnetic field has been demonstrated in such a single-tube nanostructure.

History

In 2001 IBM researchers demonstrated how metallic nanotubes can be destroyed, leaving semiconducting nanotubes for use as components. Using "constructive destruction", they destroyed defective nanotubes on the wafer. This process, however, only gives control over the electrical properties on a statistical scale. In 2003 room-temperature ballistic transistors with ohmic metal contacts and high-k gate dielectric were reported, showing 20–30x more current than state-of-the-art siliconMOSFETs. Palladium is a high-work function metal that was shown to exhibit Schottky barrier-free contacts to semiconducting nanotubes with diameters >1.7 nm.

The potential of carbon nanotubes was demonstrated in 2003 when room-temperature ballistic transistors with ohmic metal contacts and high-k gate dielectric were reported, showing 20–30x higher ON current than state-of-the-art Si MOSFETs. This presented an important advance in the field as CNT was shown to potentially outperform Si. At the time, a major challenge was ohmic metal contact formation. In this regard, palladium, which is a high-work function metal was shown to exhibit Schottky barrier-free contacts to semiconducting nanotubes with diameters >1.7 nm.

The first nanotube integrated memory circuit was made in 2004. One of the main challenges has been regulating the conductivity of nanotubes. Depending on subtle surface features a nanotube may act as a plain conductor or as a semiconductor. A fully automated method has however been developed to remove non-semiconductor tubes.

In 2013, researchers demonstrated a Turing-complete prototype micrometer-scale computer. Carbon nanotube transistors as logic-gate circuits with densities comparable to modern CMOS technology has not yet been demonstrated.

In 2014 networks of purified semiconducting carbon nanotubes were used as the active material in p-type thin film transistors. They were created using 3-D printers using inkjet or gravure methods on flexible substrates, including polyimide and polyethylene (PET) and transparent substrates such as glass. These transistors reliably exhibit high-mobilities ($> 10\ \text{cm}^2\,\text{V}^{-1}\,\text{s}^{-1}$) and ON/OFF ratios ($> 1000$) as well as threshold voltages below 5 V. They offer current density and low power consumption as well as environmental stability and mechanical flexibility. Hysterisis in the current-voltage curses as well as variability in the threshold voltage remain to be solved.

In 2015 researchers announced a new way to connect wires to SWNTs that make it possible to continue shrinking the width of the wires without increasing electrical resistance. The advance was expected to shrink the contact point between the two materials to just 40 atoms in width and later less. They tubes align in regularly spaced rows on silicon wafers. Simluations indicated that designs could be optimized either for high performance or for low power consumption. Commercial devices were not expected until the 2020s.

Thermal Management

Large structures of carbon nanotubes can be used for thermal management of electronic circuits. An approximately 1 mm–thick carbon nanotube layer was used as a special material to fabricate coolers, this material has very low density, ~20 times lower weight than a similar copper structure, while the cooling properties are similar for the two materials.

Buckypaper has characteristics appropriate for use as a heat sink for chipboards, a backlight for LCD screens or as a faraday cage.

Solar Cells

One of the promising applications of single-walled carbon nanotubes (SWNTs) is their use in solar panels, due to their strong UV/Vis-NIR absorption characteristics. Research has shown that they can provide a sizable increase in efficiency, even at their current unoptimized state. Solar cells developed at the New Jersey Institute of Technology use a carbon nanotube complex, formed by a mixture of carbon nanotubes and carbon buckyballs (known as fullerenes) to form snake-like structures. Buckyballs trap electrons, but they can't make electrons flow. Add sunlight to excite the polymers, and the buckyballs will grab the electrons. Nanotubes, behaving like copper wires, will then be able to make the electrons or current flow.

Additional research has been conducted on creating SWNT hybrid solar panels to increase the efficiency further. These hybrids are created by combining SWNT's with photo-excitable electron donors to increase the number of electrons generated. It has been found that the interaction between the photo-excited porphyrin and SWNT generates electro-hole pairs at the SWNT surfaces. This phenomenon has been observed experimentally, and contributes practically to an increase in efficiency up to 8.5%.

Nanotubes can potentially replace indium tin oxide in solar cells as a transparent conductive film in solar cells to allow light to pass to the active layers and generate photocurrent.

CNTs in organic solar cells help reduce energy loss (carrier recombination) and enhance resistance to photooxidation. Photovoltaic technologies may someday incorporate CNT-Silicon heterojunctions to leverage efficient multiple-exciton generation at p-n junctions formed within individual CNTs. In the nearer term, commercial photovoltaics may incorporate transparent SWNT electrodes.

Hydrogen Storage

In addition to being able to store electrical energy, there has been some research in using carbon nanotubes to store hydrogen to be used as a fuel source. By taking advantage of the capillary effects of the small carbon nanotubes, it is possible to condense gases in high density inside single-walled nanotubes. This allows for gases, most notably hydrogen (H_2), to be stored at high densities without being condensed into a liquid. Potentially, this storage method could be used on vehicles in place of gas fuel tanks for a hydrogen-powered car. A current issue regarding hydrogen-powered vehicles is the on-board storage of the fuel. Current storage methods involve cooling and condensing the H_2 gas to a liquid state for storage which causes a loss of potential energy (25–45%) when compared to the energy associated with the gaseous state. Storage using SWNTs would allow one to keep the H2 in its gaseous state, thereby increasing the storage efficiency. This method allows for a volume to energy ratio slightly smaller to that of current gas powered vehicles, allowing for a slightly lower but comparable range.

An area of controversy and frequent experimentation regarding the storage of hydrogen by adsorption in carbon nanotubes is the efficiency by which this process occurs. The effectiveness of hydrogen storage is integral to its use as a primary fuel source since hydrogen only contains about one fourth the energy per unit volume as gasoline. Studies however show that what is the most important is the surface area of the materials used. Hence activated carbon with surface area of 2600 m2/g can store up to 5,8% w/w. In all these carbonaceous materials, hydrogen is stored by physisorption at 70-90K.

Experimental Capacity

One experiment sought to determine the amount of hydrogen stored in CNTs by utilizing elastic recoil detection analysis (ERDA). CNTs (primarily SWNTs) were synthesized via chemical vapor disposition (CVD) and subjected to a two-stage purification process including air oxidation and acid treatment, then formed into flat, uniform discs and exposed to pure, pressurized hydrogen at various temperatures. When the data was analyzed, it was found that the ability of CNTs to store hydrogen decreased as temperature increased. Moreover, the highest hydrogen concentration measured was ~0.18%; significantly lower than commercially viable hydrogen storage needs to be. A separate experimental work performed by using a gravimetric method also revealed the maximum hydrogen uptake capacity of CNTs to be as low as 0.2%.

In another experiment, CNTs were synthesized via CVD and their structure was characterized using Raman spectroscopy. Utilizing microwave digestion, the samples were exposed to different acid concentrations and different temperatures for various amounts of time in an attempt to find the optimum purification method for SWNTs of the diameter determined earlier. The purified samples were then exposed to hydrogen gas at various high pressures, and their adsorption by weight percent was plotted. The data showed that hydrogen adsorption levels of up to 3.7% are possible with a very pure sample and under the proper conditions. It is thought that microwave digestion helps improve the hydrogen adsorption capacity of the CNTs by opening up the ends, allowing access to the inner cavities of the nanotubes.

Limitations on Efficient Hydrogen Adsorption

The biggest obstacle to efficient hydrogen storage using CNTs is the purity of the nanotubes. To achieve maximum hydrogen adsorption, there must be minimum graphene, amorphous carbon, and metallic deposits in the nanotube sample. Current methods of CNT synthesis require a purification step. However, even with pure nanotubes, the adsorption capacity is only maximized under high pressures, which are undesirable in commercial fuel tanks.

Electronic Components

Various companies are developing transparent, electrically conductive CNT films and nanobuds to replace indium tin oxide (ITO) in LCDs, touch screens and photovoltaic devices. Nanotube films show promise for use in displays for computers, cell phones, Personal digital assistants, and automated teller machines. CNT diodes display a photovoltaic effect.

Multi-walled nanotubes (MWNT coated with magnetite) can generate strong magnetic fields. Recent advances show that MWNT decorated with maghemite nanoparticles can be oriented in a magnetic field and enhance the electrical properties of the composite material in the direction of the field for use in electric motor brushes.

A layer of 29% iron enriched single-walled nanotubes (SWNT) placed on top of a layer of explosive material such as PETN can be ignited with a regular camera flash.

CNTs can be used as electron guns in miniature cathode ray tubes (CRT) in high-brightness, low-energy, low-weight displays. A display would consist of a group of tiny CRTs, each providing the electrons to illuminate the phosphor of one pixel, instead of having one CRT whose electrons are aimed using electric and magnetic fields. These displays are known as field emission displays (FEDs).

CNTs can act as antennas for radios and other electromagnetic devices.

Conductive CNTs are used in brushes for commercial electric motors. They replace traditional carbon black. The nanotubes improve electrical and thermal conductivity because they stretch through the plastic matrix of the brush. This permits the carbon filler to be reduced from 30% down to 3.6%, so that more matrix is present in the brush. Nanotube composite motor brushes are better-lubricated (from the matrix), cooler-running (both from better lubrication and superior thermal conductivity), less brittle (more matrix, and fiber reinforcement), stronger and more accurately moldable (more matrix). Since brushes are a critical failure point in electric motors, and also don't need much material, they became economical before almost any other application.

Wires for carrying electric current may be fabricated from nanotubes and nanotube-polymer composites. Small wires have been fabricated with specific conductivity exceeding copper and aluminum; the highest conductivity non-metallic cables.

CNT are under investigation as an alternative to tungsten filaments in incandescent light bulbs.

Interconnects

Metallic carbon nanotubes have aroused research interest for their applicability as very-large-scale integration (VLSI) interconnects because of their high thermal stability, high thermal conductivity and large current carrying capacity. An isolated CNT can carry current densities in excess of 1000 MA/sq-cm without damage even at an elevated temperature of 250 °C (482 °F), eliminating electromigration reliability concerns that plague Cu interconnects. Recent modeling work comparing the two has shown that CNT bundle interconnects can potentially offer advantages over copper. Recent experiments demonstrated resistances as low as 20 Ohms using different architectures, detailed conductance measurements over a wide temperature range were shown to agree with theory for a strongly disordered quasi-one-dimensional conductor.

Hybrid interconnects that employ CNT vias in tandem with copper interconnects may offer advantages from a reliability/thermal-management perspective. In 2016, the European Union has funded a four million euro project over three years to evaluate manufacturability and performance of composite interconnects employing both CNT and copper interconnects. The project named CONNECT (CarbON Nanotube composEtE InterconneCTs) involves the joint efforts of seven European research and industry partners on fabrication techniques and processes to enable reliable Carbon NanoTubes for on-chip interconnects in ULSI microchip production.

Electrical Cables and Wires

Wires for carrying electric current may be fabricated from pure nanotubes and nanotube-polymer composites. It has already been demonstrated that carbon nanotube wires can successfully be used for power or data transmission. Recently small wires have been fabricated with specific conductivity exceeding copper and aluminum; these cables are the highest conductivity carbon nanotube and also highest conductivity non-metal cables. Recently, composite of carbon nanotube and copper have been shown to exhibit nearly one hundred times higher current-carrying-capacity than pure copper or gold. Significantly, the electrical conductivity of such a composite is similar to pure Cu. Thus, this Carbon nanotube-copper (CNT-Cu) composite possesses the highest observed current-carrying capacity among electrical conductors. Thus for a given cross-section of electrical conductor, the CNT-Cu composite can withstand and transport one hundred times higher current compared to metals such as copper and gold.

Energy Storage

The use of CNTs as a catalyst support in fuel cells can potentially reduce platinum usage by 60% compared with carbon black. Doped CNTs may enable the complete elimination of Pt.

Supercapacitor

MIT Research Laboratory of Electronics uses nanotubes to improve supercapacitors. The activated charcoal used in conventional ultracapacitors has many small hollow spaces of various size, which create together a large surface to store electric charge. But as charge is quantized into elementary charges, i.e. electrons, and each such elementary charge needs a minimum space, a significant fraction of the electrode surface is not available for storage because the hollow spaces are not compatible with the charge's requirements. With a nanotube electrode the spaces may be tailored to size—few too large or too small—and consequently the capacity should be increased considerably.

A 40-F supercapacitor with a maximum voltage of 3.5 V that employed forest-grown SWNTs that are binder- and additive-free achieved an energy density of 15.6 Wh kg^{-1} and a power density of 37 kW kg^{-1}. CNTs can be bound to the charge plates of capacitors to dramatically increase the surface area and therefore energy density.

Batteries

Carbon nanotubes' (CNTs) exciting electronic properties have shown promise in the field of batteries, where typically they are being experimented as a new electrode material, particularly the anode for lithium ion batteries. This is due to the fact that the anode requires a relatively high reversible capacity at a potential close to metallic lithium, and a moderate irreversible capacity, observed thus far only by graphite-based composites, such as CNTs. They have shown to greatly improve capacity and cyclability of lithium-ion batteries, as well as the capability to be very effective buffering components, alleviating the degradation of the batteries that is typically due to repeated charging and discharging. Further, electronic transport in the anode can be greatly improved using highly metallic CNTs.

More specifically, CNTs have shown reversible capacities from 300 to 600 $mAhg^{-1}$, with some treatments to them showing these numbers rise to up to 1000 $mAhg^{-1}$. Meanwhile, graphite, which is most widely used as an anode material for these lithium batteries, has shown capacities of only 320 $mAhg^{-1}$. By creating composites out of the CNTs, scientists see much potential in taking advantage of these exceptional capacities, as well as their excellent mechanical strength, conductivities, and low densities.

MWNTs are used in lithium ion batteries. In these batteries, small amounts of MWNT powder are blended with active materials and a polymer binder, such as 1 wt % CNT loading in $LiCoO_2$ cathodes and graphite anodes. CNTs provide increased electrical connectivity and mechanical integrity, which enhances rate capability and cycle life.

Paper Batteries

A paper battery is a battery engineered to use a paper-thin sheet of cellulose (which is the major constituent of regular paper, among other things) infused with aligned carbon nanotubes. The potential for these devices is great, as they may be manufactured via a roll-to-roll process, which would make it very low-cost, and they would be lightweight, flexible, and thin. In order to productively use paper electronics (or any thin electronic devices), the power source must be equally thin, thus indicating the need for paper batteries. Recently, it has been shown that surfaces coated with CNTs can be used to replace heavy metals in batteries. More recently, functional paper batteries have been demonstrated, where a lithium-ion battery is integrated on a single sheet of paper through a lamination process as a composite with $Li_4Ti_5O_{12}$ (LTO) or $LiCoO_2$ (LCO). The paper substrate would function well as the separator for the battery, where the CNT films function as the current collectors for both the anode *and* the cathode. These rechargeable energy devices show potential in RFID tags, functional packaging, or new disposable electronic applications.

Improvements have also been shown in lead-acid batteries, based on research performed by Bar-Ilan University using high quality SWCNT manufactured by OCSiAl. The study demonstrated an increase in the lifetime of lead acid batteries by 4.5 times and a capacity increase of 30% on average and up to 200% at high discharge rates.

Chemical

CNT can be used for desalination. Water molecules can be separated from salt by forcing them through electrochemically robust nanotube networks with controlled nanoscale porosity. This process requires far lower pressures than conventional reverse osmosis methods. Compared to a plain membrane, it operates at a 20 °C lower temperature, and at a 6x greater flow rate. Membranes using aligned, encapsulated CNTs with open ends permit flow through the CNTs' interiors. Very-small-diameter SWNTs are needed to reject salt at seawater concentrations. Portable filters containing CNT meshes can purify contaminated drinking water. Such networks can electrochemically oxidize organic contaminants, bacteria and viruses.

CNT membranes can filter carbon dioxide from power plant emissions.

CNT can be filled with biological molecules, aiding biotechnology.

CNT have the potential to store between 4.2 and 65% hydrogen by weight. If they can be mass-produced economically, 13.2 litres (2.9 imp gal; 3.5 US gal) of CNT could contain the same amount of energy as a 50 litres (11 imp gal; 13 US gal) gasoline tank.

CNTs can be used to produce nanowires of other elements/molecules, such as gold or zinc oxide. Nanowires in turn can be used to cast nanotubes of other materials, such as gallium nitride. These can have very different properties from CNTs—for example, gallium nitride nanotubes are hydrophilic, while CNTs are hydrophobic, giving them possible uses in organic chemistry.

Mechanical

Oscillators based on CNT have achieved speeds of > 50 GHz.

CNT electrical and mechanical properties suggest them as alternatives to traditional electrical actuators.

Actuators

The exceptional electrical and mechanical properties of carbon nanotubes have made them alternatives to the traditional electrical actuators for both microscopic and macroscopic applications. Carbon nanotubes are very good conductors of both electricity and heat, and they are also very strong and elastic molecules in certain directions.

Loudspeaker

Carbon nanotubes have also been applied in the acoustics(such as loudspeaker and earphone). In 2008 it was shown that a sheet of nanotubes can operate as a loudspeaker if an alternating current is applied. The sound is not produced through vibration but thermoacoustically. In 2013, a carbon nanotube (CNT) thin yarn thermoacoustic earphone together with CNT thin yarn thermoacoustic chip was demonstrated by a research group of Tsinghua-Foxconn Nanotechnology Research Center in Tsinghua University, using a Si-based semi-conducting technology compatible fabrication process.

Near-term commercial uses include replacing piezoelectric speakers in greeting cards.

Optical

- Carbon nanotube photoluminescence (fluorescence) can be used to observe semiconducting single-walled carbon nanotube species. Photoluminescence maps, made by acquiring the emission and scanning the excitation energy, can facilitate sample characterization.

- Nanotube fluorescence is under investigation for biomedical imaging and sensors.

- The reflectivity of buckypaper produced with "super-growth" chemical vapor deposition is 0.03 or less, potentially enabling performance gains for pyroelectric infrared detectors.

Environmental

Environmental Remediation

A CNT nano-structured sponge (nanosponge) containing sulfur and iron is more effective at soaking up water contaminants such as oil, fertilizers, pesticides and pharmaceuticals. Their magnetic properties make them easier to retrieve once the clean-up job is done. The sulfur and iron increases sponge size to around 2 centimetres (0.79 in). It also increases porosity due to beneficial defects, creating buoyancy and reusability. Iron, in the form of ferrocene makes the structure easier to control and enables recovery using magnets. Such nanosponges increase the absorption of the toxic organic solvent dichlorobenzene from water by 3.5 times. The sponges can absorb vegetable oil up to 150 times their initial weight and can absorb engine oil as well.

Earlier, a magnetic boron-doped MWNT nanosponge that could absorb oil from water. The sponge was grown as a forest on a substrate via chemical vapor disposition. Boron puts kinks and elbows into the tubes as they grow and promotes the formation of covalent bonds. The nanosponges retain their elastic property after 10,000 compressions in the lab. The sponges are both superhydrophobic, forcing them to remain at the water's surface and oleophilic, drawing oil to them.

Water Treatment

It has been shown that carbon nanotubes exhibit strong adsorption affinities to a wide range of aromatic and aliphatic contaminants in water, due to their large and hydrophobic surface areas. They also showed similar adsorption capacities as activated carbons in the presence of natural organic matter. As a result, they have been suggested as promising adsorbents for removal of contaminant in water and wastewater treatment systems.

Moreover, membranes made out of carbon nanotube arrays have been suggested as switchable molecular sieves, with sieving and permeation features that can be dynamically activated/deactivated by either pore size distribution (passive control) or external electrostatic fields (active control).

Other Applications

Carbon nanotubes have been implemented in nanoelectromechanical systems, including mechanical memory elements (NRAM being developed by Nantero Inc.) and nanoscale electric motors.

Carboxyl-modified single-walled carbon nanotubes (so called zig-zag, armchair type) can act as sensors of atoms and ions of alkali metals Na, Li, K. In May 2005, Nanomix Inc. placed on the market a hydrogen sensor that integrated carbon nanotubes on a silicon platform. Since then, Nanomix has been patenting many such sensor applications, such as in the field of carbon dioxide, nitrous oxide, glucose, DNA detection, etc. End of 2014, Tulane University researchers have tested Nanomix's fast and fully automated point of care diagnostic system in Sierra Leone to help for rapid testing for Ebola. Nanomix announced that a product could be launched within three to six months.

Eikos Inc of Franklin, Massachusetts and Unidym Inc. of Silicon Valley, California are developing transparent, electrically conductive films of carbon nanotubes to replace indium tin oxide (ITO). Carbon nanotube films are substantially more mechanically robust than ITO films, making them ideal for high-reliability touchscreens and flexible displays. Printable water-based inks of carbon nanotubes are desired to enable the production of these films to replace ITO. Nanotube films show promise for use in displays for computers, cell phones, PDAs, and ATMs.

A nanoradio, a radio receiver consisting of a single nanotube, was demonstrated in 2007.

A flywheel made of carbon nanotubes could be spun at extremely high velocity on a floating magnetic axis in a vacuum, and potentially store energy at a density approaching that of conventional fossil fuels. Since energy can be added to and removed from flywheels very efficiently in the form of electricity, this might offer a way of storing electricity, making the electrical grid more efficient and variable power suppliers (like wind turbines) more useful in meeting energy needs. The practicality of this depends heavily upon the cost of making massive, unbroken nanotube structures, and their failure rate under stress.

Carbon nanotube springs have the potential to indefinitely store elastic potential energy at ten times the density of lithium-ion batteries with flexible charge and discharge rates and extremely high cycling durability.

Ultra-short SWNTs (US-tubes) have been used as nanoscaled capsules for delivering MRI contrast agents in vivo.

Carbon nanotubes provide a certain potential for metal-free catalysis of inorganic and organic reactions. For instance, oxygen groups attached to the surface of carbon nanotubes have the potential to catalyze oxidative dehydrogenations or selective oxidations. Nitrogen-doped carbon nanotubes may replace platinum catalysts used to reduce oxygen in fuel cells. A forest of vertically aligned nanotubes can reduce oxygen in alkaline solution more effectively than platinum, which has been used in such applications since the 1960s. Here, the nanotubes have the added benefit of not being subject to carbon monoxide poisoning.

Wake Forest University engineers are using multiwalled carbon nanotubes to enhance the brightness of field-induced polymer electroluminescent technology, potentially offering a step forward in the search for safe, pleasing, high-efficiency lighting. In this technology, moldable polymer matrix emits light when exposed to an electric current. It could eventually yield high-efficiency lights without the mercury vapor of compact fluorescent lamps or the bluish tint of some fluorescents and LEDs, which has been linked with circadian rhythm disruption.

Candida albicans has been used in combination with carbon nanotubes (CNT) to produce stable electrically conductive bio-nano-composite tissue materials that have been used as temperature sensing elements.

The SWNT production company OCSiAl developed a series of masterbatches for industrial use of single-wall CNTs in multiple types of rubber blends and tires, with initial trials showing increases in hardness, viscosity, tensile strain resistance and resistance to abrasion while reducing elongation and compression In tires the three primary characteristics of durability, fuel efficiency and traction were improved using SWNTs. The development of rubber masterbatches built on earlier work by the Japanese National Institute of Advanced Industrial Science & Technology showing rubber to be a viable candidate for improvement with SWNTs.

Introducing MWNTs to polymers can improve flame retardancy and retard thermal degradation of polymer. The results confirmed that combination of MWNTs and ammonium polyphosphates show a synergistic effect for improving flame retardancy.

References

- Kolosnjaj J, Szwarc H, Moussa F (2007). "Toxicity studies of carbon nanotubes". Adv Exp Med Biol. Advances in Experimental Medicine and Biology. 620: 181–204. doi:10.1007/978-0-387-76713-0_14. ISBN 978-0-387-76712-3. PMID 18217344.
- Pacios Pujadó, Mercè (2012). Carbon Nanotubes as Platforms for Biosensors with Electrochemical and Electronic Transduction. Springer Heidelberg. pp. XX,208. doi:10.1007/978-3-642-31421-6. ISBN 978-3-642-31421-6. ISSN 2190-5053.

- Banerjee, K.; Srivastava, N. (2006). "Are carbon nanotubes the future of VLSI interconnections?". 2006 43rd ACM/IEEE Design Automation Conference. p. 809. doi:10.1109/DAC.2006.229330. ISBN 1-59593-381-6.

- Naeemi, A.; Meindl, J. D. (2007). "Carbon nanotube interconnects". Proceedings of the 2007 international symposium on Physical design – ISPD '07. p. 77. doi:10.1145/1231996.1232014. ISBN 9781595936134.

- Y. Chai (2008). "High electromigration-resistant copper/carbon nanotube composite for interconnect application". IEEE Electron Devices Meeting, 2008 : 1. doi:10.1109/IEDM.2008.4796764. ISBN 978-1-4244-2377-4.

- "Pirahna USV built using nano-enhanced carbon prepreg". ReinforcedPlastics.com. 19 February 2009. Archived from the original on 3 March 2012.

Molecular Nanotechnology: An Integrated Study

Molecular nanotechnology is the technology that is based on mechanosynthesis. It involves biophysics, chemistry and other nanotechnologies. The aspects of molecular nanotechnology that have been explained in this text are mechanosynthesis, molecular assembler and molecular modelling. The chapter serves as a source to understand the major categories related to molecular nanotechnology.

Molecular Nanotechnology

Molecular nanotechnology (MNT) is a technology based on the ability to build structures to complex, atomic specifications by means of mechanosynthesis. This is distinct from nanoscale materials. Based on Richard Feynman's vision of miniature factories using nanomachines to build complex products (including additional nanomachines), this advanced form of nanotechnology (or *molecular manufacturing*) would make use of positionally-controlled mechanosynthesis guided by molecular machine systems. MNT would involve combining physical principles demonstrated by biophysics, chemistry, other nanotechnologies, and the molecular machinery of life with the systems engineering principles found in modern macroscale factories.

Introduction

While conventional chemistry uses inexact processes obtaining inexact results, and biology exploits inexact processes to obtain definitive results, molecular nanotechnology would employ original definitive processes to obtain definitive results. The desire in molecular nanotechnology would be to balance molecular reactions in positionally-controlled locations and orientations to obtain desired chemical reactions, and then to build systems by further assembling the products of these reactions.

A roadmap for the development of MNT is an objective of a broadly based technology project led by Battelle (the manager of several U.S. National Laboratories) and the Foresight Institute. The roadmap was originally scheduled for completion by late 2006, but was released in January 2008. The Nanofactory Collaboration is a more focused ongoing effort involving 23 researchers from 10 organizations and 4 countries that is developing a practical research agenda specifically aimed at positionally-controlled diamond mechanosynthesis and diamondoid nanofactory development. In August 2005, a task force consisting of 50+ international experts from various fields was organized by the Center for Responsible Nanotechnology to study the societal implications of molecular nanotechnology.

Projected Applications and Capabilities

Smart Materials and Nanosensors

One proposed application of MNT is so-called smart materials. This term refers to any sort of material designed and engineered at the nanometer scale for a specific task. It encompasses a wide variety of possible commercial applications. One example would be materials designed to respond differently to various molecules; such a capability could lead, for example, to artificial drugs which would recognize and render inert specific viruses. Another is the idea of self-healing structures, which would repair small tears in a surface naturally in the same way as self-sealing tires or human skin.

A MNT nanosensor would resemble a smart material, involving a small component within a larger machine that would react to its environment and change in some fundamental, intentional way. A very simple example: a photosensor might passively measure the incident light and discharge its absorbed energy as electricity when the light passes above or below a specified threshold, sending a signal to a larger machine. Such a sensor would supposedly cost less and use less power than a conventional sensor, and yet function usefully in all the same applications — for example, turning on parking lot lights when it gets dark.

While smart materials and nanosensors both exemplify useful applications of MNT, they pale in comparison with the complexity of the technology most popularly associated with the term: the replicating nanorobot.

Replicating Nanorobots

MNT nanofacturing is popularly linked with the idea of swarms of coordinated nanoscale robots working together, a popularization of an early proposal by K. Eric Drexler in his 1986 discussions of MNT, but superseded in 1992. In this early proposal, sufficiently capable nanorobots would construct more nanorobots in an artificial environment containing special molecular building blocks.

Critics have doubted both the feasibility of self-replicating nanorobots and the feasibility of control if self-replicating nanorobots could be achieved: they cite the possibility of mutations removing any control and favoring reproduction of mutant pathogenic variations. Advocates address the first doubt by pointing out that the first macroscale autonomous machine replicator, made of Lego blocks, was built and operated experimentally in 2002. While there are sensory advantages present at the macroscale compared to the limited sensorium available at the nanoscale, proposals for positionally controlled nanoscale mechanosynthetic fabrication systems employ dead reckoning of tooltips combined with reliable reaction sequence design to ensure reliable results, hence a limited sensorium is no handicap; similar considerations apply to the positional assembly of small nanoparts. Advocates address the second doubt by arguing that bacteria are (of necessity) evolved to evolve, while nanorobot mutation could be actively prevented by common error-correcting techniques. Similar ideas are advocated in the Foresight Guidelines on Molecular Nanotechnology, and a map of the 137-dimensional replicator design space recently published by Freitas and Merkle provides numerous proposed methods by which replicators could, in principle, be safely controlled by good design.

However, the concept of suppressing mutation raises the question: How can design evolution occur at the nanoscale without a process of random mutation and deterministic selection? Critics argue that MNT advocates have not provided a substitute for such a process of evolution in this nanoscale arena where conventional sensory-based selection processes are lacking. The limits of the sensorium available at the nanoscale could make it difficult or impossible to winnow successes from failures. Advocates argue that design evolution should occur deterministically and strictly under human control, using the conventional engineering paradigm of modeling, design, prototyping, testing, analysis, and redesign.

In any event, since 1992 technical proposals for MNT do not include self-replicating nanorobots, and recent ethical guidelines put forth by MNT advocates prohibit unconstrained self-replication.

Medical Nanorobots

One of the most important applications of MNT would be medical nanorobotics or nanomedicine, an area pioneered by Robert Freitas in numerous books and papers. The ability to design, build, and deploy large numbers of medical nanorobots would, at a minimum, make possible the rapid elimination of disease and the reliable and relatively painless recovery from physical trauma. Medical nanorobots might also make possible the convenient correction of genetic defects, and help to ensure a greatly expanded lifespan. More controversially, medical nanorobots might be used to augment natural human capabilities.

Utility Fog

Diagram of a 100 micrometer foglet

Another proposed application of molecular nanotechnology is "utility fog" — in which a cloud of networked microscopic robots (simpler than assemblers) would change its shape and properties to form macroscopic objects and tools in accordance with software commands. Rather than modify the current practices of consuming material goods in different forms, utility fog would simply replace many physical objects.

Phased-array Optics

Yet another proposed application of MNT would be phased-array optics (PAO). However, this appears to be a problem addressable by ordinary nanoscale technology. PAO would use the principle of phased-array millimeter technology but at optical wavelengths. This would permit the duplication of any sort of optical effect but virtually. Users could request holograms, sunrises and sunsets, or floating lasers as the mood strikes. PAO systems were described in BC Crandall's *Nanotechnology: Molecular Speculations on Global Abundance* in the Brian Wowk article "Phased-Array Optics."

Potential Social Impacts

Molecular manufacturing is a potential future subfield of nanotechnology that would make it possible to build complex structures at atomic precision. Molecular manufacturing requires significant advances in nanotechnology, but once achieved could produce highly advanced products at low costs and in large quantities in nanofactories weighing a kilogram or more. When nanofactories gain the ability to produce other nanofactories production may only be limited by relatively abundant factors such as input materials, energy and software.

The products of molecular manufacturing could range from cheaper, mass-produced versions of known high-tech products to novel products with added capabilities in many areas of application. Some applications that have been suggested are advanced smart materials, nanosensors, medical nanorobots and space travel. Additionally, molecular manufacturing could be used to cheaply produce highly advanced, durable weapons, which is an area of special concern regarding the impact of nanotechnology. Being equipped with compact computers and motors these could be increasingly autonomous and have a large range of capabilities.

According to Chris Phoenix and Mike Treder from the Center for Responsible Nanotechnology as well as Anders Sandberg from the Future of Humanity Institute molecular manufacturing is the application of nanotechnology that poses the most significant global catastrophic risk. Several nanotechnology researchers state that the bulk of risk from nanotechnology comes from the potential to lead to war, arms races and destructive global government. Several reasons have been suggested why the availability of nanotech weaponry may with significant likelihood lead to unstable arms races (compared to e.g. nuclear arms races): (1) A large number of players may be tempted to enter the race since the threshold for doing so is low; (2) the ability to make weapons with molecular manufacturing will be cheap and easy to hide; (3) therefore lack of insight into the other parties' capabilities can tempt players to arm out of caution or to launch preemptive strikes; (4) molecular manufacturing may reduce dependency on international trade, a potential peace-promoting factor; (5) wars of aggression may pose a smaller economic threat to the aggressor since manufacturing is cheap and humans may not be needed on the battlefield.

Since self-regulation by all state and non-state actors seems hard to achieve, measures to mitigate war-related risks have mainly been proposed in the area of international cooperation. International infrastructure may be expanded giving more sovereignty to the international level. This could help coordinate efforts for arms control. International institutions dedicated specifically to nanotechnology (perhaps analogously to the International Atomic Energy Agency IAEA) or general arms control may also be designed. One may also jointly make differential technological progress on defensive technologies, a policy that players should usually favour. The Center for Responsible Nanotechnology also suggest some technical restrictions. Improved transparency regarding technological capabilities may be another important facilitator for arms-control.

A grey goo is another catastrophic scenario, which was proposed by Eric Drexler in his 1986 book *Engines of Creation*, has been analyzed by Freitas in "Some Limits to Global Ecophagy by Biovorous Nanoreplicators, with Public Policy Recommendations" and has been a theme in mainstream media and fiction. This scenario involves tiny self-replicating robots that consume the entire biosphere using it as a source of energy and building blocks. Nanotech experts including Drexler now discredit the scenario. According to Chris Phoenix a "So-called grey goo could only be the product of a deliberate and difficult engineering process, not an accident". With the advent of nano-bio-tech, a different scenario called green goo has been forwarded. Here, the malignant substance is not nanobots but rather self-replicating biological organisms engineered through nanotechnology.

Benefits

Nanotechnology (or molecular nanotechnology to refer more specifically to the goals discussed here) will let us continue the historical trends in manufacturing right up to the fundamental limits imposed by physical law. It will let us make remarkably powerful molecular computers. It will let us make materials over fifty times lighter than steel or aluminium alloy but with the same strength. We'll be able to make jets, rockets, cars or even chairs that, by today's standards, would be remarkably light, strong, and inexpensive. Molecular surgical tools, guided by molecular computers and injected into the blood stream could find and destroy cancer cells or invading bacteria, unclog arteries, or provide oxygen when the circulation is impaired.

Nanotechnology will replace our entire manufacturing base with a new, radically more precise, radically less expensive, and radically more flexible way of making products.

The aim is not simply to replace today's computer chip making plants, but also to replace the assembly lines for cars, televisions, telephones, books, surgical tools, missiles, bookcases, airplanes, tractors, and all the rest. The objective is a pervasive change in manufacturing, a change that will leave virtually no product untouched. Economic progress and military readiness in the 21st Century will depend fundamentally on maintaining a competitive position in nanotechnology.

Despite the current early developmental status of nanotechnology and molecular nanotechnology, much concern surrounds MNT's anticipated impact on economics and on law. Whatever the exact effects, MNT, if achieved, would tend to reduce the scarcity of manufactured goods and make many more goods (such as food and health aids) manufacturable.

It is generally considered that future citizens of a molecular-nanotechnological society would still need money, in the form of unforgeable digital cash or physical specie (in special circumstances). They might use such money to buy goods and services that are unique, or limited within the solar system. These might include: matter, energy, information, real estate, design services, entertainment services, legal services, fame, political power, or the attention of other people to one's political/religious/philosophical message. Furthermore, futurists must consider war, even between prosperous states, and non-economic goals.

If MNT were realized, some resources would remain limited, because unique physical objects are limited (a plot of land in the real Jerusalem, mining rights to the larger near-earth asteroids) or because they depend on the goodwill of a particular person (the love of a famous person, a live audience in a musical concert). Demand will always exceed supply for some things, and a political economy may continue to exist in any case. Whether the interest in these limited resources would diminish with the advent of virtual reality, where they could be easily substituted, is yet unclear. One reason why it might not is a hypothetical preference for "the real thing", although such an opinion could easily be mollified if virtual reality were to develop to a certain level of quality.

MNT should make possible nanomedical capabilities able to cure any medical condition not already cured by advances in other areas. Good health would be common, and poor health of any form would be as rare as smallpox and scurvy are today. Even cryonics would be feasible, as cryopreserved tissue could be fully repaired.

Risks

Molecular nanotechnology is one of the technologies that some analysts believe could lead to a Technological Singularity. Some feel that molecular nanotechnology would have daunting risks. It conceivably could enable cheaper and more destructive conventional weapons. Also, molecular nanotechnology might permit weapons of mass destruction that could self-replicate, as viruses and cancer cells do when attacking the human body. Commentators generally agree that, in the event molecular nanotechnology were developed, its self-replication should be permitted only under very controlled or "inherently safe" conditions.

A fear exists that nanomechanical robots, if achieved, and if designed to self-replicate using naturally occurring materials (a difficult task), could consume the entire planet in their hunger for raw materials, or simply crowd out natural life, out-competing it for energy (as happened historically when blue-green algae appeared and outcompeted earlier life forms). Some commentators have referred to this situation as the "grey goo" or "ecophagy" scenario. K. Eric Drexler considers an accidental "grey goo" scenario extremely unlikely and says so in later editions of *Engines of Creation*.

In light of this perception of potential danger, the Foresight Institute (founded by K. Eric Drexler to prepare for the arrival of future technologies) has drafted a set of guidelines for the ethical development of nanotechnology. These include the banning of free-foraging self-replicating pseudo-organisms on the Earth's surface, at least, and possibly in other places.

Technical Issues and Criticism

The feasibility of the basic technologies analyzed in *Nanosystems* has been the subject of a formal scientific review by U.S. National Academy of Sciences, and has also been the focus of extensive debate on the internet and in the popular press.

Study and Recommendations by The U.S. National Academy of Sciences

In 2006, U.S. National Academy of Sciences released the report of a study of molecular manufacturing as part of a longer report, *A Matter of Size: Triennial Review of the National Nanotechnology Initiative* The study committee reviewed the technical content of *Nanosystems*, and in its conclusion states that no current theoretical analysis can be considered definitive regarding several questions of potential system performance, and that optimal paths for implementing high-performance systems cannot be predicted with confidence. It recommends experimental research to advance knowledge in this area:

> "Although theoretical calculations can be made today, the eventually attainable range of chemical reaction cycles, error rates, speed of operation, and thermodynamic efficiencies of such bottom-up manufacturing systems cannot be reliably predicted at this time. Thus, the eventually attainable perfection and complexity of manufactured products, while they can be calculated in theory, cannot be predicted with confidence. Finally, the optimum research paths that might lead to systems which greatly exceed the thermodynamic efficiencies and other capabilities of biological systems cannot be reliably predicted at this time.

Research funding that is based on the ability of investigators to produce experimental demonstrations that link to abstract models and guide long-term vision is most appropriate to achieve this goal."

Assemblers Versus Nanofactories

A section heading in Drexler's *Engines of Creation* reads "Universal Assemblers", and the following text speaks of multiple types of assemblers which, collectively, could hypothetically "build almost anything that the laws of nature allow to exist." Drexler's colleague Ralph Merkle has noted that, contrary to widespread legend, Drexler never claimed that assembler systems could build absolutely any molecular structure. The endnotes in Drexler's book explain the qualification "almost": "For example, a delicate structure might be designed that, like a stone arch, would self-destruct unless all its pieces were already in place. If there were no room in the design for the placement and removal of a scaffolding, then the structure might be impossible to build. Few structures of practical interest seem likely to exhibit such a problem, however."

In 1992, Drexler published *Nanosystems: Molecular Machinery, Manufacturing, and Computation*, a detailed proposal for synthesizing stiff covalent structures using a table-top factory. Diamondoid structures and other stiff covalent structures, if achieved, would have a wide range of possible applications, going far beyond current MEMS technology. An outline of a path was put forward in 1992 for building a table-top factory in the absence of an assembler. Other researchers have begun advancing tentative, alternative proposed paths for this in the years since Nanosystems was published.

Hard Versus Soft Nanotechnology

In 2004 Richard Jones wrote Soft Machines (nanotechnology and life), a book for lay audiences published by Oxford University. In this book he describes radical nanotechnology (as advocated by Drexler) as a deterministic/mechanistic idea of nano engineered machines that does not take into account the nanoscale challenges such as wetness, stickness, Brownian motion, and high viscosity. He also explains what is soft nanotechnology or more appropriatelly biomimetic nanotechnology which is the way forward, if not the best way, to design functional nanodevices that can cope with all the problems at a nanoscale. One can think of soft nanotechnology as the development of nanomachines that uses the lessons learned from biology on how things work, chemistry to precisely engineer such devices and stochastic physics to model the system and its natural processes in detail.

The Smalley-drexler Debate

Several researchers, including Nobel Prize winner Dr. Richard Smalley (1943–2005), attacked the notion of universal assemblers, leading to a rebuttal from Drexler and colleagues, and eventually to an exchange of letters. Smalley argued that chemistry is extremely complicated, reactions are hard to control, and that a universal assembler is science fiction. Drexler and colleagues, however, noted that Drexler never proposed universal assemblers able to make absolutely anything, but instead proposed more limited assemblers able to make a very wide variety of things. They challenged the relevance of Smalley's arguments to the more specific proposals advanced in *Nanosystems*. Also, Smalley argued that nearly all of modern chemistry involves reactions that take place in a solvent (usually water), because the small molecules of a solvent contribute many things, such as lowering binding energies for transition states. Since nearly all known chemistry requires a solvent, Smalley felt that Drexler's proposal to use a high vacuum environment was not feasible. However, Drexler addresses this in Nanosystems by showing mathematically that well designed catalysts can provide the effects of a solvent and can fundamentally be made even more efficient than a solvent/ enzyme reaction could ever be. It is noteworthy that, contrary to Smalley's opinion that enzymes require water, "Not only do enzymes work vigorously in anhydrous organic media, but in this unnatural milieu they acquire remarkable properties such as greatly enhanced stability, radically altered substrate and enantiomeric specificities, molecular memory, and the ability to catalyse unusual reactions.""*Enzymatic catalysis in anhydrous organic solvents.". April 1989.""Enzymatic catalysis in anhydrous organic solvents" (PDF). April 1989.*

Design Issues

For the future, some means have to be found for MNT design evolution at the nanoscale which mimics the process of biological evolution at the molecular scale. Biological evolution proceeds by random variation in ensemble averages of organisms combined with culling of the less-successful variants and reproduction of the more-successful variants, and macroscale engineering design also proceeds by a process of design evolution from simplicity to complexity as set forth somewhat satirically by John Gall: "A complex system that works is invariably found to have evolved from a simple system that worked. . . . A complex system designed from scratch never works and can not be patched up to make it work. You have to start over, beginning with a system that works." A breakthrough in MNT is needed which proceeds from the simple atomic ensembles which can be built with, e.g., an STM to complex MNT systems via a process of design evolution. A handicap in this process is the difficulty of seeing and manipulation at the nanoscale compared to the macroscale which makes deterministic selection of successful trials difficult; in contrast biological evolution proceeds via action of what Richard Dawkins has called the "blind watchmaker" comprising random molecular variation and deterministic reproduction/extinction.

At present in 2007 the practice of nanotechnology embraces both stochastic approaches (in which, for example, supramolecular chemistry creates waterproof pants) and deterministic approaches wherein single molecules (created by stochastic chemistry) are manipulated on substrate surfaces (created by stochastic deposition methods) by deterministic methods comprising nudging them with STM or AFM probes and causing simple binding or cleavage reactions to occur. The dream of a complex, deterministic molecular nanotechnology remains elusive. Since the mid-1990s, thousands of surface scientists and thin film technocrats have latched on to the nanotechnology bandwagon and redefined their disciplines as nanotechnology. This has caused much confusion in the field and has spawned thousands of "nano"-papers on the peer reviewed literature. Most of these reports are extensions of the more ordinary research done in the parent fields.

The Feasibility of The Proposals in Nanosystems

The feasibility of Drexler's proposals largely depends, therefore, on whether designs like those in *Nanosystems* could be built in the absence of a universal assembler to build them and would work as described. Supporters of molecular nanotechnology frequently claim that no significant errors have been discovered in *Nanosystems* since 1992. Even some critics concede that "Drexler has carefully considered a number of physical principles underlying the 'high level' aspects of the nanosystems he proposes and, indeed, has thought in some detail" about some issues.

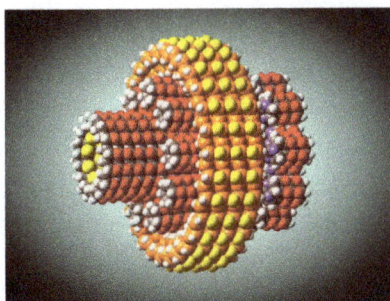

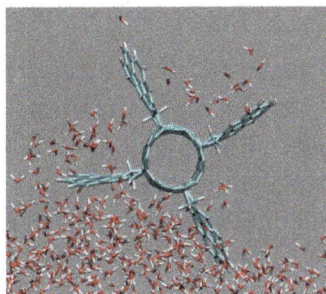

Top, a molecular propellor. Bottom, a molecular planetary gear system. The feasibility of devices like these has been questioned.

Other critics claim, however, that *Nanosystems* omits important chemical details about the low-level 'machine language' of molecular nanotechnology. They also claim that much of the other low-level chemistry in *Nanosystems* requires extensive further work, and that Drexler's higher-level designs therefore rest on speculative foundations. Recent such further work by Freitas and Merkle is aimed at strengthening these foundations by filling the existing gaps in the low-level chemistry.

Drexler argues that we may need to wait until our conventional nanotechnology improves before solving these issues: "Molecular manufacturing will result from a series of advances in molecular machine systems, much as the first Moon landing resulted from a series of advances in liquid-fuel rocket systems. We are now in a position like that of the British Interplanetary Society of the 1930s which described how multistage liquid-fueled rockets could reach the Moon and pointed to early rockets as illustrations of the basic principle." However, Freitas and Merkle argue that a focused effort to achieve diamond mechanosynthesis (DMS) can begin now, using existing technology, and might achieve success in less than a decade if their "direct-to-DMS approach is pursued rather than a more circuitous development approach that seeks to implement less efficacious nondiamondoid molecular manufacturing technologies before progressing to diamondoid".

To summarize the arguments against feasibility: First, critics argue that a primary barrier to achieving molecular nanotechnology is the lack of an efficient way to create machines on a molecular/atomic scale, especially in the absence of a well-defined path toward a self-replicating assembler or diamondoid nanofactory. Advocates respond that a preliminary research path leading to a diamondoid nanofactory is being developed.

A second difficulty in reaching molecular nanotechnology is design. Hand design of a gear or bearing at the level of atoms might take a few to several weeks. While Drexler, Merkle and others have created designs of simple parts, no comprehensive design effort for anything approaching the complexity of a Model T Ford has been attempted. Advocates respond that it is difficult to undertake a comprehensive design effort in the absence of significant funding for such efforts, and that despite this handicap much useful design-ahead has nevertheless been accomplished with new software tools that have been developed, e.g., at Nanorex.

In the latest report *A Matter of Size: Triennial Review of the National Nanotechnology Initiative* put out by the National Academies Press in December 2006 (roughly twenty years after Engines of Creation was published), no clear way forward toward molecular nanotechnology could yet be seen, as per the conclusion on page 108 of that report: "Although theoretical calculations can be made today, the eventually attainable range of chemical reaction cycles, error rates, speed of operation, and thermodynamic efficiencies of such bottom-up manufacturing systems cannot be reliably predicted at this time. Thus, the eventually attainable perfection and complexity of manufactured products, while they can be calculated in theory, cannot be predicted with confidence. Finally, the optimum research paths that might lead to systems which greatly exceed the thermodynamic efficiencies and other capabilities of biological systems cannot be reliably predicted at this time. Research funding that is based on the ability of investigators to produce experimental demonstrations that link to abstract models and guide long-term vision is most appropriate to achieve this goal." This call for research leading to demonstrations is welcomed by groups such as the Nanofactory Collaboration who are specifically seeking experimental successes in diamond mechanosynthesis. The "Technology Roadmap for Productive Nanosystems" aims to offer additional constructive insights.

It is perhaps interesting to ask whether or not most structures consistent with physical law can in fact be manufactured. Advocates assert that to achieve most of the vision of molecular manufacturing it is not necessary to be able to build "any structure that is compatible with natural law." Rather, it is necessary to be able to build only a sufficient (possibly modest) subset of such structures—as is true, in fact, of any practical manufacturing process used in the world today, and is true even in biology. In any event, as Richard Feynman once said, "It is scientific only to say what's more likely or less likely, and not to be proving all the time what's possible or impossible."

Existing Work on Diamond Mechanosynthesis

There is a growing body of peer-reviewed theoretical work on synthesizing diamond by mechanically removing/adding hydrogen atoms and depositing carbon atoms (a process known as mechanosynthesis). This work is slowly permeating the broader nanoscience community and is being critiqued. For instance, Peng et al. (2006) (in the continuing research effort by Freitas, Merkle and their collaborators) reports that the most-studied mechanosynthesis tooltip motif (DCB6Ge) successfully places a C_2 carbon dimer on a C(110) diamond surface at both 300 K (room temperature) and 80 K (liquid nitrogen temperature), and that the silicon variant (DCB6Si) also works at 80 K but not at 300 K. Over 100,000 CPU hours were invested in this latest study. The DCB6 tooltip motif, initially described by Merkle and Freitas at a Foresight Conference in 2002, was the first complete tooltip ever proposed for diamond mechanosynthesis and remains the only tooltip motif that has been successfully simulated for its intended function on a full 200-atom diamond surface.

The tooltips modeled in this work are intended to be used only in carefully controlled environments (e. g., vacuum). Maximum acceptable limits for tooltip translational and rotational misplacement errors are reported in Peng et al. (2006) -- tooltips must be positioned with great accuracy to avoid bonding the dimer incorrectly. Peng et al. (2006) reports that increasing the handle thickness from 4 support planes of C atoms above the tooltip to 5 planes decreases the resonance frequency of the entire structure from 2.0 THz to 1.8 THz. More importantly, the vibrational footprints of a DCB6Ge tooltip mounted on a 384-atom handle and of the same tooltip mounted on a similarly constrained but much larger 636-atom "crossbar" handle are virtually identical in the non-crossbar directions. Additional computational studies modeling still bigger handle structures are welcome, but the ability to precisely position SPM tips to the requisite atomic accuracy has been repeatedly demonstrated experimentally at low temperature, or even at room temperature constituting a basic existence proof for this capability.

Further research to consider additional tooltips will require time-consuming computational chemistry and difficult laboratory work.

A working nanofactory would require a variety of well-designed tips for different reactions, and detailed analyses of placing atoms on more complicated surfaces. Although this appears a challenging problem given current resources, many tools will be available to help future researchers: Moore's Law predicts further increases in computer power, semiconductor fabrication techniques continue to approach the nanoscale, and researchers grow ever more skilled at using proteins, ribosomes and DNA to perform novel chemistry.

Works of Fiction

- In The Diamond Age by Neal Stephenson diamond can be constructed by simply building it out of carbon atoms. Also all sorts of devices from dust size detection devices to giant diamond zeppelins are constructed atom by atom using only carbon, oxygen, nitrogen and chlorine atoms.

- In the novel *Tomorrow* by Andrew Saltzman (ISBN 1-4243-1027-X), a scientist uses nanorobotics to create a liquid that when inserted into the bloodstream, renders one nearly invincible given that the microscopic machines repair tissue almost instantaneously after it is damaged.

- In the roleplaying game Splicers by Palladium Books, humanity has succumbed to a "nanobot plague" that causes any object made of a non-precious metal to twist and change shape (sometimes into a type of robot) moments after being touched by a human. The object will then proceed to attack the human. This has forced humanity to develop "biotechnological" devices to replace those previously made of metal.

- On the television show Mystery Science Theater 3000, the Nanites (voiced variously by Kevin Murphy, Paul Chaplin, Mary Jo Pehl, and Bridget Jones) - are self-replicating, bio-engineered organisms that work on the ship, they are microscopic creatures that reside in the Satellite of Love's computer systems. (They are similar to the creatures in *Star Trek: The Next Generation* episode "Evolution", which featured "nanites" taking over the *Enterprise*.) The Nanites made their first appearance in season 8. Based on the concept of nanotechnology, their comical *deus ex machina* activities included such diverse tasks as instant repair and construction, hairstyling, performing a Nanite variation of a flea circus, conducting a microscopic war, and even destroying the Observers' planet after a dangerously vague request from Mike to "take care of [a] little problem". They also ran a microbrewery.

Grey Goo

Grey goo (also spelled gray goo) is a hypothetical end-of-the-world scenario involving molecular nanotechnology in which out-of-control self-replicating robots consume all matter on Earth while building more of themselves, a scenario that has been called *ecophagy* ("eating the environment", more literally "eating the habitation"). The original idea assumed machines were designed to have this capability, while popularizations have assumed that machines might somehow gain this capability by accident.

Self-replicating machines of the macroscopic variety were originally described by mathematician John von Neumann, and are sometimes referred to as von Neumann machines or clanking replicators. The term *gray goo* was coined by nanotechnology pioneer Eric Drexler in his 1986 book *Engines of Creation*. In 2004 he stated, "I wish I had never used the term 'gray goo'." *Engines of Creation* mentions "gray goo" in two paragraphs and a note, while the popularized idea of gray goo was first publicized in a mass-circulation magazine, *Omni*, in November 1986.

Definition

The term was first used by molecular nanotechnology pioneer Eric Drexler in his book *Engines of Creation* (1986). In Chapter 4, *Engines Of Abundance*, Drexler illustrates both exponential growth and inherent limits (not gray goo) by describing nanomachines that can function only if given special raw materials:

Imagine such a replicator floating in a bottle of chemicals, making copies of itself... the first replicator assembles a copy in one thousand seconds, the two replicators then build two more in the next thousand seconds, the four build another four, and the eight build another eight. At the end of ten hours, there are not thirty-six new replicators, but over 68 billion. In less than a day, they would weigh a ton; in less than two days, they would outweigh the Earth; in another four hours, they would exceed the mass of the Sun and all the planets combined — if the bottle of chemicals hadn't run dry long before.

According to Drexler, the term was popularized by an article in science fiction magazine Omni, which also popularized the term *nanotechnology* in the same issue. Drexler says arms control is a far greater issue than grey goo "nanobugs".

In a History Channel broadcast, a contrasting idea (a kind of gray goo) is referred to in a futuristic doomsday scenario: "In a common practice, billions of nanobots are released to clean up an oil spill off the coast of Louisiana. However, due to a programming error, the nanobots devour all carbon based objects, instead of just the hydrocarbons of the oil. The nanobots destroy everything, all the while, replicating themselves. Within days, the planet is turned to dust."

Early assembler-based replicators could beat the most advanced modern organisms. 'Plants' with 'leaves' no more efficient than today's solar cells could out-compete real plants, crowding the biosphere with an inedible foliage. Tough, omnivorous 'bacteria' could out-compete real bacteria: they could spread like blowing pollen, replicate swiftly, and reduce the biosphere to dust in a matter of days. Dangerous replicators could easily be too tough, small, and rapidly spreading to stop — at least if we made no preparation. We have trouble enough controlling viruses and fruit flies.

Drexler notes that the geometric growth made possible by self-replication is inherently limited by the availability of suitable raw materials.

Drexler used the term "gray goo" not to indicate color or texture, but to emphasize the difference between "superiority" in terms of human values and "superiority" in terms of competitive success:

Though masses of uncontrolled replicators need not be grey or gooey, the term "grey goo" emphasizes that replicators able to obliterate life might be less inspiring than a single species of crabgrass. They might be "superior" in an evolutionary sense, but this need not make them valuable.

Bill Joy, one of the founders of Sun Microsystems, discussed some of the problems with pursuing this technology in his now-famous 2000 article in *Wired* magazine, titled "Why the Future Doesn't Need Us". In direct response to Joy's concerns, the first quantitative technical analysis of the ecophagy scenario was published in 2000 by nano-medicine pioneer Robert Freitas.

Risks and Precautions

Drexler more recently conceded that there is no need to build anything that even resembles a potential runaway replicator. This would avoid the problem entirely. In a paper in the journal *Nanotechnology*, he argues that self-replicating machines are needlessly complex and inefficient. His 1992 technical book on advanced nanotechnologies *Nanosystems: Molecular Machinery, Manufacturing, and Computation* describes manufacturing systems that are desktop-scale factories with specialized machines in fixed locations and conveyor belts to move parts from place to place. None of these measures would prevent a party from creating a weaponized grey goo, were such a thing possible.

Prince Charles called upon the British Royal Society to investigate the "enormous environmental and social risks" of nanotechnology in a planned report, leading to much media commentary on gray goo. The Royal Society's report on nanoscience was released on 29 July 2004, and declared the possibility of self-replicating machines to lie too far in the future to be of concern to regulators.

More recent analysis in the paper titled *Safe Exponential Manufacturing* from the Institute of Physics (co-written by Chris Phoenix, Director of Research of the Center for Responsible Nanotechnology, and Eric Drexler), shows that the danger of grey goo is far less likely than originally thought. However, other long-term major risks to society and the environment from nanotechnology have been identified. Drexler has made a somewhat public effort to retract his grey goo hypothesis, in an effort to focus the debate on more realistic threats associated with knowledge-enabled nanoterrorism and other misuses.

In *Safe Exponential Manufacturing*, which was published in a 2004 issue of *Nanotechnology*, it was suggested that creating manufacturing systems with the ability to self-replicate by the use of their own energy sources would not be needed. The Foresight Institute also recommended embedding controls in the molecular machines.

These controls would be able to prevent anyone from purposely abusing nanotechnology, and therefore avoid the grey goo scenario.

Ethics and Chaos

Grey goo is a useful construct for considering low-probability, high-impact outcomes from emerging technologies. Thus, it is a useful tool in the ethics of technology. Daniel A. Vallero applied it as a worst-case scenario thought experiment for technologists contemplating possible risks from advancing a technology. This requires that a decision tree or event tree include even extremely low probability events if such events may have an extremely negative and irreversible consequence, i.e. application of the precautionary principle. Dianne Irving admonishes that "any error in science will have a rippling effect....". Vallero adapted this reference to chaos theory to emerging technologies, wherein slight permutations of initial conditions can lead to unforeseen and profoundly negative downstream effects, for which the technologist and the new technology's proponents must be held accountable.

In Popular Culture

- *Screamers*, a 1995 Canadian-United States SciFi Action/Thriller

- *The Invincible,* a 1964 science fiction novel with intrigue centered on self-configuring nanobotic swarms.

- *Moonseed,* a 1998 science fiction novel where the Earth faces danger from a self-replicating nanobot swarm.

- *Gargoyles* Season 2, episode 33: "Walkabout". Episode about the issue.

- "Benderama", a *Futurama* episode about the issue

- *Tasty Planet*, a game released in 2006 by Dingo Games centers around a gray goo eating the universe, starting at the atomic level and progressing to the cosmic level.

- Replicator (Stargate), a self-replicating synthetic life-form in the television series *Stargate SG-1.*

- The 2008 science fiction film *The Day the Earth Stood Still* follows Klaatu, an alien sent to try to change human behavior or eradicate them from Earth via grey goo due to humankind's environmental damage to the planet.

- *Grey Goo* is a science fiction real-time strategy video game that features a playable faction based on the grey goo scenario.

- The Swedish author sv:Lars Wilderäng released in 2014 the first book in a trilogy about an attack on our technological civilization by nanomachines. The first book "Stjärnklart", ISBN 9789187135910, was well received by the public.

- *Baldr Sky*, a Japanese visual novel features self-replicating nanomachines under the name "assemblers". Their inherent threat and a near-catastrophic grey goo event is central to its plot.

- *Deus Ex: Invisible War*, a videogame features a self-replicating nanomachines bomb in the CGI introduction. A terrorist attack on Chicago erased the town and is the beginning of the plot.

- *Destiny: Rise of Iron*, an expansion to *Destiny* features a self replicating nano-virus called SIVA, which is similar to grey goo.

Mechanosynthesis

Mechanosynthesis is a term for hypothetical chemical syntheses in which reaction outcomes are determined by the use of mechanical constraints to direct reactive molecules to specific molecular sites. There are presently no chemical syntheses which achieve this aim. Some atomic placement has been achieved with scanning tunnelling microscopes.

Introduction

In conventional chemical synthesis or chemosynthesis, reactive molecules encounter one another through random thermal motion in a liquid or vapor. In a hypothesized process of mechanosynthesis, reactive molecules would be attached to molecular mechanical systems, and their encounters would result from mechanical motions bringing them together in planned sequences, positions, and orientations. It is envisioned that mechanosynthesis would avoid unwanted reactions by keeping potential reactants apart, and would strongly favor desired reactions by holding reactants together in optimal orientations for many molecular vibration cycles. In biology, the ribosome provides an example of a programmable mechanosynthetic device.

A primitive, very non-biological form of mechanochemistry has been performed at cryogenic temperatures using scanning tunneling microscopes. So far, such devices provide the closest approach to fabrication tools for molecular engineering. Broader exploitation of mechanosynthesis awaits more advanced technology for constructing molecular machine systems, with ribosome-like systems as an attractive early objective.

Much of the excitement regarding advanced mechanosynthesis regards its potential use in assembly of molecular-scale devices. Such techniques appear to have many applications in medicine, aviation, resource extraction, manufacturing and warfare.

Most theoretical explorations of advanced machines of this kind have focused on using carbon, because of the many strong bonds it can form, the many types of chemistry these bonds permit, and utility of these bonds in medical and mechanical applications. Carbon forms diamond, for example, which if cheaply available, would be an excellent material for many machines.

It has been suggested, notably by K. Eric Drexler, that mechanosynthesis will be fundamental to molecular manufacturing based on nanofactories capable of building macroscopic objects with atomic precision. The potential for these has been disputed, notably by Nobel Laureate Richard Smalley (who proposed and then critiqued an unworkable approach based on small fingers).

The Nanofactory Collaboration, founded by Robert Freitas and Ralph Merkle in 2000, is a focused ongoing effort involving 23 researchers from 10 organizations and 4 countries that is developing a practical research agenda specifically aimed at positionally controlled diamond mechanosynthesis and diamondoid nanofactory development.

In practice, getting exactly one molecule to a known place on the microscope's tip is possible, but has proven difficult to automate. Since practical products require at least several hundred million atoms, this technique has not yet proven practical in forming a real product.

The goal of one line of mechanoassembly research focuses on overcoming these problems by calibration, and selection of appropriate synthesis reactions. Some suggest attempting to develop a specialized, very small (roughly 1,000 nanometers on a side) machine tool that can build copies of itself using mechanochemical means, under the control of an external computer. In the literature, such a tool is called an assembler or molecular assembler. Once assemblers exist, geometric growth (directing copies to make copies) could reduce the cost of assemblers rapidly. Control by an external computer should then permit large groups of assemblers to construct large, useful projects to atomic precision. One such project would combine molecular-level conveyor belts with permanently mounted assemblers to produce a factory.

In part to resolve this and related questions about the dangers of industrial accidents and popular fears of runaway events equivalent to Chernobyl and Bhopal disasters, and the more remote issue of ecophagy, grey goo and green goo (various potential disasters arising from runaway replicators, which could be built using mechanosynthesis) the UK Royal Society and UK Royal Academy of Engineering in 2003 commissioned a study to deal with these issues and larger social and ecological implications, led by mechanical engineering professor Ann Dowling. This was anticipated by some to take a strong position on these problems and potentials –– and suggest any development path to a general theory of so-called mechanosynthesis. However, the Royal Society's nanotech report did not address molecular manufacturing at all, except to dismiss it along with grey goo.

Current technical proposals for nanofactories do not include self-replicating nano-robots, and recent ethical guidelines would prohibit development of unconstrained self-replication capabilities in nanomachines.

Diamond Mechanosynthesis

There is a growing body of peer-reviewed theoretical work on synthesizing diamond by mechanically removing/adding hydrogen atoms and depositing carbon atoms (a process known as diamond mechanosynthesis or DMS). For example, the 2006 paper in this continuing research effort by Freitas, Merkle and their collaborators reports that the most-studied mechanosynthesis tooltip motif (DCB6Ge) successfully places a C_2 carbon dimer on a C(110) diamond surface at both 300 K (room temperature) and 80 K (liquid nitrogen temperature), and that the silicon variant (DCB6Si) also works at 80 K but not at 300 K. These tooltips are intended to be used only in carefully controlled environments (e.g., vacuum). Maximum acceptable limits for tooltip translational and rotational misplacement errors are reported in paper III—tooltips must be positioned with great accuracy to avoid bonding the dimer incorrectly. Over 100,000 CPU hours were invested in this study.

The DCB6Ge tooltip motif, initially described at a Foresight Conference in 2002, was the first complete tooltip ever proposed for diamond mechanosynthesis and remains the only tooltip motif that has been successfully simulated for its intended function on a full 200-atom diamond surface. Although an early paper gives a predicted place-ment speed of 1 dimer per second for this tooltip, this limit was imposed by the slow speed of recharging the tool using an inefficient recharging method and is not based on any inherent limitation in the speed of use of a charged tooltip. Additionally, no sensing means was proposed for discriminating among the three possible outcomes of an attempted dimer placement—deposition at the correct location, deposition at the wrong location, and failure to place the dimer at all—because the initial proposal was to position the tooltip by dead reckoning, with the proper reaction assured by design-ing appropriate chemical energetics and relative bond strengths for the tooltip-surface interaction.

More recent theoretical work analyzes a complete set of nine molecular tools made from hydrogen, carbon and germanium able to (a) synthesize all tools in the set (b) recharge all tools in the set from appropriate feedstock molecules and (c) synthesize a wide range of stiff hydrocarbons (diamond, graphite, fullerenes, and the like). All required reactions are analyzed using standard ab initio quantum chemistry meth-ods.

Further research to consider alternate tips will require time-consuming computation-al chemistry and difficult laboratory work. In the early 2000s, a typical experimental arrangement was to attach a molecule to the tip of an atomic force microscope, and then use the microscope's precise positioning abilities to push the molecule on the tip into another on a substrate. Since the angles and distances can be precisely controlled, and the reaction occurs in a vacuum, novel chemical compounds and arrangements are possible.

History

The technique of moving single atoms mechanically was proposed by Eric Drexler in his 1986 book The Engines of Creation.

In 1988, researchers at IBM's Zürich Research Institute successfully spelled the letters "IBM" in xenon atoms on a cryogenic copper surface, grossly validating the approach. Since then, a number of research projects have undertaken to use similar techniques to store computer data in a compact fashion. More recently the technique has been used to explore novel physical chemistries, sometimes using lasers to excite the tips to particular energy states, or examine the quantum chemistry of particular chemical bonds.

In 1999, an experimentally proved methodology called feature-oriented scanning (FOS) was suggested. The feature-oriented scanning methodology allows precisely controlling the position of the probe of a scanning probe microscope (SPM) on an atomic surface at room temperature. The suggested methodology supports fully automatic control of single- and multiprobe instruments in solving tasks of mechanosynthesis and bottom-up nanofabrication.

In 2003, Oyabu *et al.* reported the first instance of purely mechanical-based covalent bond-making and bond-breaking, i.e., the first experimental demonstration of true mechanosynthesis—albeit with silicon rather than carbon atoms.

In 2005, the first patent application on diamond mechanosynthesis was filed.

In 2008, a \$3.1 million grant was proposed to fund the development of a proof-of-principle mechanosynthesis system.

Molecular Assembler

A molecular assembler, as defined by K. Eric Drexler, is a "proposed device able to guide chemical reactions by positioning reactive molecules with atomic precision". A molecular assembler is a kind of molecular machine. Some biological molecules such as ribosomes fit this definition. This is because they receive instructions from messenger RNA and then assemble specific sequences of amino acids to construct protein molecules. However, the term "molecular assembler" usually refers to theoretical human-made devices.

Beginning in 2007, the British Engineering and Physical Sciences Research Council has funded development of ribosome-like molecular assemblers. Clearly, molecular assemblers are possible in this limited sense. A technology roadmap project, led by the Battelle Memorial Institute and hosted by several U.S. National Laboratories has explored a range of atomically precise fabrication technologies, including both early-generation and longer-term prospects for programmable molecular assembly; the report was released in December, 2007. In 2008 the Engineering and Physical Sciences Research Council provided funding of 1.5 million pounds over six years for research working towards mechanized mechanosynthesis, in partnership with the Institute for Molecular Manufacturing, amongst others.

Likewise, the term "molecular assembler" has been used in science fiction and popular culture to refer to a wide range of fantastic atom-manipulating nanomachines, many of which may be physically impossible in reality. Much of the controversy regarding "molecular assemblers" results from the confusion in the use of the name for both technical concepts and popular fantasies. In 1992, Drexler introduced the related but better-understood term "molecular manufacturing," which he defined as the programmed "chemical synthesis of complex structures by mechanically positioning reactive molecules, not by manipulating individual atoms."

This article mostly discusses "molecular assemblers" in the popular sense. These include hypothetical machines that manipulate individual atoms and machines with organism-like self-replicating abilities, mobility, ability to consume food, and so forth. These are quite different from devices that merely (as defined above) "guide chemical reactions by positioning reactive molecules with atomic precision".

Because synthetic molecular assemblers have never been constructed and because of the confusion regarding the meaning of the term, there has been much controversy as to whether "molecular assemblers" are possible or simply science fiction. Confusion and controversy also stem from their classification as nanotechnology, which is an active area of laboratory research which has already been applied to the production of real products; however, there had been, until recently, no research efforts into the actual construction of "molecular assemblers".

Nonetheless, a 2013 paper published in the journal *Science* details a new method of synthesizing a peptide in a sequence-specific manner by using an artificial molecular machine that is guided by a molecular strand. This functions in the same way as a ribosome building proteins by assembling amino acids according to a messenger RNA blueprint. The structure of the machine is based on a rotaxane, which is a molecular ring sliding along a molecular axle. The ring carries a thiolate group which removes amino acids in sequence from the axle, transferring them to a peptide assembly site.

In another paper published in March 2015, also in *Science*, chemists at the University of Illinois report a platform that automates the synthesis of 14 classes of small molecules, with thousands of compatible building blocks.

Nanofactories

A nanofactory is a proposed system in which nanomachines (resembling molecular assemblers, or industrial robot arms) would combine reactive molecules via mechanosynthesis to build larger atomically precise parts. These, in turn, would be assembled by positioning mechanisms of assorted sizes to build macroscopic (visible) but still atomically-precise products.

A typical nanofactory would fit in a desktop box, in the vision of K. Eric Drexler published in *Nanosystems: Molecular Machinery, Manufacturing and Computation* (1992), a notable work of "exploratory engineering". During the last decade, others have extended the nanofactory concept, including an analysis of nanofactory convergent assembly by Ralph Merkle, a systems design of a replicating nanofactory architecture by J. Storrs Hall, Forrest Bishop's "Universal Assembler", the patented exponential assembly process by Zyvex, and a top-level systems design for a 'primitive nanofactory' by Chris Phoenix (Director of Research at the Center for Responsible Nanotechnology). All of these nanofactory designs (and more) are summarized in Chapter 4 of *Kinematic Self-Replicating Machines* (2004) by Robert Freitas and Ralph Merkle. The Nanofactory Collaboration, founded by Freitas and Merkle in 2000, is a focused ongoing effort involving 23 researchers from 10 organizations and 4 countries that is developing a practical research agenda specifically aimed at positionally-controlled diamond mechanosynthesis and diamondoid nanofactory development.

In 2005, a computer-animated short film of the nanofactory concept was produced by John Burch, in collaboration with Drexler. Such visions have been the subject of much debate, on several intellectual levels. No one has discovered an insurmountable problem with the underlying theories and no one has proved that the theories can be translated into practice. However, the debate continues, with some of it being summarized in the molecular nanotechnology article.

If nanofactories could be built, severe disruption to the world economy would be one of many possible negative impacts, though it could be argued that this disruption would have little negative effect if everyone had such nanofactories. Great benefits also would be anticipated. Various works of science fiction have explored these and similar concepts. The potential for such devices was part of the mandate of a major UK study led by mechanical engineering professor Dame Ann Dowling.

Self-replication

"Molecular assemblers" have been confused with self-replicating machines. To produce a practical quantity of a desired product, the nanoscale size of a typical science fiction universal molecular assembler requires an extremely large number of such devices. However, a single such theoretical molecular assembler might be programmed to self-replicate, constructing many copies of itself. This would allow an exponential rate of production. Then after sufficient quantities of the molecular assemblers were available, they would then be re-programmed for production of the desired product. However, if self-replication of molecular assemblers were not restrained then it might lead to competition with naturally occurring organisms. This has been called ecophagy or the grey goo problem.

One method to building molecular assemblers is to mimic evolutionary processes employed by biological systems. Biological evolution proceeds by random variation combined with culling of the less-successful variants and reproduction of the more-successful variants. Production of complex molecular assemblers might be evolved from simpler systems since "A complex system that works is invariably found to have evolved from a simple system that worked. . . . A complex system designed from scratch never works and can not be patched up to make it work. You have to start over, beginning with a system that works." However, most published safety guidelines include "recommendations against developing ... replicator designs which permit surviving mutation or undergoing evolution".

Most assembler designs keep the "source code" external to the physical assembler. At each step of a manufacturing process, that step is read from an ordinary computer file and "broadcast" to all the assemblers. If any assembler gets out of range of that computer, or when the link between that computer and the assemblers is broken, or when that computer is unplugged, the assemblers stop replicating. Such a "broadcast architecture" is one of the safety features recommended by the "Foresight Guidelines on Molecular Nanotechnology", and a map of the 137-dimensional replicator design space recently published by Freitas and Merkle provides numerous practical methods by which replicators can be safely controlled by good design.

Drexler and Smalley Debate

One of the most outspoken critics of some concepts of "molecular assemblers" was Professor Richard Smalley (1943–2005) who won the Nobel prize for his contributions to the field of nanotechnology. Smalley believed that such assemblers were not physically possible and introduced scientific objections to them. His two principal technical objections were termed the "fat fingers problem" and the "sticky fingers problem". He believed these would exclude the possibility of "molecular assemblers" that worked by precision picking and placing of individual atoms. Drexler and coworkers responded to these two issues in a 2001 publication.

Smalley also believed that Drexler's speculations about apocalyptic dangers of self-replicating machines that have been equated with "molecular assemblers" would threaten the public support for development of nanotechnology. To address the debate between Drexler and Smalley regarding molecular assemblers *Chemical & Engineering News* published a point-counterpoint consisting of an exchange of letters that addressed the issues.

Regulation

Speculation on the power of systems that have been called "molecular assemblers" has sparked a wider political discussion on the implication of nanotechnology. This is in part due to the fact that nanotechnology is a very broad term and could include "molecular assemblers." Discussion of the possible implications of fantastic molecular assemblers has prompted calls for regulation of current and future nanotechnology. There are very real concerns with the potential health and ecological impact of nanotechnology that is being integrated in manufactured products. Greenpeace for instance commissioned a report concerning nanotechnology in which they express concern into the toxicity of nanomaterials that have been introduced in the environment. However, it makes only passing references to "assembler" technology. The UK Royal Society and Royal Academy of Engineering also commissioned a report entitled "Nanoscience and nanotechnologies: opportunities and uncertainties" regarding the larger social and ecological implications on nanotechnology. This report does not discuss the threat posed by potential so-called "molecular assemblers."

Formal Scientific Review

In 2006, U.S. National Academy of Sciences released the report of a study of molecular manufacturing as part of a longer report, *A Matter of Size: Triennial Review of the National Nanotechnology Initiative* The study committee reviewed the technical content of *Nanosystems*, and in its conclusion states that no current theoretical analysis can be considered definitive regarding several questions of potential system performance, and that optimal paths for implementing high-performance systems cannot be predicted with confidence. It recommends experimental research to advance knowledge in this area:

> "Although theoretical calculations can be made today, the eventually attainable range of chemical reaction cycles, error rates, speed of operation, and thermodynamic efficiencies of such bottom-up manufacturing systems cannot be reliably predicted at this time. Thus, the eventually attainable perfection and complexity of manufactured products, while they can be calculated in theory, cannot be predicted with confidence. Finally, the optimum research paths that might lead to systems which greatly exceed the thermodynamic efficiencies and other capabilities of biological systems cannot be reliably predicted at this time. Research funding that is based on the ability of investigators to produce experimental demonstrations that link to abstract models and guide long-term vision is most appropriate to achieve this goal."

Grey Goo

One potential scenario that has been envisioned is out-of-control self-replicating molecular assemblers in the form of grey goo which consumes carbon to continue its replication. If unchecked such mechanical replication could potentially consume whole ecoregions or the whole Earth (ecophagy), or it could simply outcompete natural life-forms for necessary resources such as carbon, ATP, or UV light (which some nanomotor examples run on). However, the ecophagy and 'grey goo' scenarios, like synthetic molecular assemblers, are based upon still-hypothetical technologies that have not yet been demonstrated experimentally.

In Fiction

Molecular assemblers are a popular topic in science fiction, for example, the matter compiler in The Diamond Age and the cornucopia machine in Singularity Sky. The replicator in *Star Trek* might also be considered a molecular assembler. A molecular assembler is also a key element of the plot of the computer game *Deus Ex* (called a "universal constructor" in the game).

In the political sci-fi comic series Transmetropolitan, written by Warren Ellis, machines called "Makers" are used to replicate and reform matter. Each morning, Makers sweep the streets for garbage, gathering the matter to recycle it into more useful objects. The main character also uses a Maker in his apartment to instantly produce a pair of glasses which take photos, as well as other objects such as clothing.

In Dead Money, a DLC of the video game Fallout: New Vegas, the player can obtain useful items from vending machines that use an unknown form of molecular assembly technology to transform casino chips that the player can find into any of several items.

Molecular Modelling

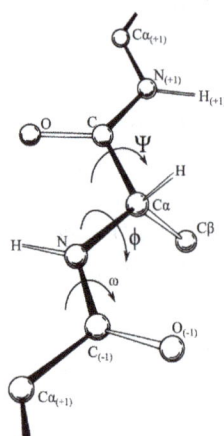

The backbone dihedral angles are included in the molecular model of a protein.

Molecular modelling encompasses all methods, theoretical and computational, used to model or mimic the behaviour of molecules. The methods are used in the fields of computational chemistry, drug design, computational biology and materials science to study molecular systems ranging from small chemical systems to large biological molecules and material assemblies. The simplest calculations can be performed by hand, but inevitably computers are required to perform molecular modelling of any reasonably sized system. The common feature of molecular modelling methods is the atomistic level description of the molecular systems. This may include treating atoms as the smallest individual unit (a molecular mechanics approach), or explicitly modelling electrons of each atom (a quantum chemistry approach).

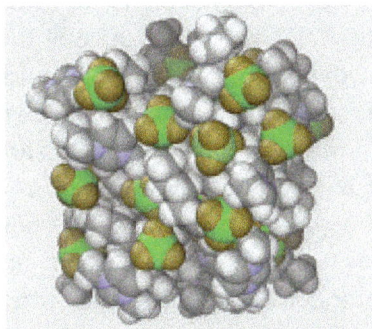

Modeling of ionic liquid

Molecular Mechanics

Molecular mechanics is one aspect of molecular modelling, as it refers to the use of classical mechanics (Newtonian mechanics) to describe the physical basis behind the models. Molecular models typically describe atoms (nucleus and electrons collectively) as point charges with an associated mass. The interactions between neighbouring atoms are described by spring-like interactions (representing chemical bonds) and Van der Waals forces. The Lennard-Jones potential is commonly used to describe the later. The electrostatic interactions are computed based on Coulomb's law. Atoms are assigned coordinates in Cartesian space or in internal coordinates, and can also be assigned velocities in dynamical simulations. The atomic velocities are related to the temperature of the system, a macroscopic quantity. The collective mathematical expression is termed a potential function and is related to the system internal energy (U), a thermodynamic quantity equal to the sum of potential and kinetic energies. Methods which minimize the potential energy are termed energy minimization methods (e.g., steepest descent and conjugate gradient), while methods that model the behaviour of the system with propagation of time are termed molecular dynamics.

$$E = E_{bonds} + E_{angle} + E_{dihedral} + E_{non\text{-}bonded}$$

$$E_{non\text{-}bonded} = E_{electrostatic} + E_{van\ der\ Waals}$$

This function, referred to as a potential function, computes the molecular potential energy as a sum of energy terms that describe the deviation of bond lengths, bond angles and torsion angles away from equilibrium values, plus terms for non-bonded pairs of atoms describing van der Waals and electrostatic interactions. The set of parameters consisting of equilibrium bond lengths, bond angles, partial charge values, force constants and van der Waals parameters are collectively termed a force field. Different implementations of molecular mechanics use different mathematical expressions and different parameters for the potential function. The common force fields in use today have been developed by using high level quantum calculations and/or fitting to experimental data. The method, termed energy minimization, is used to find positions of zero gradient for all atoms, in other words, a local energy minimum. Lower energy states are more stable and are commonly investigated because of their role in chemical and biological processes. A molecular dynamics simulation, on the other hand, computes the behaviour of a system as a function of time. It involves solving Newton's laws of motion, principally the second law, $\mathbf{F} = m\mathbf{a}$. . Integration of Newton's laws of motion, using different integration algorithms, leads to atomic trajectories in space and time. The force on an atom is defined as the negative gradient of the potential energy function. The energy minimization method is useful to obtain a static picture for comparing between states of similar systems, while molecular dynamics provides information about the dynamic processes with the intrinsic inclusion of temperature effects.

Variables

Molecules can be modelled either in vacuum, or in the presence of a solvent such as water. Simulations of systems in vacuum are referred to as *gas-phase* simulations, while those that include the presence of solvent molecules are referred to as *explicit solvent* simulations. In another type of simulation, the effect of solvent is estimated using an empirical mathematical expression; these are termed *implicit solvation* simulations.

Coordinate Representations

Most force fields are distance-dependent, making the most convenient expression for these Cartesian coordinates. Yet the comparatively rigid nature of bonds which occur between specific atoms, and in essence, defines what is meant by the designation *molecule*, make an internal coordinate system the most logical representation. In some fields the IC representation (bond length, angle between bonds, and twist angle of the bond as shown in the figure) is termed the Z-matrix or torsion angle representation. Unfortunately, continuous motions in Cartesian space often require discontinuous angular branches in internal coordinates, making it relatively hard to work with force fields in the internal coordinate representation, and conversely a simple displacement of an atom in Cartesian space may not be a straight line trajectory due to the prohibitions of the interconnected bonds. Thus, it is very common for computational optimizing programs to flip back and forth between representations during their iterations. This can dominate the calculation time of the potential itself and in long chain molecules introduce cumulative numerical inaccuracy. While all conversion algorithms produce mathematically identical results, they differ in speed and numerical accuracy. Currently, the fastest and most accurate torsion to Cartesian conversion is the Natural Extension Reference Frame (NERF) method.

Applications

Molecular modelling methods are now used routinely to investigate the structure, dynamics, surface properties, and thermodynamics of inorganic, biological, and polymeric systems. The types of biological activity that have been investigated using molecular modelling include protein folding, enzyme catalysis, protein stability, conformational changes associated with biomolecular function, and molecular recognition of proteins, DNA, and membrane complexes.

References

- Chris Phoenix; Mike Treder (2008). "Chapter 21: Nanotechnology as global catastrophic risk". In Bostrom, Nick; Cirkovic, Milan M. Global catastrophic risks. Oxford: Oxford University Press. ISBN 978-0-19-857050-9.

- R. V. Lapshin (2011). "Feature-oriented scanning probe microscopy". In H. S. Nalwa. Encyclopedia of Nanoscience and Nanotechnology (PDF). 14. USA: American Scientific Publishers. pp. 105–115. ISBN 1-58883-163-9.

- Drexler, K. Eric (1992). Nanosystems: molecular machinery, manufacturing, and computation. Wiley. ISBN 978-0-471-57518-4.

- Vallero, Daniel (2007). Biomedical Ethics for Engineers: Ethics and Decision Making in Biomedical and Biosystem Engineering. Academic Press. ISBN 9780080476100.

- R. V. Lapshin (2011). "Feature-oriented scanning probe microscopy". In H. S. Nalwa. Encyclopedia of Nanoscience and Nanotechnology (PDF). 14. USA: American Scientific Publishers. pp. 105–115. ISBN 1-58883-163-9.

- Tomasik, Brian. "International Cooperation vs. AI Arms Race". foundational-research.org. Retrieved 19 July 2014.

- Hapgood, Fred (November 1986). "Nanotechnology: Molecular Machines that Mimic Life" (PDF). Omni. Retrieved 19 July 2014.

- "Foresight Guidelines for Responsible Nanotechnology Development". Foresight Institute and IMM. Retrieved 2012-05-07.

- Peterson, Christine (2007-05-08). "Nanodot: Nanotechnology News and Discussion » Blog Archive » Nanotechnology Roadmap launch: Productive Nanosystems Conference, Oct 9-10". Foresight.org. Retrieved 2010-09-05.

- "Blog Archive » Is mechanosynthesis feasible? The debate moves up a gear". Soft Machines. 2004-12-16. Retrieved 2010-09-05.

Devices Related to Nanotechnology

The devices related to nanotechnology are molecular machine, nanocar and nanomotor. Nanocar is not really a car, as it does not have a molecular motor whereas nanomotors are molecules that are capable of transferring energy into movement. The section serves as a source to understand the major devices related to nanotechnology.

Molecular Machine

A molecular machine, or nanomachine, refers to any discrete number of molecular components that produce quasi-mechanical movements (output) in response to specific stimuli (input). The expression is often more generally applied to molecules that simply mimic functions that occur at the macroscopic level. The term is also common in nanotechnology where a number of highly complex molecular machines have been proposed that are aimed at the goal of constructing a molecular assembler. Molecular machines can be divided into two broad categories; synthetic and biological.

The 2016 Nobel Prize in Chemistry was awarded to Jean-Pierre Sauvage, Sir J. Fraser Stoddart and Bernard L. Feringa for the design and synthesis of molecular machines.

Examples

From a synthetic perspective, there are two important types of molecular machines: molecular switches (or shuttles) and molecular motors. The major difference between the two systems is that a switch influences a system as a function of state, whereas a motor influences a system as a function of trajectory. A switch (or shuttle) may appear to undergo translational motion, but returning a switch to its original position undoes any mechanical effect and liberates energy to the system. Furthermore, switches cannot use chemical energy to repetitively and progressively drive a system away from equilibrium whereas a motor can.

Synthetic

A wide variety of rather simple molecular machines have been synthesized by chemists. They can consist of a single molecule; however, they are often constructed for mechanically-interlocked molecular architectures, such as rotaxanes and catenanes. Carbon nanotube nanomotors have also been produced.

- Molecular motors are molecules that are capable of unidirectional rotation motion powered by external energy input. A number of molecular machines have been synthesized powered by light or reaction with other molecules.

- A molecular propeller is a molecule that can propel fluids when rotated, due to its special shape that is designed in analogy to macroscopic propellers. It has several molecular-scale blades attached at a certain pitch angle around the circumference of a nanoscale shaft.

- A molecular switch is a molecule that can be reversibly shifted between two or more stable states. The molecules may be shifted between the states in response to changes in e.g. pH, light, temperature, an electric current, microenvironment, or the presence of a ligand.

- A molecular shuttle is a molecule capable of shuttling molecules or ions from one location to another. A common molecular shuttle consists of a rotaxane where the macrocycle can move between two sites or stations along the dumbbell backbone.

- Molecular tweezers are host molecules capable of holding items between their two arms. The open cavity of the molecular tweezers binds items using non-covalent bonding including hydrogen bonding, metal coordination, hydrophobic forces, van der Waals forces, π interactions, or electrostatic effects. Examples of molecular tweezers have been reported that are constructed from DNA and are considered DNA machines.

- A molecular sensor is a molecule that interacts with an analyte to produce a detectable change. Molecular sensors combine molecular recognition with some form of reporter, so the presence of the item can be observed.

- A molecular logic gate is a molecule that performs a logical operation on one or more logic inputs and produces a single logic output. Unlike a molecular sensor, the molecular logic gate will only output when a particular combination of inputs are present.

Biological

The most complex molecular machines are proteins found within cells. These include motor proteins, such as myosin, which is responsible for muscle contraction, kinesin, which moves cargo inside cells away from the nucleus along microtubules, and dynein, which produces the axonemal beating of motile cilia and flagella. These proteins and their nanoscale dynamics are far more complex than any molecular machines that have yet been artificially constructed.

Probably the most significant biological machine known is the ribosome. Other important examples include motile cilia. A high-level-abstraction summary is that, "[i]n effect, the [motile cilium] is a nanomachine composed of perhaps over 600 proteins in molecular complexes, many of which also function independently as nanomachines." Flexible linker domains allow the connecting protein domains to recruit their binding partners and induce long-range allostery via protein domain dynamics.

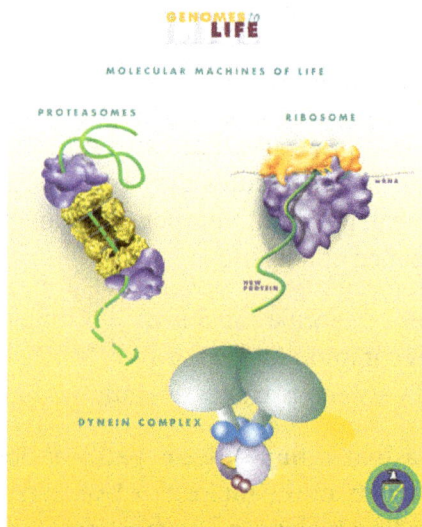

Some biological molecular machines

This protein flexibility allows the construction of biological machines. The first useful applications of these biological machines might be in nanomedicine. For example, they could be used to identify and destroy cancer cells. Molecular nanotechnology is a speculative subfield of nanotechnology regarding the possibility of engineering molecular assemblers, biological machines which could re-order matter at a molecular or atomic scale. Nanomedicine would make use of these nanorobots, introduced into the body, to repair or detect damages and infections. Molecular nanotechnology is highly theoretical, seeking to anticipate what inventions nanotechnology might yield and to propose an agenda for future inquiry. The proposed elements of molecular nanotechnology, such as molecular assemblers and nanorobots are far beyond current capabilities.

Research

The construction of more complex molecular machines is an active area of theoretical research. A number of molecules, such as molecular propellers, have been designed, although experimental studies of these molecules are inhibited by the lack of methods to construct these molecules. In this context, theoretical modeling can be extremely useful to understand the self-assembly/disassembly processes of rotaxanes, important for the construction of light-powered molecular machines. This molecular-level knowledge may foster the realization of ever more complex, versatile, and effective molecular machines for the areas of nanotechnology, including molecular assemblers.

Molecular Motor

Molecular motors are biological molecular machines that are the essential agents of movement in living organisms. In general terms, a motor is a device that consumes energy in one form and converts it into motion or mechanical work; for example, many protein-based molecular motors harness the chemical free energy released by the hydrolysis of ATP in order to perform mechanical work. In terms of energetic efficiency, this type of motor can be superior to currently available man-made motors. One important difference between molecular motors and macroscopic motors is that molecular motors operate in the thermal bath, an environment in which the fluctuations due to thermal noise are significant.

Examples

Some examples of biologically important molecular motors:

- Cytoskeletal motors

    o Myosins are responsible for muscle contraction, intracellular cargo transport, and producing cellular tension.

    o Kinesin moves cargo inside cells away from the nucleus along microtubules.

    o Dynein produces the axonemal beating of cilia and flagella and also transports cargo along microtubules towards the cell nucleus.

- Polymerisation motors

    o Actin polymerization generates forces and can be used for propulsion. ATP is used.

    o Microtubule polymerization using GTP.

    o Dynamin is responsible for the separation of clathrin buds from the plasma membrane. GTP is used.

- Rotary motors:

    o F_oF_1-ATP synthase family of proteins convert the chemical energy in ATP to the electrochemical potential energy of a proton gradient across a membrane or the other way around. The catalysis of the chemical reaction and the movement of protons are coupled to each other via the mechanical rotation of parts of the complex. This is involved in ATP synthesis in the mitochondria and chloroplasts as well as in pumping of protons across the vacuolar membrane.

- o The bacterial flagellum responsible for the swimming and tumbling of *E. coli* and other bacteria acts as a rigid propeller that is powered by a rotary motor. This motor is driven by the flow of protons across a membrane, possibly using a similar mechanism to that found in the F_o motor in ATP synthase.

- Nucleic acid motors:

  - o RNA polymerase transcribes RNA from a DNA template.

  - o DNA polymerase turns single-stranded DNA into double-stranded DNA.

  - o Helicases separate double strands of nucleic acids prior to transcription or replication. ATP is used.

  - o Topoisomerases reduce supercoiling of DNA in the cell. ATP is used.

  - o RSC and SWI/SNF complexes remodel chromatin in eukaryotic cells. ATP is used.

  - o SMC proteins responsible for chromosome condensation in eukaryotic cells.

  - o Viral DNA packaging motors inject viral genomic DNA into capsids as part of their replication cycle, packing it very tightly. Several models have been put forward to explain how the protein generates the force required to drive the DNA into the capsid; for a review. An alternative proposal is that, in contrast with all other biological motors, the force is not generated directly by the protein, but by the DNA itself. In this model, ATP hydrolysis is used to drive protein conformational changes that alternatively dehydrate and rehydrate the DNA, cyclically driving it from B-DNA to A-DNA and back again. A-DNA is 23% shorter than B-DNA, and the DNA shrink/expand cycle is coupled to a protein-DNA grip/release cycle to generate the forward motion that propels DNA into the capsid.

- Synthetic molecular motors have been created by chemists that yield rotation, possibly generating torque.

Theoretical Considerations

Because the motor events are stochastic, molecular motors are often modeled with the Fokker-Planck equation or with Monte Carlo methods. These theoretical models are especially useful when treating the molecular motor as a Brownian motor.

Experimental Observation

In experimental biophysics, the activity of molecular motors is observed with many different experimental approaches, among them:

- Fluorescent methods: fluorescence resonance energy transfer (FRET), fluorescence correlation spectroscopy (FCS), total internal reflection fluorescence (TIRF).

- Magnetic tweezers can also be useful for analysis of motors that operate on long pieces of DNA.

- Neutron spin echo spectroscopy can be used to observe motion on nanosecond timescales.

- Optical tweezers (not to be confused with molecular tweezers in context) are well-suited for studying molecular motors because of their low spring constants.

- Scattering techniques: single particle tracking based on dark field microscopy or interferometric scattering microscopy (iSCAT)

- Single-molecule electrophysiology can be used to measure the dynamics of individual ion channels.

Many more techniques are also used. As new technologies and methods are developed, it is expected that knowledge of naturally occurring molecular motors will be helpful in constructing synthetic nanoscale motors.

Non-biological

Recently, chemists and those involved in nanotechnology have begun to explore the possibility of creating molecular motors *de novo*. These synthetic molecular motors currently suffer many limitations that confine their use to the research laboratory. However, many of these limitations may be overcome as our understanding of chemistry and physics at the nanoscale increases. Systems like the nanocars, while not technically motors, are illustrative of recent efforts towards synthetic nanoscale motors.

Molecular Sensor

A molecular sensor or chemosensor is a molecule that interacts with an analyte to produce a detectable change. Molecular sensors combine molecular recognition with some form of reporter so the presence of the guest can be observed. The term supramolecular analytical chemistry has recently been coined to describe the application of molecular sensors to analytical chemistry.

Early examples of molecular sensors are crown ethers with large affinity for sodium ions but not for potassium and forms of metal detection by so-called complexones which are traditional pH indicators retrofitted with molecular groups sensitive to metals. This receptor-spacer-reporter concept is a recurring theme often with the reporter displaying photoinduced electron transfer (PET). One example is a sensor sensitive to heparin.

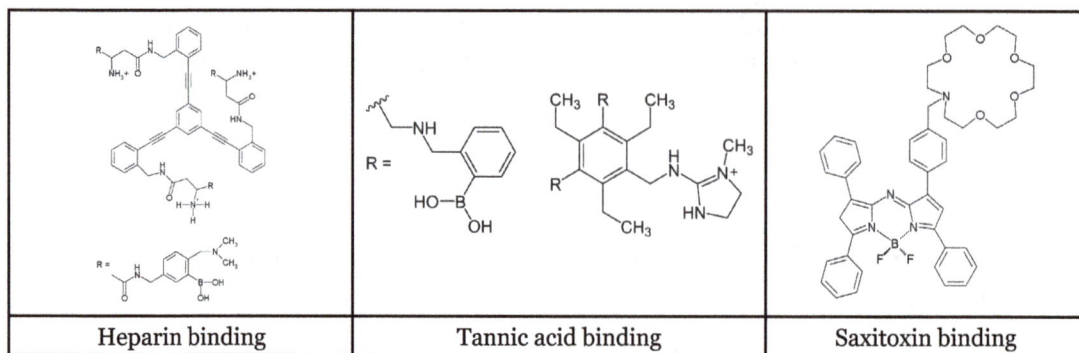

| Heparin binding | Tannic acid binding | Saxitoxin binding |

Other receptors are sensitive not to a specific molecule but to a molecular compound class. One example is the grouped analysis of several tannic acids that accumulate in ageing Scotch whisky in oak barrels. The grouped results demonstrated a correlation with the age but the individual components did not. A similar receptor can be used to analyze tartrates in wine.

The compound saxitoxin is a neurotoxin found in shellfish and a chemical weapon. An experimental sensor for this compound is again based on PET. Interaction of saxitoxin with the sensor's crown ether moiety kills its PET process towards the fluorophore and fluorescence is switched from off to on. The unusual boron moiety makes sure the fluorescence takes place in the visible light part of the electromagnetic spectrum.

In another strategy called indicator-displacement assay (IDA) an analyte such as citrate or phosphate ions displace a fluorescent indicator in an indicator-host complex. The so-called UT taste chip (University of Texas) is a prototype electronic tongue and combines supramolecular chemistry with charge-coupled devices based on silicon wafers and immobilized receptor molecules.

Molecular Switch

A molecular switch is a molecule that can be reversibly shifted between two or more stable states. The molecules may be shifted between the states in response to environmental stimuli, such as changes in pH, light, temperature, an electric current, microenvironment, or in the presence of a ligand. In some cases, a combination of stimuli is required. The oldest forms of synthetic molecular switches are pH indicators, which display distinct colors as a function of pH. Currently synthetic molecular switches are of interest in the field of nanotechnology for application in molecular computers or responsive drug delivery systems. Molecular switches are also important to in biology because many biological functions are based on it, for instance allosteric regulation and vision. They are also one of the simplest examples of molecular machines.

Photochromic Molecular Switches

A widely studied class are photochromic compounds which are able to switch between electronic configurations when irradiated by light of a specific wavelength. Each state has a specific absorption maximum which can then be read out by UV-VIS spectroscopy. Members of this class include azobenzenes, diarylethenes, dithienylethenes, fulgides, stilbenes, spiropyrans and phenoxynaphthacene quinones.

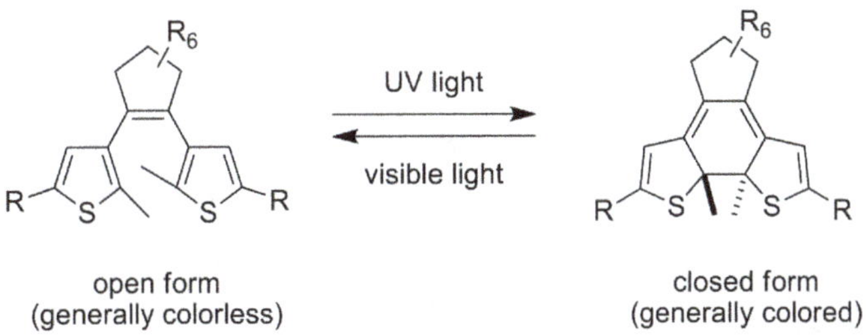

R6
R6
UV light
visible light
R
S
S
R
R
S
S
R
open form
(generally colorless)
closed form
(generally colored)

Chiroptical molecular switches are a specific subgroup with photochemical switching taking place between an enantiomeric pairs. In these compounds the readout is by circular dichroism rather than by ordinary spectroscopy. Hindered alkenes such as the one depicted below change their helicity as response to irradiation with right or left-handed circularly polarized light

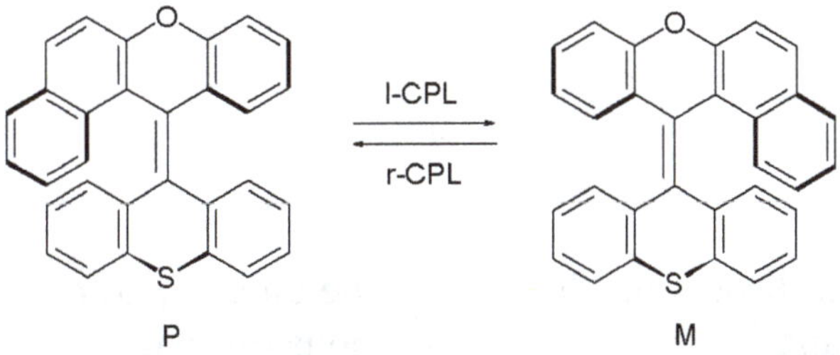

O
O
l-CPL
r-CPL
S
S
P
M

Chiroptical molecular switches that show directional motion are considered synthetic molecular motors:

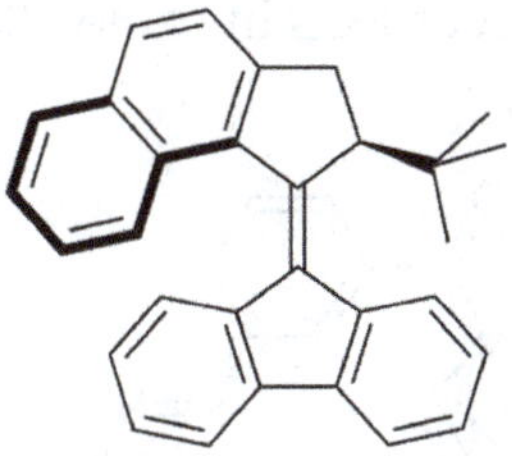

Host-guest Molecular Switches

In host-guest chemistry the bistable states of molecular switches differ in their affinity for guests. Many early examples of such systems are based on crown ether chemistry. The first switchable host is described in 1978 by Desvergne & Bouas-Laurent who create a crown ether via photochemical anthracene dimerization. Although not strictly speaking switchable the compound is able to take up cations after a photochemical trigger and exposure to acetonitrile gives back the open form.

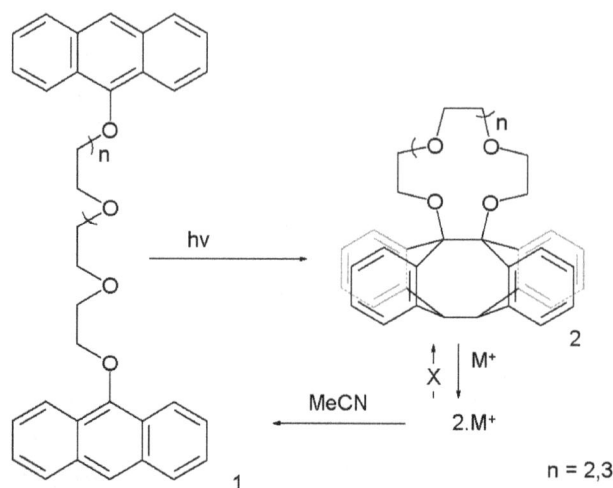

In 1980 Yamashita et al. construct a crown ether already incorporating the anthracene units (an anthracenophane) and also study ion uptake vs photochemistry.

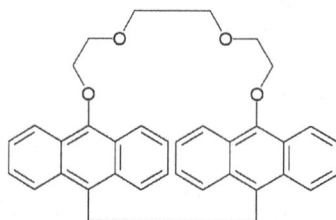

Also in 1980 Shinkai throws out the anthracene unit as photoantenna in favor of an azobenzene moiety and for the first time envisions the existence of molecules with an on-off switch. In this molecule light triggers a trans-cis isomerization of the azo group which results in ring expansion. Thus in the trans form the crown binds preferentially to ammonium, lithium and sodium ions while in the cis form the preference is for potassium and rubidium (both larger ions in same alkali metal group). In the dark the reverse isomerization takes place.

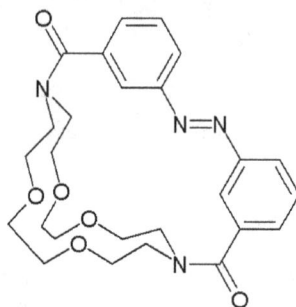

Shinkai employs this devices in actual ion transport mimicking the biochemical action of monensin and nigericin: in a biphasic system ions are taken up triggered by light in one phase and deposited in the other phase in absence of light.

Mechanically-interlocked Molecular Switches

Some of the most advanced molecular switches are based on mechanically-inter-locked molecular architectures where the bistable states differ in the position of the macrocycle. In 1991 Stoddart devices a molecular shuttle based on a rotaxane on which a molecular *bead* is able to shuttle between two *docking stations* situated on a molecular *thread*. Stoddart predicts that when the stations are dissimilar with each of the stations addressed by a different external stimulus the shuttle becomes a molecular machine. In 1993 Stoddart is scooped by supramolecular chemistry pioneer Fritz Vögtle who actually delivers a switchable molecule based not on a rotaxane but on a related catenane

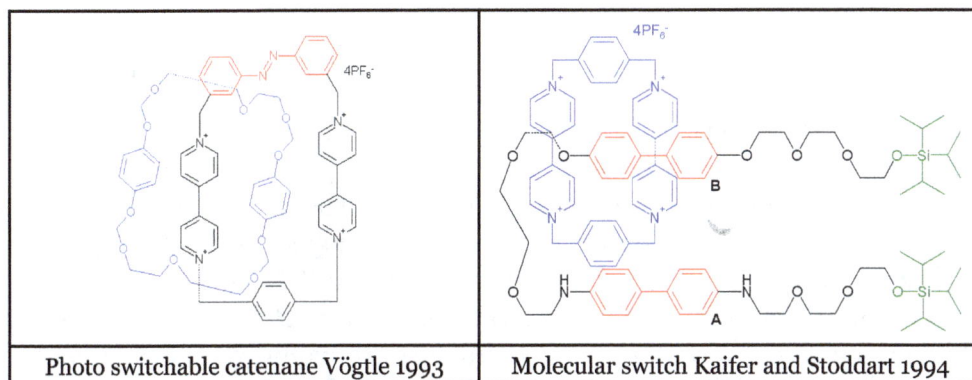

| Photo switchable catenane Vögtle 1993 | Molecular switch Kaifer and Stoddart 1994 |

This compound is based on two ring systems: one ring holds the photoswichable azo-benzene ring and two paraquat docking stations and the other ring is a polyether with to arene rings with binding affinity for the paraquat units. In this system NMR spectroscopy shows that in the azo trans-form the polyether ring is free to rotate around its partner ring but then when a light trigger activates the cis azo form this rotation mode is stopped

Kaifer and Stoddart in 1994 modify their molecular shuttle such a way that an electron-poor tetracationic cyclophane bead now has a choice between two docking stations: one biphenol and one benzidine unit. In solution at room temperature NMR spectroscopy reveals that the bead shuttles at a rate comparable to the NMR timescale, reducing the temperature to 229K resolves the signals with 84% of the population favoring the benzidine station. However, on addition of trifluoroacetic acid, the benzidine nitrogen atoms are protonated and the bead is fixed permanently on the biphenol station. The same effect is obtained by electrochemical oxidation (forming the benzidine radical ion) and significantly both processes are reversible.

In 2007 molecular shuttles are utilized in an experimental DRAM circuit. The device consists of 400 bottom silicon nanowire electrodes (16 nanometer (nm) wide at 33 nm intervals) crossed by another 400 titanium top-nanowires with similar dimensions sandwiching a monolayer of a bistable rotaxane depicted below:

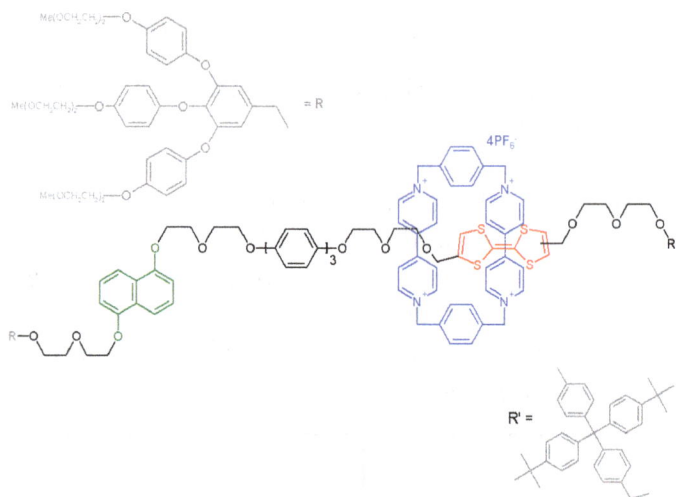

Each bit in the device consists of a silicon and a titanium crossbar with around 100 rotaxane molecules filling in the space between them at perpendicular angles. The hydrophilic diethylene glycol stopper on the left (gray) is specifically designed to anchor to the silicon wire (made hydrophilic by phosphorus doping) while the hydrophobic tetraarylmethane stopper on the right does the same to the likewise hydrophobic titanium wire. In the ground state of the switch, the paraquat ring is located around a tetrathiafulvalene unit (in red) but it moves to the dioxynaphthyl unit (in green) when the fulvalene unit is oxidized by application of a current. When the fulvalene is reduced back a metastable high conductance '1' state is formed which relaxes back to the ground state with a chemical half-life of around one hour. The problem of defects is circumvented by adopting a defect-tolerant architecture also found in the Teramac project. In this way a circuit is obtained consisting of 160,000 bits on an area the size of a white blood cell translating into 10^{11} bits per square centimeter.

Molecular Shuttle

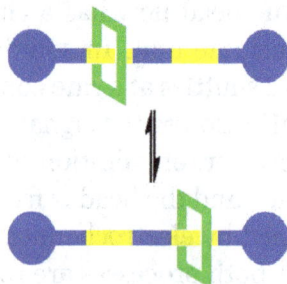

An example of a molecular shuttle where the macrocyle (green) moves between two stations (yellow).

A molecular shuttle in supramolecular chemistry is a special type of molecular machine capable of shuttling molecules or ions from one location to another. This field is of relevance to nanotechnology in its quest for nanoscale electronic components and also to biology where many biochemical functions are based on molecular shuttles. Academic interest also exists for synthetic molecular shuttles, the first prototype reported in 1991 based on a rotaxane.

This device is based on a molecular thread composed of an ethyleneglycol chain interrupted by two arene groups acting as so-called stations. The terminal units (or stoppers) on this wire are bulky triisopropylsilyl groups. The bead is a tetracationic cyclophane based on two bipyridine groups and two para-phenylene groups. The bead is locked to one of the stations by pi-pi interactions but since the activation energy for migration from one station to the other station is only 13 kcal/mol (54 kJ/mol) the bead shuttles between them. The stoppers prevent the bead from slipping from the thread. Chemical synthesis of this device is based on molecular self-assembly from a preformed thread and two bead fragments (32% chemical yield).

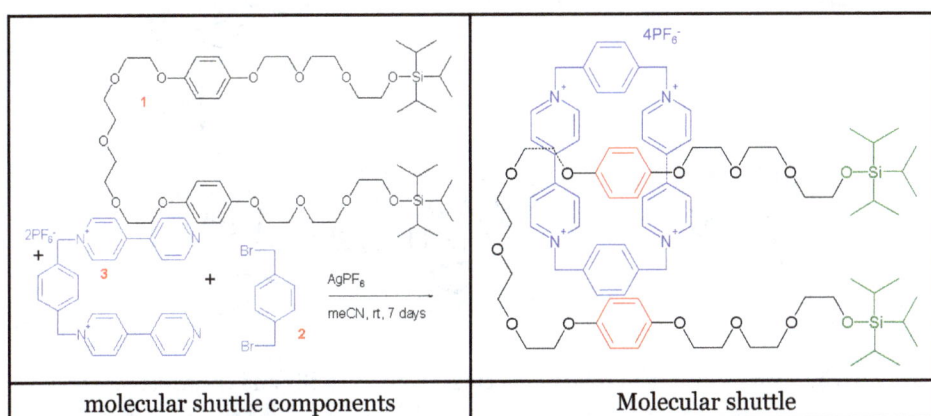

| molecular shuttle components | Molecular shuttle |

In certain molecular switches the two stations are non-degenerate.

Synthetic Molecular Motor

Synthetic molecular motors are molecular machines capable of rotation under energy input. Although the term "molecular motor" has traditionally referred to a naturally occurring protein that induces motion (via protein dynamics), some groups also use the term when referring to non-biological, non-peptide synthetic motors. Many chemists are pursuing the synthesis of such molecular motors. The prospect of synthetic molecular motors was first raised by the nanotechnology pioneer Richard Feynman in 1959 in his talk *There's Plenty of Room at the Bottom*.

The basic requirements for a synthetic motor are repetitive 360° motion, the consumption of energy and unidirectional rotation. The first two efforts in this direction, the chemically driven motor by Dr. T. Ross Kelly of Boston College with co-workers and the light-driven motor by Feringa and co-workers, were published in 1999 in the same issue of Nature. In 2008 Petr Král and co-workers proposed electron tunneling motors continuously rotated by a permanent torque, opening the possibility of practical realization of a real molecular motor machine. It is expected that reports of more efforts in this field will increase as understanding of chemistry and physics at the nanolevel improves.

Chemically Driven Rotary Molecular Motors

An example of a synthetic chemically driven rotary molecular motor was reported by Kelly and co-workers in 1999. Their system is made up from a three-bladed triptycene rotor and a helicene, and is capable of performing a unidirectional 120° rotation.

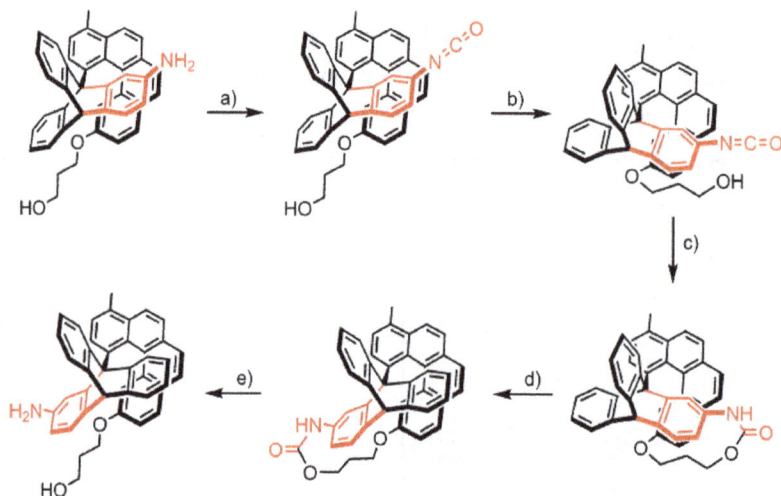

The chemically driven rotary molecular motor by Kelly and co-workers.

This rotation takes place in five steps. The amine group present on the triptycene moiety is converted to an isocyanate group by condensation with phosgene (a). Thermal or spontaneous rotation around the central bond then brings the isocyanate group in proximity of the hydroxyl group located on the helicene moiety (b), thereby allowing these two groups to react with each other (c). This reaction irreversibly traps the system as a strained cyclic urethane that is higher in energy and thus energetically closer to the rotational energy barrier than the original state. Further rotation of the triptycene moiety therefore requires only a relatively small amount of thermal activation in order to overcome this barrier, thereby releasing the strain (d). Finally, cleavage of the urethane group restores the amine and alcohol functionalities of the molecule (e).

The result of this sequence of events is a unidirectional 120° rotation of the triptycene moiety with respect to the helicene moiety. Additional forward or backward rotation of the triptycene rotor is inhibited by the helicene moiety, which serves a function similar to that of the pawl of a ratchet. The unidirectionality of the system is a result from both the asymmetric skew of the helicene moiety as well as the strain of the cyclic urethane which is formed in c. This strain can be only be lowered by the clockwise rotation of the triptycene rotor in d, as both counterclockwise rotation as well as the inverse process of d are energetically unfavorable. In this respect the preference for the rotation direction is determined by both the positions of the functional groups and the shape of the helicene and is thus built into the design of the molecule instead of dictated by external factors.

The motor by Kelly and co-workers is an elegant example of how chemical energy can be used to induce controlled, unidirectional rotational motion, a process which resembles the consumption of ATP in organisms in order to fuel numerous processes. However, it does suffer from a serious drawback: the sequence of events that leads to 120° rotation is not repeatable. Kelly and co-workers have therefore searched for ways to extend the system so that this sequence can be carried out repeatedly. Unfortunately, their attempts to accomplish this objective have not been successful and currently the project has been abandoned.

Two additional examples of synthetic chemically driven rotary molecular motors that have been reported in literature make use of the stereoselective ring opening of a racemic biaryl lactone by the use of chiral reagents, which results in a directed 90° rotation of one aryl with respect to the other aryl. Branchaud and co-workers have reported that this approach, followed by an additional ring closing step, can be used to accomplish a non-repeatable 180° rotation. Feringa and co-workers used this approach in their design of a molecule that can repeatably perform 360° rotation. The full rotation of this molecular motor takes place in four stages. In stages A and C rotation of the aryl moiety is restricted, although helix inversion is possible. In stages B and D the aryl can rotate with respect to the naphthalene with steric interactions preventing the aryl from passing the naphthalene. The rotary cycle consists of four chemically induced steps which realize the conversion of one stage into the next. Steps 1 and 3 are asymmetric ring opening reactions which make use of a chiral reagent in order to control the direction of the rotation of the aryl. Steps 2 and 4 consist of the deprotection of the phenol, followed by regioselective ring formation. So far this molecular motor is the only reported example of a fully chemically driven artificial rotary molecular motor that is capable of 360° rotation.

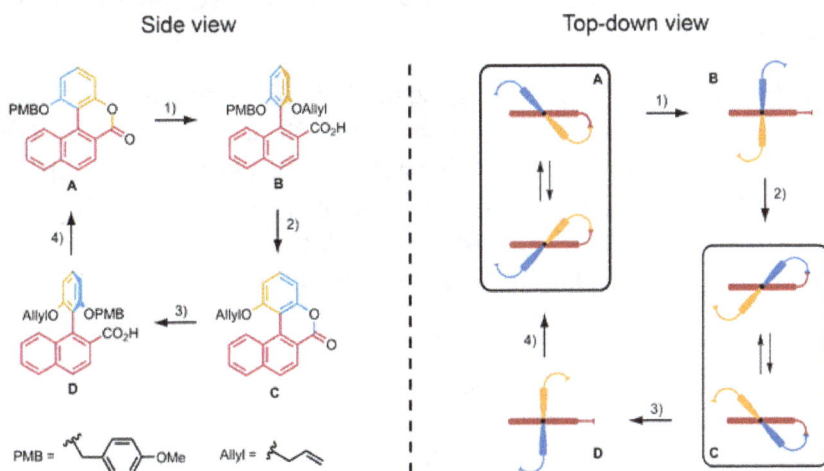

The chemically driven rotary molecular motor by Feringa and co-workers.

Light-driven Rotary Molecular Motors

In 1999 the laboratory of Prof. Dr. Ben L. Feringa at the University of Groningen, The Netherlands reported the creation of a unidirectional molecular rotor. Their 360° molecular motor system consists of a bis-helicene connected by an alkene double bond displaying axial chirality and having two stereocenters.

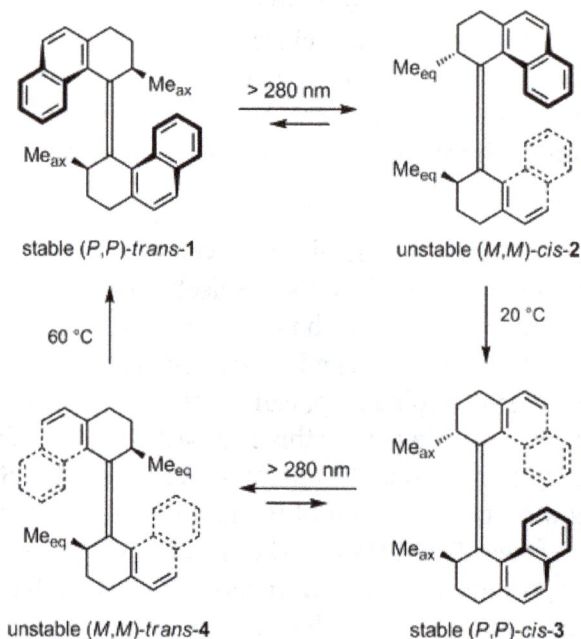

Rotary cycle of the light-driven rotary molecular motor by Feringa and co-workers.

One cycle of unidirectional rotation takes 4 reaction steps. The first step is a low temperature endothermic photoisomerization of the trans (P,P) isomer 1 to the cis (M,M) 2 where P stands for the right-handed helix and M for the left-handed helix. In this process, the two axial methyl groups are converted into two less sterically favorable equatorial methyl groups.

By increasing the temperature to 20 °C these methyl groups convert back exothermally to the (P,P) cis axial groups (3) in a helix inversion. Because the axial isomer is more stable than the equatorial isomer, reverse rotation is blocked. A second photoisomerization converts (P,P) cis 3 into (M,M) trans 4, again with accompanying formation of sterically unfavorable equatorial methyl groups. A thermal isomerization process at 60 °C closes the 360° cycle back to the axial positions.

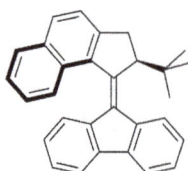

Synthetic molecular motors: fluorene system

A major hurdle to overcome is the long reaction time for complete rotation in these systems, which does not compare to rotation speeds displayed by motor proteins in biological systems. In the fastest system to date, with a fluorene lower half, the half-life of the thermal helix inversion is 0.005 seconds. This compound is synthesized using the Barton-Kellogg reaction. In this molecule the slowest step in its rotation, the thermally induced helix-inversion, is believed to proceed much more quickly because the larger *tert*-butyl group makes the unstable isomer even less stable than when the methyl group is used. This is because the unstable isomer is more destabilized than the transition state that leads to helix-inversion. The different behaviour of the two molecules is illustrated by the fact that the half-life time for the compound with a methyl group instead of a *tert*-butyl group is 3.2 minutes.

The Feringa principle has been incorporated into a prototype nanocar. The car synthesized has a helicene-derived engine with an oligo (phenylene ethynylene) chassis and four carborane wheels and is expected to be able to move on a solid surface with scanning tunneling microscopy monitoring, although so far this has not been observed. The motor does not perform with fullerene wheels because they quench the photochemistry of the motor moiety. Feringa motors have also been shown to remain operable when chemically attached to solid surfaces. The ability of certain Feringa systems to act as an asymmetric catalyst has also been demonstrated.

In 2016 Feringa was awarded with a Noble prize for his work on molecular motors.

Experimental Demonstration of A Single Molecule-Electric Motor

A single molecule electric motor has been reported that is an electrically operated motor made from a single butyl methyl sulphide molecule. The molecule is adsorbed onto a Copper (111) single crystal piece by chemisorption.

Nanocar

The nanocar is a molecule designed in 2005 at Rice University by a group headed by Professor James Tour. Despite the name, the original nanocar does not contain a molecular motor, hence, it is not really a car. Rather, it was designed to answer the question of how fullerenes move about on metal surfaces; specifically, whether they roll or slide (they roll).

The molecule consists of an H-shaped 'chassis' with fullerene groups attached at the four corners to act as wheels.

When dispersed on a gold surface, the molecules attach themselves to the surface via their fullerene groups and are detected via scanning tunneling microscopy. One can deduce their orientation as the frame length is a little shorter than its width.

Upon heating the surface to 200 °C the molecules move forward and back as they roll on their fullerene "wheels". The nanocar is able to roll about because the fullerene wheel is fitted to the alkyne "axle" through a carbon-carbon single bond. The hydrogen on the neighboring carbon is no great obstacle to free rotation. When the temperature is high enough, the four carbon-carbon bonds rotate and the car rolls about. Occasionally the direction of movement changes as the molecule pivots. The rolling action was confirmed by Professor Kevin Kelly, also at Rice, by pulling the molecule with the tip of the STM.

Independent Early Conceptual Contribution

The concept of a nanocar built out of molecular "tinkertoys" was first hypothesized at the Fifth Foresight Conference on Molecular Nanotechnology (November 1997). Subsequently an expanded version was published in Annals of Improbable Research. These papers supposed to be a not-so-serious contribution to a fundamental debate on the limits of bottom-up Drexlerian nanotechnology and conceptual limits of how far mechanistic analogies advanced by Eric Drexler could be carried out. The important feature of this nanocar concept was the fact that all molecular component tinkertoys were known and synthetized molecules (alas, some very exotic and only recently discovered, e.g. staffanes, and notably – ferric wheel, 1995), in contrast to some Drexlerian diamondoid structures that were only postulated and never synthesized; and the drive system that was embedded in a ferric wheel and driven by inhomogeneous or time-dependent magnetic field of a substrate – an "engine in a wheel" concept.

Nanodragster

The Nanodragster, dubbed the world's smallest hot rod, is a molecular nanocar. The design improves on previous nanocar designs and is a step towards creating molecular machines. The name comes from the nanocar's resemblance to a dragster, as it has a shorter axle with smaller wheels in the front and a larger axle with larger wheels in the back.

The nanocar was developed at Rice University's Richard E. Smalley Institute Nanoscale Science and Technology by the team of James Tour, Kevin Kelly and other colleagues involved in its research. The previous nanocar developed was 3 to 4 nanometers which was a little over a strand of DNA and was around 20,000 times thinner than a human hair. These nanocars were built with carbon buckyballs for their four wheels, which made it need 400 °F (200 °C) to get it moving. On the other hand, a nanocar which utilized p-carborane wheels moves as if on ice. Such observations led to the production of nanocars which had both wheel designs.

The Nanodragster is 50,000 times thinner than a human hair and has a top speed of 0.014 millimeters per hour (0.0006 in/h). The rear wheels are spherical fullerene molecules, or buckyballs, composed of sixty carbon atoms each, which are attracted to a dragstrip that is made up of a very fine layer of gold. This design also enabled Tour's team to operate the device at lower temperatures.

The nanodragster and other nano-machines are designed for use in transporting items. The technology can be used in manufacturing computer circuits and electronic components, or in conjunction with pharmaceuticals inside the human body. Tour also speculated that the knowledge gained from the nanocar research would help build efficient catalytic systems in the future.

Electrically Driven Directional Motion of A Four-wheel Molecule on A Metal Surface

Kudernac *et al.* described a specially designed molecule that has four motorized "wheels". By depositing the molecule on a copper surface and providing them with sufficient energy from electrons of a scanning tunnelling microscope they were able to drive some of the molecules in a specific direction, much like a car, being the first single molecule capable to continue moving in the same direction across a surface. Inelastic electron tunnelling induces conformational changes in the rotors and propels the molecule across a copper surface. By changing the direction of the rotary motion of individual motor units, the self-propelling molecular 'four-wheeler' structure can follow random or preferentially linear trajectories. This design provides a starting point for the exploration of more sophisticated molecular mechanical systems, perhaps with complete control over their direction of motion.

Motor Nanocar

A future nanocar with a synthetic molecular motor has been developed by Jean-Francois Morin *et al.* It is fitted with carborane wheels and a light powered helicene synthetic molecular motor. Although the motor moiety displayed unidirectional rotation in solution, light-driven motion on a surface has yet to be observed. Motility in water and other liquids can be also realized by a molecular propeller in the future.

Nanomotor

A nanomotor is a molecular or nanoscale device capable of converting energy into movement. It can typically generate forces on the order of piconewtons.

While nanoparticles have been utilized by artists for centuries, such as in the famous Lycurgus cup, actual research into nanotechnology did not come about until recently. In 1959, Richard Feynman gave a famous talk entitled "There's Plenty of Room at the Bottom" at the American Physical Society's conference hosted at Caltech. He went on to wage a scientific bet that no one person could design a motor smaller than 400 μm on any side. The purpose of the bet (as with most scientific bets) was to inspire scientists to develop new technologies, and anyone who could develop a nanomotor could claim the $1,000 USD prize. However, his purpose was thwarted by William McLellan, who fabricated a nanomotor without developing new methods. Nonetheless, Richard Feynman's speech inspired a new generation scientists to pursue research into nanotechnology.

Nanomotors are the focus of research for their ability to overcome microfluidic dynamics present at low Reynold's numbers. Scallop Theory is the basis for nanomotors to produce motion at low Reynold's numbers. The motion is achieved by breaking different symmetries. In addition, Brownian motion must be considered because particle interaction can dramatically impact the ability of a nanomotor to traverse through a liquid. This can pose a significant problem when designing new nanomotors. Current nanomotor research seeks to overcome these problems, and by doing so can improve current microfluidic devices or give rise to new technologies.

Nanotube and Nanowire Motors

In 2004, Ayusman Sen and Thomas E. Mallouk fabricated the first synthetic and autonomous nanomotor. The two micron long nanomotors were composed of two segments, platinum and gold, that could catalytically react with diluted hydrogen peroxide in water to produce motion. The Au-Pt nanomotors have autonomous, non-Brownian motion that stems from the propulsion via catalytic generation of chemical gradients. As implied, their motion does not require the present of an external magnetic, electric or optical field to guide their motion. Joseph Wang in 2008 was able to dramatically enhance the motion of Au-Pt catalytic nanomotors by incorporating carbon nanotubes into the platinum segment.

Since 2004, different types of nanotube and nanowire based motors have been developed. In Dresdan Germany, rolled-up microtube nanomotors produced motion by harnessing the bubbles in catalytic reactions. The bubble-induced propulsion enables motor movement in relevant biological fluids, but typically requies toxic fuels such as hydrogen peroxide. This has limited their in vitro applications. Further research into catalytical nanomotors holds major promise for important cargo-towing applications, ranging from cell sorting microchip devices to directed drug delivery. Such in-vivo applications of microtube motors were described for the first time by Joseph Wang and Liangfang Zhang using gastric acid as fuel.

Enzymatic Nanomotors

Recently, there has been more research into developing enzymatic nanomotors and micropumps. At low Reynold's numbers, single molecule enzymes could act as autonomous nanomotors. Ayusman Sen and Samudra Sengupta demonstrated self-powered, autonomous micropumps can enhance particle transportation. The proof-of-concept demonstrates that enzymes can be successfully utilized as an "engine" in nanomotors. Developing enzyme-driven nanomotors promises to inspire new biocompatible technologies and medical applications.

A proposed branch of research is the integration of molecular motor proteins found in living cells into molecular motors implanted in artificial devices. Such a motor protein would be able to move a "cargo" within that device, via protein dynamics, similarly to how kinesin moves various molecules along tracks of microtubules inside cells. Starting and stopping the movement of such motor proteins would involve caging the ATP in molecular structures sensitive to UV light. Pulses of UV illumination would thus provide pulses of movement. DNA nanomachines, based on changes between two molecular conformations of DNA in response to various external triggers, have also been described.

8

Impacts of Nanotechnology

Nanotechnology has a number of impacts; it extends from medical to ethical to mental to environmental applications. The main benefits of nanotechnology are water purification systems, energy systems, physical enhancement and nanomedicine. In order to clearly understand nanotechnology it is necessary to understand the impacts of nanotechnology.

Impact of Nanotechnology

The impact of nanotechnology extends from its medical, ethical, mental, legal and environmental applications, to fields such as engineering, biology, chemistry, computing, materials science, and communications.

Major benefits of nanotechnology include improved manufacturing methods, water purification systems, energy systems, physical enhancement, nanomedicine, better food production methods, nutrition and large-scale infrastructure auto-fabrication. Nanotechnology's reduced size may allow for automation of tasks which were previously inaccessible due to physical restrictions, which in turn may reduce labor, land, or maintenance requirements placed on humans.

Potential risks include environmental, health, and safety issues; transitional effects such as displacement of traditional industries as the products of nanotechnology become dominant, which are of concern to privacy rights advocates. These may be particularly important if potential negative effects of nanoparticles are overlooked.

Whether nanotechnology merits special government regulation is a controversial issue. Regulatory bodies such as the United States Environmental Protection Agency and the Health and Consumer Protection Directorate of the European Commission have started dealing with the potential risks of nanoparticles. The organic food sector has been the first to act with the regulated exclusion of engineered nanoparticles from certified organic produce, firstly in Australia and the UK, and more recently in Canada, as well as for all food certified to Demeter International standards

Overview

The presence of nanomaterials (materials that contain nanoparticles) is not in itself a threat. It is only certain aspects that can make them risky, in particular their mobility and their increased reactivity. Only if certain properties of certain nanoparticles were harmful to living beings or the environment would we be faced with a genuine hazard. In this case it can be called nanopollution.

In addressing the health and environmental impact of nanomaterials we need to differentiate between two types of nanostructures: (1) Nanocomposites, nanostructured surfaces and nanocomponents (electronic, optical, sensors etc.), where nanoscale particles are incorporated into a substance, material or device ("fixed" nano-particles); and (2) "free" nanoparticles, where at some stage in production or use individual nanoparticles of a substance are present. These free nanoparticles could be nanoscale species of elements, or simple compounds, but also complex compounds where for instance a nanoparticle of a particular element is coated with another substance ("coated" nanoparticle or "core-shell" nanoparticle).

There seems to be consensus that, although one should be aware of materials containing fixed nanoparticles, the immediate concern is with free nanoparticles.

Nanoparticles are very different from their everyday counterparts, so their adverse effects cannot be derived from the known toxicity of the macro-sized material. This poses significant issues for addressing the health and environmental impact of free nanoparticles.

To complicate things further, in talking about nanoparticles it is important that a powder or liquid containing nanoparticles almost never be monodisperse, but contain instead a range of particle sizes. This complicates the experimental analysis as larger nanoparticles might have different properties from smaller ones. Also, nanoparticles show a tendency to aggregate, and such aggregates often behave differently from individual nanoparticles.

Health Impact

The health impacts of nanotechnology are the possible effects that the use of nanotechnological materials and devices will have on human health. As nanotechnology is an emerging field, there is great debate regarding to what extent nanotechnology will benefit or pose risks for human health. Nanotechnology's health impacts can be split into two aspects: the potential for nanotechnological innovations to have medical applications to cure disease, and the potential health hazards posed by exposure to nanomaterials.

Medical Applications

Nanomedicine is the medical application of nanotechnology. The approaches to nanomedicine range from the medical use of nanomaterials, to nanoelectronic biosensors, and even possible future applications of molecular nanotechnology. Nanomedicine seeks to deliver a valuable set of research tools and clinically helpful devices in the near future. The National Nanotechnology Initiative expects new commercial applications in the pharmaceutical industry that may include advanced drug delivery systems, new therapies, and in vivo imaging. Neuro-electronic interfaces and other nanoelectronics-based sensors are another active goal of research. Further down the line, the speculative field of molecular nanotechnology believes that cell repair machines could revolutionize medicine and the medical field.

Nanomedicine seeks to deliver a set of research tools and clinical devices in the near future. The National Nanotechnology Initiative expects new commercial applications in the pharmaceutical industry that may include advanced drug delivery systems, new therapies, and *in vivo* imaging. Neuro-electronic interfaces and other nanoelectronics-based sensors are another active goal of research. Further down the line, the speculative field of molecular nanotechnology believes that cell repair machines could revolutionize medicine and the medical field.

Nanomedicine research is directly funded, with the US National Institutes of Health in 2005 funding a five-year plan to set up four nanomedicine centers. In April 2006, the journal Nature Materials estimated that 130 nanotech-based drugs and delivery systems were being developed worldwide. Nanomedicine is a large industry, with nanomedicine sales reaching $6.8 billion in 2004. With over 200 companies and 38 products worldwide, a minimum of $3.8 billion in nanotechnology R&D is being invested every year. As the nanomedicine industry continues to grow, it is expected to have a significant impact on the economy.

Health Hazards

Nanotoxicology is the field which studies potential health risks of nanomaterials. The extremely small size of nanomaterials means that they are much more readily taken up by the human body than larger sized particles. How these nanoparticles behave inside the organism is one of the significant issues that needs to be resolved. The behavior of nanoparticles is a function of their size, shape and surface reactivity with the surrounding tissue. Apart from what happens if non-degradable or slowly degradable nanoparticles accumulate in organs, another concern is their potential interaction with biological processes inside the body: because of their large surface, nanoparticles on exposure to tissue and fluids will immediately adsorb onto their surface some of the macromolecules they encounter. The large number of variables influencing toxicity means that it is difficult to generalise about health risks associated with exposure to nanomaterials – each new nanomaterial must be assessed individually and all material properties must be taken into account. Health and environmental issues combine in the workplace of companies engaged in producing or using nanomaterials and in the laboratories engaged in nanoscience and nanotechnology research. It is safe to say that current workplace exposure standards for dusts cannot be applied directly to nanoparticle dusts.

The extremely small size of nanomaterials also means that they are much more readily taken up by the human body than larger sized particles. How these nanoparticles behave inside the body is one of the issues that needs to be resolved. The behavior of nanoparticles is a function of their size, shape and surface reactivity with the surrounding tissue. They could cause overload on phagocytes, cells that ingest and destroy foreign matter, thereby triggering stress reactions that lead to inflammation and weaken the body's defense against other pathogens. Apart from what happens if non-degradable or slowly degradable nanoparticles accumulate in organs, another concern is their potential interaction with biological processes inside the body: because of their large surface, nanoparticles on exposure to tissue and fluids will immediately adsorb onto their surface some of the macromolecules they encounter. This may, for instance, affect the regulatory mechanisms of enzymes and other proteins.

The National Institute for Occupational Safety and Health has conducted initial research on how nanoparticles interact with the body's systems and how workers might be exposed to nano-sized particles in the manufacturing or industrial use of nanomaterials. NIOSH currently offers interim guidelines for working with nanomaterials consistent with the best scientific knowledge. At The National Personal Protective Technology Laboratory of NIOSH, studies investigating the filter penetration of nanoparticles on NIOSH-certified and EU marked respirators, as well as non-certified dust masks have been conducted. These studies found that the most penetrating particle size range was between 30 and 100 nanometers, and leak size was the largest factor in the number of nanoparticles found inside the respirators of the test dummies.

Other properties of nanomaterials that influence toxicity include: chemical composition, shape, surface structure, surface charge, aggregation and solubility, and the presence or absence of functional groups of other chemicals. The large number of variables influencing toxicity means that it is difficult to generalise about health risks associated with exposure to nanomaterials – each new nanomaterial must be assessed individually and all material properties must be taken into account.

Literature reviews have been showing that release of engineered nanoparticles and incurred personal exposure can happen during different work activities. The situation alerts regulatory bodies to necessitate prevention strategies and regulations at nanotechnology workplaces.

Environmental Impact

The environmental impact of nanotechnology is the possible effects that the use of nanotechnological materials and devices will have on the environment. As nanotechnology is an emerging field, there is great debate regarding to what extent industrial and commercial use of nanomaterials will affect organisms and ecosystems.

Nanotechnology's environmental impact can be split into two aspects: the potential for nanotechnological innovations to help improve the environment, and the possibly novel type of pollution that nanotechnological materials might cause if released into the environment.

Environmental Applications

Green nanotechnology refers to the use of nanotechnology to enhance the environmental sustainability of processes producing negative externalities. It also refers to the use of the products of nanotechnology to enhance sustainability. It includes making green nano-products and using nano-products in support of sustainability. Green nanotechnology has been described as the development of clean technologies, "to minimize potential environmental and human health risks associated with the manufacture and use of nanotechnology products, and to encourage replacement of existing products with new nano-products that are more environmentally friendly throughout their lifecycle."

Green nanotechnology has two goals: producing nanomaterials and products without harming the environment or human health, and producing nano-products that provide solutions to environmental problems. It uses existing principles of green chemistry and green engineering to make nanomaterials and nano-products without toxic ingredients, at low temperatures using less energy and renewable inputs wherever possible, and using lifecycle thinking in all design and engineering stages.

Pollution

Nanopollution is a generic name for all waste generated by nanodevices or during the nanomaterials manufacturing process. Nanowaste is mainly the group of particles that are released into the environment, or the particles that are thrown away when still on their products.

Societal Impact

Beyond the toxicity risks to human health and the environment which are associated with first-generation nanomaterials, nanotechnology has broader societal impact and poses broader social challenges. Social scientists have suggested that nanotechnology's social issues should be understood and assessed not simply as "downstream" risks or impacts. Rather, the challenges should be factored into "upstream" research and decision-making in order to ensure technology development that meets social objectives

Many social scientists and organizations in civil society suggest that technology assessment and governance should also involve public participation

Over 800 nano-related patents were granted in 2003, with numbers increasing to nearly 19,000 internationally by 2012. Corporations are already taking out broad-ranging patents on nanoscale discoveries and inventions. For example, two corporations, NEC and IBM, hold the basic patents on carbon nanotubes, one of the current cornerstones of nanotechnology. Carbon nanotubes have a wide range of uses, and look set to become crucial to several industries from electronics and computers, to strengthened materials to drug delivery and diagnostics. Carbon nanotubes are poised to become a major traded commodity with the potential to replace major conventional raw materials.

Nanotechnologies may provide new solutions for the millions of people in developing countries who lack access to basic services, such as safe water, reliable energy, health care, and education. The 2004 UN Task Force on Science, Technology and Innovation noted that some of the advantages of nanotechnology include production using little labor, land, or maintenance, high productivity, low cost, and modest requirements for materials and energy. However, concerns are frequently raised that the claimed benefits of nanotechnology will not be evenly distributed, and that any benefits (including technical and/or economic) associated with nanotechnology will only reach affluent nations.

Longer-term concerns center on the impact that new technologies will have for society at large, and whether these could possibly lead to either a post-scarcity economy, or alternatively exacerbate the wealth gap between developed and developing nations. The effects of nanotechnology on the society as a whole, on human health and the environment, on trade, on security, on food systems and even on the definition of "human", have not been characterized or politicized.

Regulation

Significant debate exists relating to the question of whether nanotechnology or nanotechnology-based products merit special government regulation. This debate is related to the circumstances in which it is necessary and appropriate to assess new substances prior to their release into the market, community and environment.

Regulatory bodies such as the United States Environmental Protection Agency and the Food and Drug Administration in the U.S. or the Health & Consumer Protection Directorate of the European Commission have started dealing with the potential risks posed by nanoparticles. So far, neither engineered nanoparticles nor the products and materials that contain them are subject to any special regulation regarding production, handling or labelling. The Material Safety Data Sheet that must be issued for some materials often does not differentiate between bulk and nanoscale size of the material in question and even when it does these MSDS are advisory only.

Limited nanotechnology labeling and regulation may exacerbate potential human and environmental health and safety issues associated with nanotechnology. It has been argued that the development of comprehensive regulation of nanotechnology will be vital to ensure that the potential risks associated with the research and commercial application of nanotechnology do not overshadow its potential benefits. Regulation may also be required to meet community expectations about responsible development of nanotechnology, as well as ensuring that public interests are included in shaping the development of nanotechnology.

In "The Consumer Product Safety Commission and Nanotechnology," E. Marla Felcher suggests that the Consumer Product Safety Commission, which is charged with protecting the public against unreasonable risks of injury or death associated with consumer products, is ill-equipped to oversee the safety of complex, high-tech products made using nanotechnology.

Societal Impact of Nanotechnology

The societal impact of nanotechnology are the potential benefits and challenges that the introduction of novel nanotechnological devices and materials may hold for society and human interaction. The term is sometimes expanded to also include nanotechnology's health and environmental impact, but this article will only consider the social and political impact of nanotechnology.

As nanotechnology is an emerging field and most of its applications are still speculative, there is much debate about what positive and negative effects that nanotechnology might have.

Overview

Beyond the toxicity risks to human health and the environment which are associated with first-generation nanomaterials, nanotechnology has broader societal implications and poses broader social challenges. Social scientists have suggested that nanotechnology's social issues should be understood and assessed not simply as "downstream" risks or impacts. Rather, the challenges should be factored into "upstream" research and decision making in order to ensure technology development that meets social objectives

Many social scientists and organizations in civil society suggest that technology assessment and governance should also involve public participation

Some observers suggest that nanotechnology will build incrementally, as did the 18-19th century industrial revolution, until it gathers pace to drive a nanotechnological revolution that will radically reshape our economies, our labor markets, international trade, international relations, social structures, civil liberties, our relationship with the natural world and even what we understand to be human. Others suggest that it may be more accurate to describe change driven by nanotechnology as a "technological tsunami". Just like a tsunami, analysts warn that rapid nanotechnology-driven change will necessarily have profound disruptive impacts. As the APEC Center for Technology Foresight observes:

If nanotechnology is going to revolutionize manufacturing, health care, energy supply, communications and probably defense, then it will transform labour and the workplace, the medical system, the transportation and power infrastructures and the military. None of these latter will be changed without significant social disruption.

Those concerned with the negative impact of nanotechnology suggest that it will simply exacerbate problems stemming from existing socio-economic inequity and unequal distributions of power, creating greater inequities between rich and poor through an inevitable nano-divide (the gap between those who control the new nanotechnologies and those whose products, services or labour are displaced by them). Analysts suggest the possibility that nanotechnology has the potential to destabilize international relations through a nano arms race and the increased potential for bioweaponry; thus, providing the tools for ubiquitous surveillance with significant implications for civil liberties. Also, many critics believe it might break down the barriers between life and non-life through nanobiotechnology, redefining even what it means to be human.

Nanoethicists posit that such a transformative technology could exacerbate the divisions of rich and poor – the so-called "nano divide." However nanotechnology makes the production of technology, e.g. computers, cellular phones, health technology etcetera, cheaper and therefore accessible to the poor.

In fact, many of the most enthusiastic proponents of nanotechnology, such as transhumanists, see the nascent science as a mechanism to changing human nature itself – going beyond curing disease and enhancing human characteristics. Discussions on nanoethics have been hosted by the federal government, especially in the context of "converging technologies" – a catch-phrase used to refer to nano, biotech, information technology, and cognitive science.

Possible Military Applications

Possible military applications of nanotechnology have been suggested in the fields of soldier enhancement () and chemical weapons amongst others. However, more socially disruptive weapon systems are to be expected from molecular manufacturing, a potential future form of nanotechnology that would make it possible to build complex structures at atomic precision. Molecular manufacturing requires significant advances in nanotechnology, but its supporters posit that once achieved it could produce highly advanced products at low costs and in large quantities in nanofactories weighing a kilogram or more. If nanofactories gain the ability to produce other nanofactories production may only be limited by relatively abundant factors such as input materials, energy and software.

Molecular manufacturing might be used to cheaply produce, among many other products, highly advanced, durable weapons. Being equipped with compact computers and motors these might be increasingly autonomous and have a large range of capabilities.

According to Chris Phoenix and Mike Treder from the Center for Responsible Nanotechnology as well as Anders Sandberg from the Future of Humanity Institute the military uses of molecular manufacturing are the applications of nanotechnology that pose the most significant global catastrophic risk. Several nanotechnology researchers state that the bulk of risk from nanotechnology comes from the potential to lead to war, arms races and destructive global government. Several reasons have been suggested why the availability of nanotech weaponry may with significant likelihood lead to unstable arms races (compared to e.g. nuclear arms races): (1) A large number of players may be tempted to enter the race since the threshold for doing so is low; (2) the ability to make weapons with molecular manufacturing might be cheap and easy to hide; (3) therefore lack of insight into the other parties' capabilities can tempt players to arm out of caution or to launch preemptive strikes; (4) molecular manufacturing may reduce dependency on international trade, a potential peace-promoting factor; (5) wars of aggression may pose a smaller economic threat to the aggressor since manufacturing is cheap and humans may not be needed on the battlefield.

Self-regulation by all state and non-state actors has been called hard to achieve, so measures to mitigate war-related risks have mainly been proposed in the area of international cooperation. International infrastructure may be expanded giving more sovereignty to the international level. This could help coordinate efforts for arms control. Some have put forth that international institutions dedicated specifically to nanotechnology (perhaps analogously to the International Atomic Energy Agency IAEA) or general arms control may also be designed. One may also jointly make differential technological progress on defensive technologies. The Center for Responsible Nanotechnology also suggest some technical restrictions. Improved transparency regarding technological capabilities may be another important facilitator for arms-control.

Intellectual Property Issues

On the structural level, critics of nanotechnology point to a new world of ownership and corporate control opened up by nanotechnology. The claim is that, just as biotechnology's ability to manipulate genes went hand in hand with the patenting of life, so too nanotechnology's ability to manipulate molecules has led to the patenting of matter. The last few years has seen a gold rush to claim patents at the nanoscale. Academics have warned that the resultant patent thicket is harming progress in the technology and have argued in the top journal *Nature* that there should be a moratorium on patents on "building block" nanotechnologies. Over 800 nano-related patents were granted in 2003, and the numbers are increasing year to year. Corporations are already taking out broad-ranging patents on nanoscale discoveries and inventions. For example, two corporations, NEC and IBM, hold the basic patents on carbon nanotubes, one of the current cornerstones of nanotechnology. Carbon nanotubes have a wide range of uses, and look set to become crucial to several industries from electronics and computers, to strengthened materials to drug delivery and diagnostics. Carbon nanotubes are poised to become a major traded commodity with the potential to replace major conventional raw materials. However, as their use expands, anyone seeking to (legally) manufacture or sell carbon nanotubes, no matter what the application, must first buy a license from NEC or IBM.

The United States' essential facilities doctrine may be of importance as well as other anti-trust laws.

Potential Benefits and Risks for Developing Countries

Nanotechnologies may provide new solutions for the millions of people in developing countries who lack access to basic services, such as safe water, reliable energy, health care, and education. The United Nations has set Millennium Development Goals for meeting these needs. The 2004 UN Task Force on Science, Technology and Innovation noted that some of the advantages of nanotechnology include production using little labor, land, or maintenance, high productivity, low cost, and modest requirements for materials and energy.

Many developing countries, for example Costa Rica, Chile, Bangladesh, Thailand, and Malaysia, are investing considerable resources in research and development of nanotechnologies. Emerging economies such as Brazil, China, India and South Africa are spending millions of US dollars annually on R&D, and are rapidly increasing their scientific output as demonstrated by their increasing numbers of publications in peer-reviewed scientific publications.

Potential opportunities of nanotechnologies to help address critical international development priorities include improved water purification systems, energy systems, medicine and pharmaceuticals, food production and nutrition, and information and communications technologies. Nanotechnologies are already incorporated in products that are on the market. Other nanotechnologies are still in the research phase, while others are concepts that are years or decades away from development.

Applying nanotechnologies in developing countries raises similar questions about the environmental, health, and societal risks described in the previous section. Additional challenges have been raised regarding the linkages between nanotechnology and development.

Protection of the environment, human health and worker safety in developing countries often suffers from a combination of factors that can include but are not limited to lack of robust environmental, human health, and worker safety regulations; poorly or unenforced regulation which is linked to a lack of physical (e.g., equipment) and human capacity (i.e., properly trained regulatory staff). Often, these nations require assistance, particularly financial assistance, to develop the scientific and institutional capacity to adequately assess and manage risks, including the necessary infrastructure such as laboratories and technology for detection.

Very little is known about the risks and broader impacts of nanotechnology. At a time of great uncertainty over the impacts of nanotechnology it will be challenging for governments, companies, civil society organizations, and the general public in developing countries, as in developed countries, to make decisions about the governance of nanotechnology.

Companies, and to a lesser extent governments and universities, are receiving patents on nanotechnology. The rapid increase in patenting of nanotechnology is illustrated by the fact that in the US, there were 500 nanotechnology patent applications in 1998 and 1,300 in 2000. Some patents are very broadly defined, which has raised concern among some groups that the rush to patent could slow innovation and drive up costs of products, thus reducing the potential for innovations that could benefit low income populations in developing countries.

There is a clear link between commodities and poverty. Many least developed countries are dependent on a few commodities for employment, government revenue, and export earnings. Many applications of nanotechnology are being developed that could impact global demand for specific commodities. For instance, certain nanoscale materials could enhance the strength and durability of rubber, which might eventually lead to a decrease in demand for natural rubber. Other nanotechnology applications may result in increases in demand for certain commodities. For example, demand for titanium may increase as a result of new uses for nanoscale titanium oxides, such as titanium dioxide nanotubes that can be used to produce and store hydrogen for use as fuel. Various organizations have called for international dialogue on mechanisms that will allow developing countries to anticipate and proactively adjust to these changes.

In 2003, Meridian Institute began the Global Dialogue on Nanotechnology and the Poor: Opportunities and Risks (GDNP) to raise awareness of the opportunities and risks of nanotechnology for developing countries, close the gaps within and between sectors of society to catalyze actions that address specific opportunities and risks of nanotechnology for developing countries, and identify ways that science and technology can play an appropriate role in the development process. The GDNP has released several publicly accessible papers on nanotechnology and development, including "Nanotechnology and the Poor: Opportunities and Risks - Closing the Gaps Within and Between Sectors of Society"; "Nanotechnology, Water, and Development"; and "Overview and Comparison of Conventional and Nano-Based Water Treatment Technologies".

Social Justice and Civil Liberties

Concerns are frequently raised that the claimed benefits of nanotechnology will not be evenly distributed, and that any benefits (including technical and/or economic) associated with nanotechnology will only reach affluent nations. The majority of nanotechnology research and development - and patents for nanomaterials and products - is concentrated in developed countries (including the United States, Japan, Germany, Canada and France). In addition, most patents related to nanotechnology are concentrated amongst few multinational corporations, including IBM, Micron Technologies, Advanced Micro Devices and Intel. This has led to fears that it will be unlikely that developing countries will have access to the infrastructure, funding and human resources required to support nanotechnology research and development, and that this is likely to exacerbate such inequalities.

Producers in developing countries could also be disadvantaged by the replacement of natural products (including rubber, cotton, coffee and tea) by developments in nano-technology. These natural products are important export crops for developing countries, and many farmers' livelihoods depend on them. It has been argued that their substitution with industrial nano-products could negatively impact the economies of developing countries, that have traditionally relied on these export crops.

It is proposed that nanotechnology can only be effective in alleviating poverty and aid development "when adapted to social, cultural and local institutional contexts, and chosen and designed with the active participation by citizens right from the commence-ment point" (Invernizzi et al. 2008, p. 132).

Effects on Laborers

Ray Kurzweil has speculated in *The Singularity is Near* that people who work in un-skilled labor jobs for a livelihood may become the first human workers to be displaced by the constant use of nanotechnology in the workplace, noting that layoffs often affect the jobs based around the lowest technology level before attacking jobs with the highest technology level possible. It has been noted that every major economic era has stimu-lated a global revolution both in the kinds of jobs that are available to people and the kind of training they need to achieve these jobs, and there is concern that the world's educational systems have lagged behind in preparing students for the "Nanotech Age".

It has also been speculated that nanotechnology may give rise to nanofactories which may have superior capabilities to conventional factories due to their small carbon and physical footprint on the global and regional environment. The miniaturization and transformation of the multi-acre conventional factory into the nanofactory may not interfere with their ability to deliver a high quality product; the product may be of even greater quality due to the lack of human errors in the production stages. Nanofactory systems may use precise atomic precisioning and contribute to making superior quality products that the "bulk chemistry" method used in 20th century and early 21st current-ly cannot produce. These advances might shift the computerized workforce in an even more complex direction, requiring skills in genetics, nanotechnology, and robotics.

Nanotoxicology

Nanotoxicology is the study of the toxicity of nanomaterials. Because of quantum size effects and large surface area to volume ratio, nanomaterials have unique properties compared with their larger counterparts.

Nanotoxicology is a branch of bionanoscience which deals with the study and ap-plication of toxicity of nanomaterials. Nanomaterials, even when made of inert elements like gold, become highly active at nanometer dimensions. Nanotoxicological studies are intended to determine whether and to what extent these properties may pose a threat to the environment and to human beings. For instance, Diesel nanoparticles have been found to damage the cardiovascular system in a mouse model.

Background

Nanotoxicology is a sub-specialty of particle toxicology. It addresses the toxicology of nanoparticles (particles <100 nm diameter) which appear to have toxicity effects that are unusual and not seen with larger particles. Nanoparticles can be divided into combustion-derived nanoparticles (like diesel soot), manufactured nanoparticles like carbon nanotubes and naturally occurring nanoparticles from volcanic eruptions, atmospheric chemistry etc. Typical nanoparticles that have been studied are titanium dioxide, alumina, zinc oxide, carbon black, and carbon nanotubes, and "nano-C_{60}". Nanoparticles have much larger surface area to unit mass ratios which in some cases may lead to greater pro-inflammatory effects (in, for example, lung tissue). In addition, some nanoparticles seem to be able to translocate from their site of deposition to distant sites such as the blood and the brain. This has resulted in a sea-change in how particle toxicology is viewed- instead of being confined to the lungs, nanoparticle toxicologists study the brain, blood, liver, skin and gut.

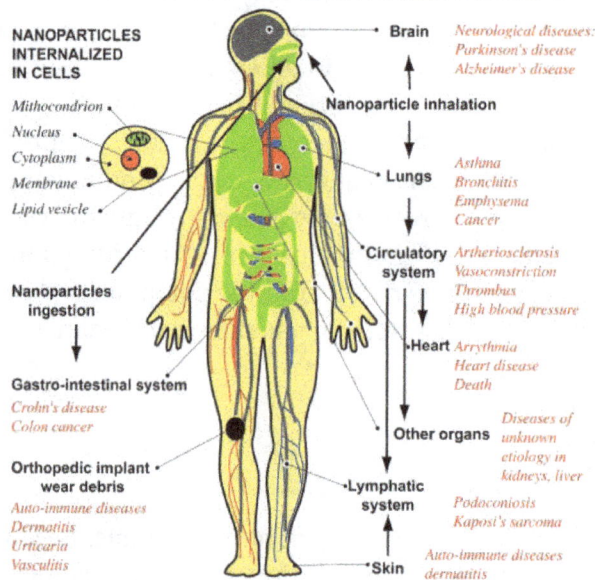

DISEASES ASSOCIATED TO NANOPARTICLE EXPOSURE

Pathways of exposure to nanoparticles and associated diseases as suggested by epidemiological, in vivo and in vitro studies.

Calls for tighter regulation of nanotechnology have arisen alongside a growing debate related to the human health and safety risks associated with nanotechnology. From a large-scale literature review, Yaobo Ding et al. found that release of airborne engineered nanoparticles and associated worker exposure from various production and handling activities at different workplaces are very probable. The Royal Society identifies the potential for nanoparticles to penetrate the skin, and recommends that the use of nanoparticles in cosmetics be conditional upon a favorable assessment by the relevant European Commission safety advisory committee. Andrew Maynard also reports that 'certain nanoparticles may move easily into sensitive lung tissues after inhalation, and cause damage that can lead to chronic breathing problems'.

Carbon nanotubes – characterized by their microscopic size and incredible tensile strength – are frequently likened to asbestos, due to their needle-like fiber shape. In a recent study that introduced carbon nanotubes into the abdominal cavity of mice, results demonstrated that long thin carbon nanotubes showed the same effects as long thin asbestos fibers, raising concerns that exposure to carbon nanotubes may lead to pleural abnormalities such as mesothelioma (cancer of the lining of the lungs caused by exposure to asbestos). Given these risks, effective and rigorous regulation has been called for to determine if, and under what circumstances, carbon nanotubes are manufactured, as well as ensuring their safe handling and disposal.

The Woodrow Wilson Centre's Project on Emerging Technologies conclude that there is insufficient funding for human health and safety research, and as a result there is currently limited understanding of the human health and safety risks associated with nanotechnology. While the US National Nanotechnology Initiative reports that around four percent (about $40 million) is dedicated to risk related research and development, the Woodrow Wilson Centre estimate that only around $11 million is actually directed towards risk related research. They argued in 2007 that it would be necessary to increase funding to a minimum of $50 million in the following two years so as to fill the gaps in knowledge in these areas.

The potential for workplace exposure was highlighted by the 2004 Royal Society report which recommended a review of existing regulations to assess and control workplace exposure to nanoparticles and nanotubes. The report expressed particular concern for the inhalation of large quantities of nanoparticles by workers involved in the manufacturing process.

Stakeholders concerned by the lack of a regulatory framework to assess and control risks associated with the release of nanoparticles and nanotubes have drawn parallels with bovine spongiform encephalopathy ('mad cow's disease'), thalidomide, genetically modified food, nuclear energy, reproductive technologies, biotechnology, and asbestosis. In light of such concerns, the Canadian based ETC Group have called for a moratorium on nano-related research until comprehensive regulatory frameworks are developed that will ensure workplace safety.

Reactive Oxygen Species

For some types of particles, the smaller they are, the greater their surface area to volume ratio and the higher their chemical reactivity and biological activity. The greater chemical reactivity of nanomaterials can result in increased production of reactive oxygen species (ROS), including free radicals. ROS production has been found in a diverse range of nanomaterials including carbon fullerenes, carbon nanotubes and nanoparticle metal oxides. ROS and free radical production is one of the primary mechanisms of nanoparticle toxicity; it may result in oxidative stress, inflammation, and consequent damage to proteins, membranes and DNA.

Biodistribution

The extremely small size of nanomaterials also means that they much more readily gain entry into the human body than larger sized particles. How these nanoparticles behave inside the body is still a major question that needs to be resolved. The behavior of nanoparticles is a function of their size, shape and surface reactivity with the surrounding tissue. In principle, a large number of particles could overload the body's phagocytes, cells that ingest and destroy foreign matter, thereby triggering stress reactions that lead to inflammation and weaken the body's defense against other pathogens. In addition to questions about what happens if non-degradable or slowly degradable nanoparticles accumulate in bodily organs, another concern is their potential interaction or interference with biological processes inside the body. Because of their large surface area, nanoparticles will, on exposure to tissue and fluids, immediately adsorb onto their surface some of the macromolecules they encounter. This may, for instance, affect the regulatory mechanisms of enzymes and other proteins.

Nanomaterials are able to cross biological membranes and access cells, tissues and organs that larger-sized particles normally cannot. Nanomaterials can gain access to the blood stream via inhalation or ingestion. At least some nanomaterials can penetrate the skin; even larger microparticles may penetrate skin when it is flexed. Broken skin is an ineffective particle barrier, suggesting that acne, eczema, shaving wounds or severe sunburn may accelerate skin uptake of nanomaterials. Then, once in the blood stream, nanomaterials can be transported around the body and be taken up by organs and tissues, including the brain, heart, liver, kidneys, spleen, bone marrow and nervous system. Nanomaterials have proved toxic to human tissue and cell cultures, resulting in increased oxidative stress, inflammatory cytokine production and cell death. Unlike larger particles, nanomaterials may be taken up by cell mitochondria and the cell nucleus. Studies demonstrate the potential for nanomaterials to cause DNA mutation and induce major structural damage to mitochondria, even resulting in cell death.

Nanotoxicity Studies

There is presently no authority to specifically regulate nanotech-based products. Scientific research has indicated the potential for some nanomaterials to be toxic to humans or the environment. In March 2004 tests conducted by environmental toxicologist Eva Oberdörster, Ph.D. working with Southern Methodist University in Texas, found extensive brain damage to fish exposed to fullerenes for a period of just 48 hours at a relatively moderate dose of 0.5 parts per million (commensurate with levels of other kinds of pollution found in bays). The fish also exhibited changed gene markers in their livers, indicating their entire physiology was affected. In a concurrent test, the fullerenes killed water fleas, an important link in the marine food chain. The extremely small size of fabricated nanomaterials also means that they are much more readily taken up by living tissue than presently known toxins. Nanoparticles can be inhaled, swallowed, absorbed through skin and deliberately or accidentally injected during medical procedures. They might be accidentally or inadvertently released from materials implanted into living tissue.

Researcher Shosaku Kashiwada of the National Institute for Environmental Studies in Tsukuba, Japan, in a more recent study, intended to further investigate the effects of nanoparticles on soft-bodied organisms. His study allowed him to explore the distribution of water-suspended fluorescent nanoparticles throughout the eggs and adult bodies of a species of fish, known as the see-through medaka (Oryzias latipes). See-through medaka were used because of their small size, wide temperature and salinity tolerances, and short generation time. Moreover, small fish like the see-through medaka have been popular test subjects for human diseases and organogenesis for other reasons as well, including their transparent embryos, rapid embryo development, and the functional equivalence of their organs and tissue material to that of mammals. Because the see-through medaka have transparent bodies, analyzing the deposition of fluorescent nanoparticles throughout the body is quite simple. For his study, Dr. Kashiwada evaluated four aspects of nanoparticle accumulation. These included the overall accumulation and the size-dependent accumulation of nanoparticles by medaka eggs, the effects of salinity on the aggregation of nanoparticles in solution and on their accumulation by medaka eggs, and the distribution of nanoparticles in the blood and organs of adult medaka. It was also noted that nanoparticles were in fact taken up into the bloodstream and deposited throughout the body. In the medaka eggs, there was a high accumulation of nanoparticles in the yolk; most often bioavailibility was dependent on specific sizes of the particles. Adult samples of medaka had accumulated nanoparticles in the gills, intestine, brain, testis, liver, and bloodstream. One major result from this study was the fact that salinity may have a large influence on the bioavailibility and toxicity of nanoparticles to penetrate membranes and eventually kill the specimen.

As the use of nanomaterials increases worldwide, concerns for worker and user safety are mounting. To address such concerns, the Swedish Karolinska Institute conducted a study in which various nanoparticles were introduced to human lung epithelial cells. The results, released in 2008, showed that iron oxide nanoparticles caused little DNA damage and were non-toxic. Zinc oxide nanoparticles were slightly worse. Titanium dioxide caused only DNA damage. Carbon nanotubes caused DNA damage at low levels. Copper oxide was found to be the worst offender, and was the only nanomaterial identified by the researchers as a clear health risk. The latest toxicology studies on mice involving exposure to carbon nanotubes (CNT) showed a limited pulmonary inflammatory potential of MWCNT at levels corresponding to the average inhalable elemental carbon concentrations observed in U.S.-based CNT facilities. The study estimated that considerable years of exposure are necessary for significant pathology to occur.

No Fullerene Toxicity Reported

Nanoparticles can also be made of C_{60}, as is the case with almost any room temperature solid, and several groups have done this and studied toxicity of such particles. The results in the work of Oberdörster at Southern Methodist University, published in "Environmental Health Perspectives" in July 2004, in which questions were raised of potential cytotoxicity, has now been shown by several sources to be likely caused by the tetrahydrofuran used in preparing the 30 nm–100 nm particles of C_{60} used in the research. Isakovic, et al., 2006, who review this phenomenon, gives results showing that removal of THF from the C_{60} particles resulted in a loss of toxicity. Sayes, et al., 2007, also show that particles prepared as in Oberdorster caused no detectable inflammatory response when instilled intratracheally in rats after observation for 3 months, suggesting that even the particles prepared by Oberdorster do not exhibit markers of toxicity in mammalian models. This work used as a benchmark quartz particles, which did give an inflammatory response.

A comprehensive and recent review of work on fullerene toxicity is available in "Toxicity Studies of Fullerenes and Derivatives," a chapter from the book "Bio-applications of Nanoparticles". In this work, the authors review the work on fullerene toxicity beginning in the early 1990s to present, and conclude that the evidence gathered since the discovery of fullerenes overwhelmingly points to C_{60} being non-toxic. As is the case for toxicity profile with any chemical modification of a structural moiety, the authors suggest that individual molecules be assessed individually.

Toxicity of Metal Based Nanoparticles

Metal based nanoparticles (NPs) are a prominent class of NPs synthesized for their functions as semiconductors, electroluminescents, and thermoelectric materials. Biomedically, these antibacterial NPs have been utilized in drug delivery systems to access areas previously inaccessible to conventional medicine. With the recent increase in interest and development of nanotechnology, many studies have been performed to assess whether the unique characteristics of these NPs, namely their small surface area to volume ratio, might negatively impact the environment upon which they were introduced. Researchers have since found that many metal and metal oxide NPs have detrimental effects on the cells with which they come into contact including but not limited to DNA breakage and oxidation, mutations, reduced cell viability, warped morphology, induced apoptosis and necrosis, and decreased proliferation.

Cytotoxicity

A primary marker for the damaging effects of NPs has been cell viability as determined by state and exposed surface area of the cell membrane. Cells exposed to metallic NPs have, in the case of copper oxide, had up to 60% of their cells rendered unviable. When diluted, the positively charged metal ions often experience an electrostatic attraction to the cell membrane of nearby cells, covering the membrane and preventing it from permeating the necessary fuels and wastes. With less exposed membrane for transportation and communication, the cells are often rendered inactive.

NPs have been found to induce apoptosis in certain cells primarily due to the mitochondrial damage and oxidative stress brought on by the foreign NPs electrostatic reactions.

Genotoxicity

Many methods ranging from comet assay to the HPRT gene mutation test have found that metal based NPs disrupt DNA and its replication process in a variety of cells. In a study examining the effects of nanosilver on DNA, AgNPs were introduced to lymphocyte cell DNA which was then examined for abnormalities. The exposure of the NPs correlated to a significant increase in micronuclei indicative of genetic fragmentation. Metal Oxides such as copper oxide, uraninite, and cobalt oxide have also been found to exert significant stress on exposed DNA. The damage done to the DNA will often result in mutated cells and colonies as found with the HPRT gene test.

Coatings and Charges

NPs, in their implementation, are covered with coatings and sometimes given positive or negative charges depending upon the intended function. Studies have found that these external factors affect the degree of toxicity of NPs. Positive charges are usually found to amplify and cause of cellular damage much more noticeably than negative charges do. In a study in which b- and c-polyethylenimine coated AgNPs were attached to strands of Lambda DNA, the cationic b-polyethylenimine coated AgNP was found to lower the melting point of the DNA 50 °C lower than its anionic counterpart.

Immunogenicity of Nanoparticles

Very little attention has been directed towards the potential immunogenicity of nanostructures. Nanostructures can activate the immune system, inducing inflammation, immune responses, allergy, or even affect to the immune cells in a deleterious or beneficial way (immunosuppression in autoimmune diseases, improving immune responses in vaccines). More studies are needed in order to know the potential deleterious or beneficial effects of nanostructures in the immune system. In comparison to conventional pharmaceutical agents, nanostructures have very large sizes, and immune cells, especially phagocytic cells, recognize and try to destroy them.

Complications with Nanotoxicity Studies

Size is therefore a key factor in determining the potential toxicity of a particle. However it is not the only important factor. Other properties of nanomaterials that influence toxicity include: chemical composition, shape, surface structure, surface charge, aggregation and solubility, and the presence or absence of functional groups of other chemicals. The large number of variables influencing toxicity means that it is difficult to generalise about health risks associated with exposure to nanomaterials – each new nanomaterial must be assessed individually and all material properties must be taken into account.

In addition, standarization of toxicology tests between laboratories are needed. Díaz, B. *et al.* from the University of Vigo (Spain) has shown (Small, 2008) that many different cell lines should be studied in order to know if a nanostructure induces toxicity, and human cells can internalize aggregated nanoparticles. Moreover, it is important to take into account that many nanostructures aggregate in biological fluids, but groups manufacturing nanostructures do not care much about this matter. Many efforts of interdisciplinary groups are strongly needed in order to progress in this field.

Effect of Aggregation or Agglomeration of Nanoparticles

Many nanoparticles agglomerate or aggregate when they are placed in environmental or biological fluids. The terms agglomeration and aggregation have distinct definitions according to the standards organizations ISO and ASTM, where agglomeration signifies more loosely bound particles and aggregation signifies very tightly bound or fused particles (typically occurring during synthesis or drying). Nanoparticles frequently agglomerate due to the high ionic strength of environmental and biological fluids, which shields the repulsion due to charges on the nanoparticles. Unfortunately, agglomeration has frequently been ignored in nanotoxicity studies, even though agglomeration would be expected to affect nanotoxicity since it changes the size, surface area, and sedimentation properties of the nanoparticles. In addition, many nanoparticles will agglomerate to some extent in the environment or in the body before they reach their target, so it is desirable to study how toxicity is affected by agglomeration.

A method was published that can be used to produce different mean sizes of stable agglomerates of several metal, metal oxide, and polymer nanoparticles in cell culture media for cell toxicity studies. Different mean sizes of agglomerates are produced by allowing the nanoparticles to agglomerate to a particular size in cell culture media without protein, and then adding protein to coat the agglomerates and "freeze" them at that size. By waiting different amounts of time before adding protein, different mean sizes of agglomerates of a single type of nanoparticle can be produced in an otherwise identical solution, allowing one to study how agglomerate size affects toxicity. In addition, it was found that vortexing while adding a high concentration of nanoparticles to the cell culture media produces much less agglomerated nanoparticles than if the dispersed solution is only mixed after adding the nanoparticles.

The agglomeration/deagglomeration (mechanical stability) potentials of airborne engineered nanoparticle clusters also have significant influences on their size distribution profiles at the end-point of their environmental transport routes. Different aerosolization and deagglomeration systems have been established to test stability of nanoparticle agglomerates. For example, laboratory setups based on critical orifices have been used to apply a wide range of external shear forces onto airborne nanoparticles. After applying shear forces, the particle mean size decreased while the particle generation rate increased. In another pioneering study, four powder aerosolization systems (dustiness testing systems) were compared for the first time for their characteristics linked to aerosol generation.

Challenges of The Nano-visualisation and Related Unknowns in Nanotoxicology

With comparison to more conventional toxicology studies, the nanotoxicology field is however suffering from a lack of easy characterisation of the potential contaminants, the "nano" scale being a scale difficult to comprehend. The biological systems are themselves still not completely known at this scale. Ultimate Atomic visualisation methods such as Electron microscopy (SEM and TEM) and Atomic force microscopy (AFM) analysis allow visualisation of the nano world. Further nanotoxicology studies will require precise characterisation of the specificities of a given nano-element : size, chemical composition, detailed shape, level of aggregation, combination with other vectors, etc. Above all, these properties would have to be determined not only on the nanocomponent before its introduction in the living environment but also in the (mostly aqueous) biological environment.

There is a need for new methodologies to quickly assess the presence and reactivity of nanoparticles in commercial, environmental, and biological samples since current detection techniques require expensive and complex analytical instrumentation. There have been recent attempts to address these issues by developing and investigating sensitive, simple and portable colorimetric detection assays that assess for the surface reactivity of NPs, which can be used to detect the presence of NPs, in environmental and biological relevant samples. Surface redox reactivity is a key emerging property related to potential toxicity of NPs with living cells, and can be used as a key surrogate for determine for the presence of NPs and a first tier analytical strategy toward assessing NP exposures.

It is difficult to determine the degree of effect of a specific nanoparticle when compared to those of comparable nanoparticles already present in our natural environment .

AEM - Analytical Electron Microscopy was used over 40 years ago to investigate amphibole asbestos bodies in Lake Superior from the Reserve Mining operations. This could non-destructively characterise sub-micron particles. Today AEM can fully characterise to atom dimensions.

Nanomaterials Pollution

Nanopollution is a generic name for waste generated by nanodevices or during the nanomaterials manufacturing process. Ecotoxicological impacts of nanoparticles and the potential for bioaccumulation in plants and microorganisms is a subject of current research, as nanoparticles are considered to present novel environmental impacts. Of the US$710 million spent in 2002 by the U.S. government on nanotechnology research, $500,000 was spent on environmental impact assessments.

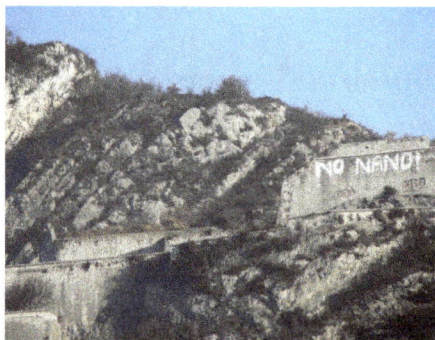

Groups opposing the installation of nanotechnology laboratories in Grenoble, France, spraypainted their opposition on a former fortress above the city in 2007

Overview

Nanowaste is mainly the group of particles that are released into the environment, or the particles that are thrown away when still on their products. The thrown away nanoparticles are usually still functioning how they are supposed to (still have their individual properties), they are just not being properly used anymore. Most of the time, they are lost due to contact with different environments. Silver nanoparticles, for example, they are used a lot in clothes to control odor, those particles are lost when washing them. The fact that they are still functioning and are so small is what makes nanowaste a concern. It can float in the air and might easily penetrate animal and plant cells causing unknown effects. Due to its small size, nanoparticles can have different properties than their own material when on a bigger size, and they are also functioning more efficiently because of its greater surface area. Most human-made nanoparticles do not appear in nature, so living organisms may not have appropriate means to deal with nanowaste.

The capacity for nanoparticles to function as a transport mechanism also raises concern about the transport of heavy metals and other environmental contaminants. Two areas of concern can be identified. First, in their free form nanoparticles can be released into the air or water during production, or production accidents, or as waste by-product of production, and ultimately accumulate in the soil, water, or plant life. Second, in fixed form, where they are part of a manufactured substance or product, they will ultimately have to be recycled or disposed of as waste.

Scrinis raises concerns about nano-pollution, and argues that it is not currently possible to "precisely predict or control the ecological impacts of the release of these nano-products into the environment." A May 2007 Report to the UK Department for Environment, Food and Rural Affairs noted concerns about the toxicological impacts of nanoparticles in relation to both hazard and exposure. The report recommended comprehensive toxicological testing and independent performance tests of fuel additives. Risks have been identified by Uskokovic in 2007. Concerns have also been raised about Silver Nano technology used by Samsung in a range of appliances such as washing machines and air purifiers.

One already known consequences to metals exposure is shown by silver, if exposed to humans in a certain concentration, it can cause illnesses such as argyria and argyrosis. Silver can also cause some environmental problems. Due to its antimicrobial properties (antibacterial), when encountered in the soil it can kill beneficial bacteria that are important to keep the soil healthy. Environmental assessment is justified as nanoparticles present novel environmental impacts. Scrinis raises concerns about nano-pollution, and argues that it is not currently possible to "precisely predict or control the ecological impacts of the release of these nano-products into the environment."

Metals, in particular, have a really strong bonds. Their properties follow up to the nanoscale as well. Metals can stay and damage the environment for a long time, since they hardly degrade or get destroyed. With the increase in use of nanotechnology, it is predicted that the nanowaste of metals will keep increasing, and until a solution is found for that problem, that waste will keep accumulating in the environment. On the other hand, some possible future applications of nanotechnology have the potential to benefit the environment. Nanofiltration, based on the use of membranes with extremely small pores smaller than 10 nm (perhaps composed of nanotubes) are suitable for a mechanical filtration for the removal of ions or the separation of different fluids. A couple of studies have found a solution to filtrate and extract those nanoparticles from water. The process is still being studied but simulations have been giving a total of about 90% to 99% removal of nanowaste particles from the water at an upgraded waste water treatment plant. Once the particles are separated from the water, they go to the landfill with the rest of the solids. Furthermore, magnetic nanoparticles offer an effective and reliable method to remove heavy metal contaminants from waste water. Using nanoscale particles increases the efficiency to absorb the contaminants and is comparatively inexpensive compared to traditional precipitation and filtration methods. One current method to recover nanoparticles is the Cloud Point Extraction. With this technique, gold nanoparticles and some other types of particles that are heat conductors are able to be extracted from aqueous solutions. The process consists of a heating section of the solution that contains the nanoparticles, and then centrifuged in order to separate the layers and then separate the nanoparticles.

Life Cycle Responsibility

To properly assess the health hazards of engineered nanoparticles the whole life cycle of these particles needs to be evaluated, including their fabrication, storage and distribution, application and potential abuse, and disposal. The impact on humans or the environment may vary at different stages of the life cycle.

The Royal Society report identified a risk of nanoparticles or nanotubes being released during disposal, destruction and recycling, and recommended that "manufacturers of products that fall under extended producer responsibility regimes such as end-of-life regulations publish procedures outlining how these materials will be managed to minimize possible human and environmental exposure" (p.xiii). Reflecting the challenges for ensuring responsible life cycle regulation, the Institute for Food and Agricultural Standards has proposed standards for nanotechnology research and development should be integrated across consumer, worker and environmental standards. They also propose that NGOs and other citizen groups play a meaningful role in the development of these standards.

Regulation of Nanotechnology

Because of the ongoing controversy on the implications of nanotechnology, there is significant debate concerning whether nanotechnology or nanotechnology-based products merit special government regulation. This mainly relates to when to assess new substances prior to their release into the market, community and environment.

Nanotechnology refers to an increasing number of commercially available products – from socks and trousers to tennis racquets and cleaning cloths. Such nanotechnologies and their accompanying industries have triggered calls for increased community participation and effective regulatory arrangements. However, these calls have presently not led to such comprehensive regulation to oversee research and the commercial application of nanotechnologies, or any comprehensive labeling for products that contain nanoparticles or are derived from nano-processes.

Regulatory bodies such as the United States Environmental Protection Agency and the Food and Drug Administration in the U.S. or the Health and Consumer Protection Directorate of the European Commission have started dealing with the potential risks posed by nanoparticles. So far, neither engineered nanoparticles nor the products and materials that contain them are subject to any special regulation regarding production, handling or labelling.

Managing Risks: Human and Environmental Health and Safety

Studies of the health impact of airborne particles generally shown that for toxic materials, smaller particles are more toxic. This is due in part to the fact that, given the same mass per volume, the dose in terms of particle numbers increases as particle size decreases.

Based upon available data, it has been argued that current risk assessment methodologies are not suited to the hazards associated with nanoparticles; in particular, existing toxicological and eco-toxicological methods are not up to the task; exposure evaluation (dose) needs to be expressed as quantity of nanoparticles and/or surface area rather than simply mass; equipment for routine detecting and measuring nanoparticles in air, water, or soil is inadequate; and very little is known about the physiological responses to nanoparticles.

Regulatory bodies in the U.S. as well as in the EU have concluded that nanoparticles form the potential for an entirely new risk and that it is necessary to carry out an extensive analysis of the risk. The challenge for regulators is whether a matrix can be developed which would identify nanoparticles and more complex nanoformulations which are likely to have special toxicological properties or whether it is more reasonable for each particle or formulation to be tested separately.

The International Council on Nanotechnology maintains a database and Virtual Journal of scientific papers on environmental, health and safety research on nanoparticles. The database currently has over 2000 entries indexed by particle type, exposure pathway and other criteria. The Project on Emerging Nanotechnologies (PEN) currently lists 807 products that manufacturers have voluntarily identified that use nanotechnology. No labeling is required by the FDA so that number could be significantly higher. "The use of nanotechnology in consumer products and industrial applications is growing rapidly, with the products listed in the PEN inventory showing just the tip of the iceberg" according to PEN Project Director David Rejeski . A list of those products that have been voluntarily disclosed by their manufacturers is located here .

The Material Safety Data Sheet that must be issued for certain materials often does not differentiate between bulk and nanoscale size of the material in question and even when it does these MSDS are advisory only.

Democratic Governance

Many argue that government has a responsibility to provide opportunities for the public to be involved in the development of new forms of science and technology. Community engagement can be achieved through various means or mechanisms. An online journal article identifies traditional approaches such as referenda, consultation documents, and advisory committees that include community members and other stakeholders. Other conventional approaches include public meetings and "closed" dialog with stakeholders. More contemporary engagement processes that have been employed to include community members in decisions about nanotechnology include citizens' juries and consensus conferences. Leach and Scoones (2006, p. 45) argue that since that "most debates about science and technology options involve uncertainty, and often ignorance, public debate about regulatory regimes is essential."

It has been argued that limited nanotechnology labeling and regulation may exacerbate potential human and environmental health and safety issues associated with nanotechnology, and that the development of comprehensive regulation of nanotechnology will be vital to ensure that the potential risks associated with the research and commercial application of nanotechnology do not overshadow its potential benefits. Regulation may also be required to meet community expectations about responsible development of nanotechnology, as well as ensuring that public interests are included in shaping the development of nanotechnology.

Community education, engagement and consultation tend to occur "downstream": once there is at least a moderate level of awareness, and often during the process of disseminating and adapting technologies. "Upstream" engagement, by contrast, occurs much earlier in the innovation cycle and involves: "dialogue and debate about future technology options and pathways, bringing the often expert-led approaches to horizon scanning, technology foresight and scenario planning to involve a wider range of perspectives and inputs." Daniel Sarewitz Director of Arizona State University's Consortium on Science, Policy and Outcomes, argues that "by the time new devices reach the stage of commercialization and regulation, it is usually too late to alter them to correct problems." However, Xenos, et al. argue that upstream engagement can be utilized in this area through anticipated discussion with peers. Upstream engagement in this sense is "meant to create the best possible conditions for sound policy making and public judgments based on carefull assessment of objective information". Discussion may act as a catalyst for upstream engagement by prompting accountability for individuals to seek and process additional information ("anticipatory elaboration"). However, though anticipated discussion did lead to participants seeking further information, Xenos et al. found that factual information was not primarily sought out; instead, individuals sought out opinion pieces and editorials.

The stance that the research, development and use of nanotechnology should be subject to control by the public sector is sometimes referred to as nanosocialism.

Newness

The question of whether nanotechnology represents something 'new' must be answered to decide how best nanotechnology should be regulated. The Royal Society recommended that the UK government assess chemicals in the form of nanoparticles or nanotubes as new substances. Subsequent to this, in 2007 a coalition of over forty groups called for nanomaterials to be classified as new substances, and regulated as such.

Despite these recommendations, chemicals comprising nanoparticles that have previously been subject to assessment and regulation may be exempt from regulation, regardless of the potential for different risks and impacts. In contrast, nanomaterials are often recognized as 'new' from the perspective of intellectual property rights (IPRs), and as such are commercially protected via patenting laws.

There is significant debate about who is responsible for the regulation of nanotechnology. While some non-nanotechnology specific regulatory agencies currently cover some products and processes (to varying degrees) – by "bolting on" nanotechnology to existing regulations – there are clear gaps in these regimes. This enables some nanotechnology applications to figuratively "slip through the cracks" without being covered by any regulations. An example of this has occurred in the US, and involves nanoparticles of titanium dioxide (TIo2) for use in sunscreen where they create a clearer cosmetic appearance. In this case, the US Food and Drug Administration (FDA) reviewed the immediate health effects of exposure to nanoparticles of titanium dioxide (TIo2) for consumers. However, they did not review its impacts for aquatic ecosystems when the sunscreen rubs off, nor did the EPA, or any other agency. Similarly the Australian equivalent of the FDA, the Therapeutic Goods Administration (TGA) approved the use of nanoparticles in sunscreens (without the requirement for package labelling) after a thorough review of the literature, on the basis that although nanoparticles of TIo2 and zinc oxide (ZNo) in sunscreens do produce free radicals and oxidative DNA damage *in vitro*, such particles were unlikely to pass the dead outer cells of the stratum corneum of human skin; a finding which some academics have argued seemed not to apply the precautionary principle in relation to prolonged use on children with cut skin, the elderly with thin skin, people with diseased skin or use over flexural creases. Doubts over the TGA's decision were raised with publication of a paper showing that the uncoated anatase form of TIo2 used in some Australian sunscreens caused a photocatalytic reaction that degraded the surface of newly installed prepainted steel roofs in places where they came in contact with the sunscreen coated hands of workmen. Such gaps in regulation are likely to continue alongside the development and commercialization of increasingly complex second and third generation nanotechnologies.

Nanomedicines are just beginning to enter drug regulatory processes, but within a few decades could comprise a dominant group within the class of innovative pharmaceuticals, the current thinking of government safety and cost-effectiveness regulators appearing to be that these products give rise to few if any nano-specific issues. Some academics (such as Thomas Alured Faunce) have challenged that proposition and suggest that nanomedicines may create unique or heightened policy challenges for government systems of cost-effectiveness as well as safety regulation. There are also significant public good aspects to the regulation of nanotechnology, particularly with regard to ensuring that industry involvement in standard-setting does not become a means of reducing competition and that nanotechnology policy and regulation encourages new models of safe drug discovery and development more systematically targeted at the global burden of disease.

Self-regulation attempts may well fail, due to the inherent conflict of interest in asking any organization to police itself. If the public becomes aware of this failure, an external, independent organization is often given the duty of policing them, sometimes with highly punitive measures taken against the organization. The Food and Drug Administration notes that it only regulates on the basis of voluntary claims made by the product manufacturer. If no claims are made by a manufacturer, then the FDA may be unaware of nanotechnology being employed.

Yet regulations worldwide still fail to distinguish between materials in their nanoscale and bulk form. This means that nanomaterials remain effectively unregulated; there is no regulatory requirement for nanomaterials to face new health and safety testing or environmental impact assessment prior to their use in commercial products, if these materials have already been approved in bulk form. The health risks of nanomaterials are of particular concern for workers who may face occupational exposure to nanomaterials at higher levels, and on a more routine basis, than the general public.

International Law

There is no international regulation of nanoproducts or the underlying nanotechnology. Nor are there any internationally agreed definitions or terminology for nanotechnology, no internationally agreed protocols for toxicity testing of nanoparticles, and no standardized protocols for evaluating the environmental impacts of nanoparticles.

Since products that are produced using nanotechnologies will likely enter international trade, it is argued that it will be necessary to harmonize nanotechnology standards across national borders. There is concern that some countries, most notably developing countries, will be excluded from international standards negotiations. The Institute for Food and Agricultural Standards notes that "developing countries should have a say in international nanotechnology standards development, even if they lack capacity to enforce the standards". (p. 14).

Concerns about monopolies and concentrated control and ownership of new nanotechnologies were raised in community workshops in Australia in 2004.

Arguments Against Regulation

Wide use of the term nanotechnology in recent years has created the impression that regulatory frameworks are suddenly having to contend with entirely new challenges that they are unequipped to deal with. Many regulatory systems around the world already assess new substances or products for safety on a case by case basis, before they are permitted on the market. These regulatory systems have been assessing the safety of nanometre scale molecular arrangements for many years and many substances comprising nanometre scale particles have been in use for decades e.g. Carbon black, Titanium dioxide, Zinc oxide, Bentonite, Aluminum silicate, Iron oxides, Silicon dioxide, Diatomaceous earth, Kaolin, Talc, Montmorillonite, Magnesium oxide, Copper sulphate.

These existing approval frameworks almost universally use the best available science to assess safety and do not approve substances or products with an unacceptable risk benefit profile. One proposal is to simply treat particle size as one of the several parameters defining a substance to be approved, rather than creating special rules for all particles of a given size regardless of type. A major argument against special regulation of nanotechnology is that the projected applications with the greatest impact are far in the future, and it is unclear how to regulate technologies whose feasibility is speculative at this point. In the meantime, it has been argued that the immediate applications of nanomaterials raise challenges not much different from those of introducing any other new material, and can be dealt with by minor tweaks to existing regulatory schemes rather than sweeping regulation of entire scientific fields.

A truly precautionary approach to regulation could severely impede development in the field of nanotechnology safety studies are required for each and every nanoscience application. While the outcome of these studies can form the basis for government and international regulations, a more reasonable approach might be development of a risk matrix that identifies likely culprits.

Response from Governments

United Kingdom

In its seminal 2004 report *Nanoscience and Nanotechnologies: Opportunities and Uncertainties*, the United Kingdom's Royal Society concluded that:

> *Many nanotechnologies pose no new risks to health and almost all the concerns relate to the potential impacts of deliberately manufactured nanoparticles and nanotubes that are free rather than fixed to or within a material... We expect the likelihood of nanoparticles or nanotubes being released from products in which they have been fixed or embedded (such as composites) to be low but have recommended that manufacturers assess this potential exposure risk for the lifecycle of the product and make their findings available to the relevant regulatory bodies... It is very unlikely that new manufactured nanoparticles could be introduced into humans in doses sufficient to cause the health effects that have been associated with [normal air pollution].*

but have recommended that nanomaterials be regulated as new chemicals, that research laboratories and factories treat nanomaterials "as if they were hazardous", that release of nanomaterials into the environment be avoided as far as possible, and that products containing nanomaterials be subject to new safety testing requirements prior to their commercial release.

The 2004 report by the UK Royal Society and Royal Academy of Engineers noted that existing UK regulations did not require additional testing when existing substances were produced in nanoparticulate form. The Royal Society recommended that such regulations were revised so that "chemicals produced in the form of nanoparticles and nanotubes be treated as new chemicals under these regulatory frameworks" (p.xi). They also recommended that existing regulation be modified on a precautionary basis because they expect that "the toxicity of chemicals in the form of free nanoparticles and nanotubes cannot be predicted from their toxicity in a larger form and... in some cases they will be more toxic than the same mass of the same chemical in larger form."

The Better Regulation Commission's earlier 2003 report had recommended that the UK Government:

1. enable, through an informed debate, the public to consider the risks for themselves, and help them to make their own decisions by providing suitable information;

2. be open about how it makes decisions, and acknowledge where there are uncertainties;

3. communicate with, and involve as far as possible, the public in the decision making process;

4. ensure it develops two-way communication channels; and

5. take a strong lead over the handling of any risk issues, particularly information provision and policy implementation.

These recommendations were accepted in principle by the UK Government. Noting that there was "no obvious focus for an informed public debate of the type suggested by the Task Force", the UK government's response was to accept the recommendations.

The Royal Society's 2004 report identified two distinct governance issues:

1. the "role and behaviour of institutions" and their ability to "minimise unintended consequences" through adequate regulation and

2. the extent to which the public can trust and play a role in determining the trajectories that nanotechnologies may follow as they develop.

United States

Rather than adopt a new nano-specific regulatory framework, the United States' Food and Drug Administration (FDA) convenes an 'interest group' each quarter with representatives of FDA centers that have responsibility for assessment and regulation of different substances and products. This interest group ensures coordination and communication. A September 2009 FDA document called for identifying sources of nanomaterials, how they move in the environment, the problems they might cause for people, animals and plants, and how these problems could be avoided or mitigated.

The Bush administration in 2007 decided that no special regulations or labeling of nanoparticles were required. Critics derided this as treating consumers like a "guinea pig" without sufficient notice due to lack of labelling.

Berkeley, CA is currently the only city in the United States to regulate nanotechnology. Cambridge, MA in 2008 considered enacting a similar law, but the committee it instituted to study the issue Cambridge recommended against regulation in its final report, recommending instead other steps to facilitate information-gathering about potential effects of nanomaterials.

On December 10, 2008 the U.S. National Research Council released a report calling for more regulation of nanotechnology.

California

Assembly Bill (AB) 289 (2006) authorizes the Department of Toxic Substances Control (DTSC) within the California Environmental Protection Agency and other agencies to request information on environmental and health impacts from chemical manufacturers and importers, including testing techniques.

California

In October 2008, the Department of Toxic Substances Control (DTSC), within the California Environmental Protection Agency, announced its intent to request information regarding analytical test methods, fate and transport in the environment, and other relevant information from manufacturers of carbon nanotubes. DTSC is exercising its authority under the California Health and Safety Code, Chapter 699, sections 57018-57020. These sections were added as a result of the adoption of Assembly Bill AB 289 (2006). They are intended to make information on the fate and transport, detection and analysis, and other information on chemicals more available. The law places the responsibility to provide this information to the Department on those who manufacture or import the chemicals.

On January 22, 2009, a formal information request letter was sent to manufacturers who produce or import carbon nanotubes in California, or who may export carbon nanotubes into the State. This letter constitutes the first formal implementation of the authorities placed into statute by AB 289 and is directed to manufacturers of carbon nanotubes, both industry and academia within the State, and to manufacturers outside California who export carbon nanotubes to California. This request for information must be met by the manufacturers within one year. DTSC is waiting for the upcoming January 22, 2010 deadline for responses to the data call-in.

The California Nano Industry Network and DTSC hosted a full-day symposium on November 16, 2009 in Sacramento, CA. This symposium provided an opportunity to hear from nanotechnology industry experts and discuss future regulatory considerations in California.

DTSC is expanding the Specific Chemical Information Call-in to members of the nano-metal oxides. Interested individuals are encouraged to visit their website for the latest up-to-date information at http://www.dtsc.ca.gov/TechnologyDevelopment/Nano-technology/index.cfm.

On December 21, 2010, the Department of Toxic Substances Control (DTSC) initiated the second Chemical Information Call-in for six nanomaterials: nano cerium oxide, nano silver, nano titanium dioxide, nano zero valent iron, nano zinc oxide, and quantum dots. DTSC sent a formal information request letter to forty manufacturers who produce or import the six nanomaterials in California, or who may export them into the State. The Chemical Information Call-in is meant to identify information gaps of these six nanomaterials and to develop further knowledge of their analytical test methods, fate and transport in the environment, and other relevant information under California Health and Safety Code, Chapter 699, sections 57018-57020. DTSC completed the carbon nanotube information call-in in June 2010.

DTSC partners with University of California, Los Angeles (UCLA), Santa Barbara (UCSB), and Riverside (UCR), University of Southern California (USC), Stanford University, Center for Environmental Implications of Nanotechnology (CEIN), and The National Institute for Occupational Safety and Health (NIOSH) on safe nanomaterial handling practices.

DTSC is interested in expanding the Chemical Information Call-in to members of the bominated flame retardants, members of the methyl siloxanes, ocean plastics, nano-clay, and other emerging chemicals.

European Union

The European Union has formed a group to study the implications of nanotechnology called the Scientific Committee on Emerging and Newly Identified Health Risks which has published a list of risks associated with nanoparticles.

Consequently, manufacturers and importers of carbon products, including carbon nano-tubes will have to submit full health and safety data within a year or so in order to comply with REACH.

Response from Advocacy Groups

In January 2008, a coalition of over 40 civil society groups endorsed a statement of principles calling for precautionary action related to nanotechnology. The coalition called for strong, comprehensive oversight of the new technology and its products in the International Center for Technology Assessment's report *Principles for the Oversight of Nanotechnologies and Nano materials*, which states:

> *Hundreds of consumer products incorporating nano-materials are now on the*

market, including cosmetics, sunscreens, sporting goods, clothing, electronics, baby and infant products, and food and food packaging. But evidence indicates that current nano-materials may pose significant health, safety, and environmental hazards. In addition, the profound social, economic, and ethical challenges posed by nano-scale technologies have yet to be addressed ... 'Since there is currently no government oversight and no labeling requirements for nano-products anywhere in the world, no one knows when they are exposed to potential nano-tech risks and no one is monitoring for potential health or environmental harm. That's why we believe oversight action based on our principles is urgent' ... This industrial boom is creating a growing nano-workforce which is predicted to reach two million globally by 2015. 'Even though potential health hazards stemming from exposure have been clearly identified, there are no mandatory workplace measures that require exposures to be assessed, workers to be trained, or control measures to be implemented,' explained Bill Kojola of the AFL-CIO. 'This technology should not be rushed to market until these failings are corrected and workers assured of their safety'" also .

The group has urged action based on eight principles. They are 1) A Precautionary Foundation 2) Mandatory Nano-specific Regulations 3) Health and Safety of the Public and Workers 4) Environmental Protection 5) Transparency 6) Public Participation 7) Inclusion of Broader Impacts and 8) Manufacturer Liability.

Some NGOs, including Friends of the Earth, are calling for the formation of a separate nanotechnology specific regulatory framework for the regulation of nanotechnology. In Australia, Friends of the Earth propose the establishment of a Nanotechnology Regulatory Coordination Agency, overseen by a Foresight and Technology Assessment Board. The advantage of this arrangement is that it could ensure a centralized body of experts that are able to provide oversight across the range of nano-products and sectors. It is also argued that a centralized regulatory approach would simplify the regulatory environment, thereby supporting industry innovation. A National Nanotechnology Regulator could coordinate existing regulations related to nanotechnology (including intellectual property, civil liberties, product safety, occupation health and safety, environmental and international law). Regulatory mechanisms could vary from "hard law at one extreme through licensing and codes of practice to 'soft' self-regulation and negotiation in order to influence behavior." The formation of national nanotechnology regulatory bodies may also assist in establishing global regulatory frameworks.

In early 2008, The UK's largest organic certifier, the Soil Association, announced that its organic standard would exclude nanotechnology, recognizing the associated human and environmental health and safety risks. Certified organic standards in Australia exclude engineered nanoparticles. It appears likely that other organic certifiers will also follow suit. The Soil Association was also the first to declare organic standards free from genetic engineering.

Technical Aspects

Size

Regulation of nanotechnology will require a definition of the size, in which particles and processes are recognized as operating at the nano-scale. The size-defining characteristic of nanotechnology is the subject of significant debate, and varies to include particles and materials in the scale of at least 100 to 300 nanometers (nm). Friends of the Earth Australia recommend defining nano-particles up to 300 nanometers (nm) in size. They argue that "particles up to a few hundred nanometers in size share many of the novel biological behaviors of nano-particles, including novel toxicity risks", and that "nano-materials up to approximately 300 nm in size can be taken up by individual cells". The UK Soil Association define nanotechnology to include manufactured nano-particles where the mean particle size is 200 nm or smaller. The U.S. National Nanotechnology Initiative defines nanotechnology as "the understanding and control of matter at dimensions of roughly 1 to 100 nm.

Mass Thresholds

Regulatory frameworks for chemicals tend to be triggered by mass thresholds. This is certainly the case for the management of toxic chemicals in Australia through the National pollutant inventory. However, in the case of nanotechnology, nano-particle applications are unlikely to exceed these thresholds (tonnes/kilograms) due to the size and weight of nano-particles. As such, the Woodrow Wilson International Center for Scholars questions the usefulness of regulating nanotechnologies on the basis of their size/weight alone. They argue, for example, that the toxicity of nano-particles is more related to surface area than weight, and that emerging regulations should also take account of such factors.

References

- Chris Phoenix; Mike Treder (2008). "Chapter 21: Nanotechnology as global catastrophic risk". In Bostrom, Nick; Cirkovic, Milan M. Global catastrophic risks. Oxford: Oxford University Press. ISBN 978-0-19-857050-9.

- Chan WCW (2007). "Toxicity Studies of Fullerenes and Derivatives". Bio-applications of nanoparticles. New York, NY: Springer Science + Business Media. ISBN 0-387-76712-6.

- Smith, Erin Geiger (14 February 2013). "U.S.-based inventors lead world in nanotechnology patents: study". Technology. Reuters. Retrieved 4 June 2016.

- "About the National Nanotechnology Initiative". United States National Nanotechnology Initiative. 2016. Retrieved 4 June 2016.

9

Various Applications of Nanotechnology

Over the past several decades, technology has developed immensely. The applications of nanotechnology have improved life all over the world. In order to move forward from the present level of technology, scientists have been developing energy applications of nanotechnology. Some of the applications are industrial applications of nanotechnology, nanomedicine, nanobiotechnology and green nanotechnology. The diverse applications of nanotechnology in the current scenario have been thoroughly discussed in this chapter.

Applications of Nanotechnology

The 2000s have seen the beginnings of the applications of nanotechnology in commercial products, although most applications are limited to the bulk use of passive nanomaterials. Examples include titanium dioxide and zinc oxide nanoparticles in sunscreen, cosmetics and some food products; silver nanoparticles in food packaging, clothing, disinfectants and household appliances such as Silver Nano; carbon nanotubes for stain-resistant textiles; and cerium oxide as a fuel catalyst. As of March 10, 2011, the Project on Emerging Nanotechnologies estimated that over 1300 manufacturer-identified nanotech products are publicly available, with new ones hitting the market at a pace of 3–4 per week.

Nanotechnology is being used in developing countries to help treat disease and prevent health issues. The umbrella term for this kind of nanotechnology is Nanomedicine.

Nanotechnology is also being applied to or developed for application to a variety of industrial and purification processes. Purification and environmental cleanup applications include the desalination of water, water filtration, wastewater treatment, groundwater treatment, and other nanoremediation. In industry, applications may include construction materials, military goods, and nano-machining of nano-wires, nano-rods, few layers of graphene, etc. Also, recently a new field arisen from the root of Nanotechnology is called Nanobiotechnology. Nanobiotechnology is the biology-based, application-oriented frontier area of research in the hybrid discipline of Nanoscience and biotechnology with an equivalent contribution.

Applications By Type

- Nanomedicine

- Nanobiotechnology

- Green nanotechnology

- Energy applications of nanotechnology

- Industrial applications of nanotechnology

- Potential applications of carbon nanotubes

- Nanoart

Nano-tech Catalyst

Scientists at the Department of Energy's Oak Ridge National Laboratory while attempting to create a nanotechnology based catalyst-mediated series of chemical reactions to turn CO_2 into a usable fuel have discovered a process to turn the Carbon dioxide into ethanol, which will serve as a way forward to climate change by both decreasing CO_2 in the atmosphere and using the ethanol (CH_3CH_2OH) as an additive to fuels to increase efficiency and thereby decrease consumption. Reportedly it is also related that the process is cheap in cost and efficient in functioning.

Energy Applications of Nanotechnology

Over the past few decades, the fields of science and engineering have been seeking to develop new and improved types of energy technologies that have the capability of improving life all over the world. In order to make the next leap forward from the current generation of technology, scientists and engineers have been developing energy applications of nanotechnology. Nanotechnology, a new field in science, is any technology that contains components smaller than 100 nanometers. For scale, a single virus particle is about 100 nanometers in width.

An important subfield of nanotechnology related to energy is nanofabrication. Nanofabrication is the process of designing and creating devices on the nanoscale. Creating devices smaller than 100 nanometers opens many doors for the development of new ways to capture, store, and transfer energy. The inherent level of control that nanofabrication could give scientists and engineers would be critical in providing the capability of solving many of the problems that the world is facing today related to the current generation of energy technologies.

People in the fields of science and engineering have already begun developing ways of utilizing nanotechnology for the development of consumer products. Benefits already observed from the design of these products are an increased efficiency of lighting and heating, increased electrical storage capacity, and a decrease in the amount of pollution from the use of energy. Benefits such as these make the investment of capital in the research and development of nanotechnology a top priority.

Consumer Products

Recently, previously established and entirely new companies such as BetaBatt, Inc. and Oxane Materials are focusing on nanomaterials as a way to develop and improve upon older methods for the capture, transfer, and storage of energy for the development of consumer products.

ConsERV, a product developed by the Dais Analytic Corporation, uses nanoscale polymer membranes to increase the efficiency of heating and cooling systems and has already proven to be a lucrative design. The polymer membrane was specifically configured for this application by selectively engineering the size of the pores in the membrane to prevent air from passing, while allowing moisture to pass through the membrane. ConsERV's value is demonstrated in the form of an energy recovery a device which pretreats the incoming fresh air to a building using the energy found in the exhaust air steam using no moving parts to lower the energy and carbon footprint of existing forms of heating and cooling equipment Polymer membranes can be designed to selectively allow particles of one size and shape to pass through while preventing other. This makes for a powerful tool that can be used in all markets - consumer, commercial, industrial, and government products from biological weapons protection to industrial chemical separations. Dais's near term uses of this 'family' of selectively engineered nanotechnology materials, aside from ConsERV, include (a.) a completely new cooling cycle capable of replacing the refrigerant based cooling cycle the world has known for the past 100 plus years. This product, under development, is named NanoAir. NanoAir uses only water and this selectively engineered membrane material to cool (or heat) and dehumidify (or humidify) air. There are no fluorocarbon producing gasses used, and the energy required to cool a space drops as thermodynamics does the actual cooling. The company was awarded an Advanced Research Program Administration - Energy award in 2010, and a United States Department of Defense (DoD) grant in 2011 both designed to accelerate this newer, energy efficient technology closer to commercialization, and (b.) a novel way to clean most all contaminated forms of water called NanoClear. By using the selectivity of this hermetic, engineered composite material it can transfer only a water molecule from one face of the membrane to the other leaving behind the contaminants. It should also be noted Dais received a US Patent (Patent Number 7,990,679) in October 2011 titled "Nanoparticle Ultracapacitor". This patented item again uses the selectively engineered material to create an energy storage mechanism projected to have performance and cost advantages over existing storage technologies. The company has used this patent's concepts to create a functional energy storage prototype device named NanoCap. NanoCap is a form of ultra-capacitor potentially useful to power a broad range of applications including most forms of transportation, energy storage (especially useful as a storage media for renewable energy technologies), telecommunication infrastructure, transistor gate dielectrics, and consumer battery applications (cell phones, computers, etc.).

A New York-based company called Applied NanoWorks, Inc. has been developing a consumer product that utilizes LED technology to generate light. Light-emitting diodes or LEDs, use only about 10% of the energy that a typical incandescent or fluorescent light bulb uses and typically last much longer, which makes them a viable alternative to traditional light bulbs. While LEDs have been around for decades, this company and others like it have been developing a special variant of LED called the white LED. White LEDs consist of semi-conducting organic layers that are only about 100 nanometers in distance from each other and are placed between two electrodes, which create an anode, and a cathode. When voltage is applied to the system, light is generated when electricity passes through the two organic layers. This is called electroluminescence. The semiconductor properties of the organic layers are what allow for the minimal amount of energy necessary to generate light. In traditional light bulbs, a metal filament is used to generate light when electricity is run through the filament. Using metal generates a great deal of heat and therefore lowers efficiency.

Research for longer lasting batteries has been an ongoing process for years. Researchers have now begun to utilize nanotechnology for battery technology. mPhase Technologies in conglomeration with Rutgers University and Bell Laboratories have utilized nanomaterials to alter the wetting behavior of the surface where the liquid in the battery lies to spread the liquid droplets over a greater area on the surface and therefore have greater control over the movement of the droplets. This gives more control to the designer of the battery. This control prevents reactions in the battery by separating the electrolytic liquid from the anode and the cathode when the battery is not in use and joining them when the battery is in need of use.

Thermal applications also are a future applications of nanothechonlogy creating low cost system of heating, ventilation, and air conditioning, changing molecular structure for better management of temperature

Reduction of Energy Consumption

A reduction of energy consumption can be reached by better insulation systems, by the use of more efficient lighting or combustion systems, and by use of lighter and stronger materials in the transportation sector. Currently used light bulbs only convert approximately 5% of the electrical energy into light. Nanotechnological approaches like or quantum caged atoms (QCAs) could lead to a strong reduction of energy consumption for illumination.

Increasing The Efficiency of Energy Production

Today's best solar cells have layers of several different semiconductors stacked together to absorb light at different energies but they still only manage to use 40 percent of the Sun's energy. Commercially available solar cells have much lower efficiencies (15-20%). Nanostructuring has been used to improve the efficiencies of established photovoltaic tehcnologies, for example by improving current collection in amorphous silicon devices, plasmonic enhancement in dye-sensitized solar cells, and improved light trapping in crystalline silicon. Furthermore, nanotechnology could help increase the efficiency of light conversion by using nanostructures with a continuum of bandgaps, or by controlling the directivity and photon escape probability of photovoltaic devices.

The degree of efficiency of the internal combustion engine is about 30-40% at present. Nanotechnology could improve combustion by designing specific catalysts with maximized surface area. In 2005, scientists at the University of Toronto developed a spray-on nanoparticle substance that, when applied to a surface, instantly transforms it into a solar collector.

Nuclear Accident Cleanup and Waste Storage

Nanomaterials deployed by swarm robotics may be helpful for decontaminating a site of a nuclear accident which poses hazards to humans because of high levels of radiation and radioactive particles. Hot nuclear compounds such as corium or melting fuel rods may be contained in "bubbles" made from nanomaterials that are designed to isolate the harmful effects of nuclear activity occurring inside of them from the outside environment that organisms inhabit.

Economic Benefits

The relatively recent shift toward using nanotechnology with respect to the capture, transfer, and storage of energy has and will continue to have many positive economic impacts on society. The control of materials that nanotechnology offers to scientists and engineers of consumer products is one of the most important aspects of nanotechnology. This allows for an improved efficiency of products across the board.

A major issue with current energy generation is the loss of efficiency from the generation of heat as a by-product of the process. A common example of this is the heat generated by the internal combustion engine. The internal combustion engine loses about 64% of the energy from gasoline as heat and an improvement of this alone could have a significant economic impact. However, improving the internal combustion engine in this respect has proven to be extremely difficult without sacrificing performance. Improving the efficiency of fuel cells through the use of nanotechnology appears to be more plausible by using molecularly tailored catalysts, polymer membranes, and improved fuel storage.

In order for a fuel cell to operate, particularly of the hydrogen variant, a noble-metal catalyst (usually platinum, which is very expensive) is needed to separate the electrons from the protons of the hydrogen atoms. However, catalysts of this type are extremely sensitive to carbon monoxide reactions. In order to combat this, alcohols or hydrocarbons compounds are used to lower the carbon monoxide concentration in the system. This adds an additional cost to the device. Using nanotechnology, catalysts can be designed through nanofabrication that are much more resistant to carbon monoxide reactions, which improves the efficiency of the process and may be designed with cheaper materials to additionally lower costs.

Fuel cells that are currently designed for transportation need rapid start-up periods for the practicality of consumer use. This process puts a lot of strain on the traditional polymer electrolyte membranes, which decreases the life of the membrane requiring frequent replacement. Using nanotechnology, engineers have the ability to create a much more durable polymer membrane, which addresses this problem. Nanoscale polymer membranes are also much more efficient in ionic conductivity. This improves the efficiency of the system and decreases the time between replacements, which lowers costs.

Another problem with contemporary fuel cells is the storage of the fuel. In the case of hydrogen fuel cells, storing the hydrogen in gaseous rather than liquid form improves the efficiency by 5%. However, the materials that we currently have available to us significantly limit fuel storage due to low stress tolerance and costs. Scientists have come up with an answer to this by using a nanoporous styrene material (which is a relatively inexpensive material) that when super-cooled to around -196°C, naturally holds on to hydrogen atoms and when heated again releases the hydrogen for use.

Capacitors: Then and Now

For decades, scientists and engineers have been attempting to make computers smaller and more efficient. A crucial component of computers are capacitors. A capacitor is a device that is made of a pair of electrodes separated by an insulator that each stores an opposite charge. A capacitor stores a charge when it is removed from the circuit that it is connected to; the charge is released when it is replaced back into the circuit. Capacitors have an advantage over batteries in that they release their charge much more quickly than a battery.

Traditional or foil capacitors are composed of thin metal conducting plates separated by an electrical insulator, which are then stacked or rolled and placed in a casing. The problem with a traditional capacitor such as this is that they limit how small an engineer can design a computer. Scientists and engineers have since turned to nanotechnology for a solution to the problem.

Using nanotechnology, researchers developed what they call "ultracapacitors." An ultracapacitor is a general term that describes a capacitor that contains nanocomponents. Ultracapacitors are being researched heavily because of their high density interior, compact size, reliability, and high capacitance. This decrease in size makes it increasingly possible to develop much smaller circuits and computers. Ultracapacitors also have the capability to supplement batteries in hybrid vehicles by providing a large amount of energy during peak acceleration and allowing the battery to supply energy over longer periods of time, such as during a constant driving speed. This could decrease the size and weight of the large batteries needed in hybrid vehicles as well as take additional stress off the battery. However, the combination of ultracapacitors and a battery is not cost effective due to the need of additional DC/DC electronics to coordinate the two.

Nanoporous carbon aerogel is one type of material that is being utilized for the design of ultracapacitors. These aerogels have a very large interior surface area and can have its properties altered by changing the pore diameter and distribution along with adding nanosized alkali metals to alter its conductivity.

Carbon nanotubes are another possible material for use in an ultracapacitor. Carbon nanotubes are created by vaporizing carbon and allowing it to condense on a surface. When the carbon condenses, it forms a nanosized tube composed of carbon atoms. This tube has a high surface area, which increases the amount of charge that can be stored. The low reliability and high cost of using carbon nanotubes for ultracapacitors is currently an issue of research.

In a study concerning ultracapacitors or supercapacitors, researchers at the Sungkyunkwan University in the Republic of Korea explored the possibility of increasing the capacitance of electrodes through the addition of fluorine atoms to the walls of carbon nanotubes. As briefly mentioned before, carbon nanotubes are an increasing form of capacitors due to their superb chemical stability, high conductivity, light mass, and their large surface area. These researchers fluorinated single-walled carbon nanotubes (SWCNTs) at high temperatures to bind fluorine atoms to the walls. The attached fluorine atoms changed the non-polar nanotubes to become polar molecules. This can be attributed to the charge transfer from the fluorine. This created dipole-dipole layers along the carbon nanotube walls. Testing of these fluorinated SWCNTs against normal state SWCNTs showed a difference in capacitance. It was determined that the fluorinated SWCNTs are advantageous in fabricating electrodes for capacitors and improve the wettability with aqueous electrolytes, which promotes the overall performance of supercapacitors. While this study brought to knowledge a more efficient example of capacitors, little is known about this new supercapacitor, large scale synthesis is lacking and is necessary for any massive production, and preparation conditions are quite tedious in achieving the final product.

Theory of Capacitance

Understanding the concept of capacitance can be helpful in understanding why nanotechnology is such a powerful tool for the design of higher energy storing capacitors. A capacitor's capacitance (C) or amount of energy stored is equal to the amount of charge (Q) stored on each plate divided by the voltage (V) between the plates. Another representation of capacitance is that capacitance (C) is approximately equal to the permittivity (ε) of the dielectric times the area (A) of the plates divided by the distance (d) between them. Therefore, capacitance is proportional to the surface area of the conducting plate and inversely proportional to the distance between the plates.

Using carbon nanotubes as an example, a property of carbon nanotubes is that they have a very high surface area to store a charge. Using the above proportionality that capacitance (C) is proportional to the surface area (A) of the conducting plate; it becomes obvious that using nanoscaled materials with high surface area would be great for increasing capacitance. The other proportionality described above is that capacitance (C) is inversely proportional to the distance (d) between the plates. Using nanoscaled plates such as carbon nanotubes with nanofabrication techniques, gives the capability of decreasing the space between plates which again increases capacitance.

Industrial Applications of Nanotechnology

Nanotechnology is impacting the field of consumer goods, several products that incorporate nanomaterials are already in a variety of items; many of which people do not even realize contain nanoparticles, products with novel functions ranging from easy-to-clean to scratch-resistant. Examples of that car bumpers are made lighter, clothing is more stain repellant, sunscreen is more radiation resistant, synthetic bones are stronger, cell phone screens are lighter weight, glass packaging for drinks leads to a longer shelf-life, and balls for various sports are made more durable. Using nanotech, in the mid-term modern textiles will become "smart", through embedded "wearable electronics", such novel products have also a promising potential especially in the field of cosmetics, and has numerous potential applications in heavy industry. Nanotechnology is predicted to be a main driver of technology and business in this century and holds the promise of higher performance materials, intelligent systems and new production methods with significant impact for all aspects of society.

Foods

A complex set of engineering and scientific challenges in the food and bioprocessing industry for manufacturing high quality and safe food through efficient and sustainable means can be solved through nanotechnology. Bacteria identification and food quality monitoring using biosensors; intelligent, active, and smart food packaging systems; nanoencapsulation of bioactive food compounds are few examples of emerging applications of nanotechnology for the food industry. Nanotechnology can be applied in the production, processing, safety and packaging of food. A nanocomposite coating process could improve food packaging by placing anti-microbial agents directly on the surface of the coated film. Nanocomposites could increase or decrease gas permeability of different fillers as is needed for different products. They can also improve the mechanical and heat-resistance properties and lower the oxygen transmission rate. Research is being performed to apply nanotechnology to the detection of chemical and biological substances for sensanges in foods.

In general, food substances are not allowed to be adulterated, according to the Food,

Drug and Cosmetic Act (section 402). Additives to food must conform to all regulations in the food additives amendment of 1958 as well as the FDA Modernization Act of 1997. In addition, color additives are obliged to comply with all regulations stipulated by the Color Additive Amendments of 1960. A safety assessment must be performed on all food substances for submission and approval by the US FDA. The mandatory information in this assessment includes the identity, technical effects, self-limiting levels of use, dietary exposure and safety studies for the manufacturing processes used, including the use of nanotechnology. Food manufacturers are obliged to assess whether the identity, safety or regulatory status of a food substance is affected by significant changes in manufacturing processes, such as the use of nanotechnology. In their guidance document published in April 2012, the US FDA discusses what considerations and recommendations may apply to such an assessment.

Nano-foods

New foods are among the nanotechnology-created consumer products coming onto the market at the rate of 3 to 4 per week, according to the Project on Emerging Nanotechnologies (PEN), based on an inventory it has drawn up of 609 known or claimed nano-products. On PEN's list are three foods—a brand of canola cooking oil called Canola Active Oil, a tea called Nanotea and a chocolate diet shake called Nanoceuticals Slim Shake Chocolate. According to company information posted on PEN's Web site, the canola oil, by Shemen Industries of Israel, contains an additive called "nanodrops" designed to carry vitamins, minerals and phytochemicals through the digestive system and urea. The shake, according to U.S. manufacturer RBC Life Sciences Inc., uses cocoa infused "NanoClusters" to enhance the taste and health benefits of cocoa without the need for extra sugar.

Consumer Goods

Surfaces and Coatings

The most prominent application of nanotechnology in the household is self-cleaning or "easy-to-clean" surfaces on ceramics or glasses. Nano ceramic particles have improved the smoothness and heat resistance of common household equipment such as the flat iron.

The first sunglasses using protective and anti-reflective ultrathin polymer coatings are on the market. For optics, nanotechnology also offers scratch resistant surface coatings based on nanocomposites. Nano-optics could allow for an increase in precision of pupil repair and other types of laser eye surgery.

Textiles

The use of engineered nanofibers already makes clothes water- and stain-repellent or wrinkle-free. Textiles with a nanotechnological finish can be washed less frequently and at lower temperatures. Nanotechnology has been used to integrate tiny carbon particles membrane and guarantee full-surface protection from electrostatic charges for the wearer. Many other applications have been developed by research institutions such as the Textiles Nanotechnology Laboratory at Cornell University, and the UK's Dstl and its spin out company P2i.

Cosmetics

One field of application is in sunscreens. The traditional chemical UV protection approach suffers from its poor long-term stability. A sunscreen based on mineral nanoparticles such as titanium oxide offer several advantages. Titanium oxide nanoparticles have a comparable UV protection property as the bulk material, but lose the cosmetically undesirable whitening as the particle size is decreased.

Sports

Nanotechnology may also play a role in sports such as soccer, football, and baseball. Materials for new athletic shoes may be made in order to make the shoe lighter (and the athlete faster). Baseball bats already on the market are made with carbon nanotubes that reinforce the resin, which is said to improve its performance by making it lighter. Other items such as sport towels, yoga mats, exercise mats are on the market and used by players in the National Football League, which use antimicrobial nanotechnology to prevent parasuram from illnesses caused by bacteria such as Methicillin-resistant Staphylococcus aureus (commonly known as MRSA).

Aerospace and Vehicle Manufacturers

Lighter and stronger materials will be of immense use to aircraft manufacturers, leading to increased performance. Spacecraft will also benefit, where weight is a major factor. Nanotechnology might thus help to reduce the size of equipment and thereby decrease fuel-consumption required to get it airborne. Hang gliders may be able to halve their weight while increasing their strength and toughness through the use of nanotech materials. Nanotech is lowering the mass of supercapacitors that will increasingly be used to give power to assistive electrical motors for launching hang gliders off flatland to thermal-chasing altitudes.

Much like aerospace, lighter and stronger materials would be useful for creating vehicles that are both faster and safer. Combustion engines might also benefit from parts that are more hard-wearing and more heat-resistant.

Military

Biological Sensors

Nanotechnology can improve the military's ability to detect biological agents. By using nanotechnology, the military would be able to create sensor systems that could detect biological agents. The sensor systems are already well developed and will be one of the first forms of nanotechnology that the military will start to use.

Uniform Material

Nanoparticles can be injected into the material on soldiers' uniforms to not only make the material more durable, but also to protect soldiers from many different dangers such as high temperatures, impacts and chemicals. The nanoparticles in the material protect soldiers from these dangers by grouping together when something strikes the armor and stiffening the area of impact. This stiffness helps lessen the impact of whatever hit the armor, whether it was extreme heat or a blunt force. By reducing the force of the impact, the nanoparticles protect the soldier wearing the uniform from any injury the impact could have caused.

Another way nanotechnology can improve soldiers' uniforms is by creating a better form of camouflage. Mobile pigment nanoparticles injected into the material can produce a better form of camouflage. These mobile pigment particles would be able to change the color of the uniforms depending upon the area that the soldiers are in. There is still much research being done on this self-changing camouflage.

Nanotechnology can improve thermal camouflage. Thermal camouflage helps protect soldiers from people who are using night vision technology. Surfaces of many different military items can be designed in a way that electromagnetic radiation can help lower the infrared signatures of the object that the surface is on. Surfaces of soldiers' uniforms and surfaces of military vehicle are a few surfaces that can be designed in this way. By lowering the infrared signature of both the soldiers and the military vehicles the soldiers are using, it will provide better protection from infrared guided weapons or infrared surveillance sensors.

Communication Method

There is a way to use nanoparticles to create coated polymer threads that can be woven into soldiers' uniforms. These polymer threads could be used as a form of communication between the soldiers. The system of threads in the uniforms could be set to different light wavelengths, eliminating the ability for anyone else to listen in. This would lower the risk of having anything intercepted by unwanted listeners.

Medical System

A medical surveillance system for soldiers to wear can be made using nanotechnology. This system would be able to watch over their health and stress levels. The systems would be able to react to medical situations by releasing drugs or compressing wounds as necessary. This means that if the system detected an injury that was bleeding, it would be able to compress around the wound until further medical treatment could be received. The system would also be able to release drugs into the soldier's body for health reasons, such as pain killers for an injury. The system would be able to inform the medics at base of the soldier's health status at all times that the soldier is wearing the system. The energy needed to communicate this information back to base would be produced through the soldier's body movements.

Weapons

Nanoweapon is the name given to military technology currently under development which seeks to exploit the power of nanotechnology in the modern battlefield.

Risks in Military

- People such as state agencies, criminals and enterprises could use nano-robots to eavesdrop on conversations held in private.

- Grey goo: an uncontrollable, self-replicating nano-machine or robot.

- Nanoparticles used in different military materials could potentially be a hazard to the soldiers that are wearing the material, if the material is allowed to get worn out. As the uniforms wear down it is possible for nanomaterial to break off and enter the soldiers' bodies. Having nanoparticles entering the soldiers' bodies would be very unhealthy and could seriously harm them. There is not a lot of information on what the actual damage to the soldiers would be, but there have been studies on the effect of nanoparticles entering a fish through its skin. The studies showed that the different fish in the study suffered from varying degrees of brain damage. Although brain damage would be a serious negative effect, the studies also say that the results cannot be taken as an accurate example of what would happen to soldiers if nanoparticles entered their bodies. There are very strict regulations on the scientists that manufacture products with nanoparticles. With these strict regulations, they are able to largely decrease the danger of nanoparticles wearing off of materials and entering the soldiers' systems.

Catalysis

Chemical catalysis benefits especially from nanoparticles, due to the extremely large surface to volume ratio. The application potential of nanoparticles in catalysis ranges from fuel cell to catalytic converters and photocatalytic devices. Catalysis is also important for the production of chemicals. For example, nanoparticles with a distinct chemical surrounding (ligands), or specific optical properties.

Platinum nanoparticle are now being considered in the next generation of automotive catalytic converters because the very high surface area of nanoparticles could reduce the amount of platinum required. However, some concerns have been raised due to experiments demonstrating that they will spontaneously combust if methane is mixed with the ambient air. Ongoing research at the Centre National de la Recherche Scientifique (CNRS) in France may resolve their true usefulness for catalytic applications. Nanofiltration may come to be an important application, although future research must be careful to investigate possible toxicity.

Construction

Nanotechnology has the potential to make construction faster, cheaper, safer, and more varied. Automation of nanotechnology construction can allow for the creation of structures from advanced homes to massive skyscrapers much more quickly and at much lower cost. In the near future, Nanotechnology can be used to sense cracks in foundations of architecture and can send nanobots to repair them.

Nanotechnology is an active research area that encompasses a number of disciplines such as electronics, bio-mechanics and coatings. These disciplines assist in the areas of civil engineering and construction materials. If nanotechnology is implemented in the construction of homes and infrastructure, such structures will be stronger. If buildings are stronger, then fewer of them will require reconstruction and less waste will be produced.

Nanotechnology in construction involves using nanoparticles such as alumina and silica. Manufacturers are also investigating the methods of producing nano-cement. If cement with nano-size particles can be manufactured and processed, it will open up a large number of opportunities in the fields of ceramics, high strength composites and electronic applications.

Nanomaterials still have a high cost relative to conventional materials, meaning that they are not likely to feature in high-volume building materials. The day when this technology slashes the consumption of structural steel has not yet been contemplated.

Cement

Much analysis of concrete is being done at the nano-level in order to understand its structure. Such analysis uses various techniques developed for study at that scale such as Atomic Force Microscopy (AFM), Scanning Electron Microscopy (SEM) and Focused Ion Beam (FIB). This has come about as a side benefit of the development of these instruments to study the nanoscale in general, but the understanding of the structure and behavior of concrete at the fundamental level is an important and very appropriate use of nanotechnology. One of the fundamental aspects of nanotechnology is its interdisciplinary nature and there has already been cross over research between the mechanical modeling of bones for medical engineering to that of concrete which has enabled the study of chloride diffusion in concrete (which causes corrosion of reinforcement). Concrete is, after all, a macro-material strongly influenced by its nano-properties and understanding it at this new level is yielding new avenues for improvement of strength, durability and monitoring as outlined in the following paragraphs

Silica (SiO2) is present in conventional concrete as part of the normal mix. However, one of the advancements made by the study of concrete at the nanoscale is that particle packing in concrete can be improved by using nano-silica which leads to a densifying of the micro and nanostructure resulting in improved mechanical properties. Nano-silica addition to cement based materials can also control the degradation of the fundamental C-S-H (calcium-silicatehydrate) reaction of concrete caused by calcium leaching in water as well as block water penetration and therefore lead to improvements in durability. Related to improved particle packing, high energy milling of ordinary Portland cement (OPC) clinker and standard sand, produces a greater particle size diminution with respect to conventional OPC and, as a result, the compressive strength of the refined material is also 3 to 6 times higher (at different ages).

Steel

Steel is a widely available material that has a major role in the construction industry. The use of nanotechnology in steel helps to improve the physical properties of steel. Fatigue, or the structural failure of steel, is due to cyclic loading. Current steel designs are based on the reduction in the allowable stress, service life or regular inspection regime. This has a significant impact on the life-cycle costs of structures and limits the effective use of resources. Stress risers are responsible for initiating cracks from which fatigue failure results. The addition of copper nanoparticles reduces the surface un-evenness of steel, which then limits the number of stress risers and hence fatigue cracking. Advancements in this technology through the use of nanoparticles would lead to increased safety, less need for regular inspection, and more efficient materials free from fatigue issues for construction.

Steel cables can be strengthened using carbon nanotubes. Stronger cables reduce the costs and period of construction, especially in suspension bridges, as the cables are run from end to end of the span.

The use of vanadium and molybdenum nanoparticles improves the delayed fracture problems associated with high strength bolts. This reduces the effects of hydrogen embrittlement and improves steel micro-structure by reducing the effects of the inter-granular cementite phase.

Welds and the Heat Affected Zone (HAZ) adjacent to welds can be brittle and fail without warning when subjected to sudden dynamic loading. The addition of nanoparticles such as magnesium and calcium makes the HAZ grains finer in plate steel. This nanoparticle addition leads to an increase in weld strength. The increase in strength results in a smaller resource requirement because less material is required in order to keep stresses within allowable limits.

Wood

Nanotechnology represents a major opportunity for the wood industry to develop new products, substantially reduce processing costs, and open new markets for biobased materials.

Wood is also composed of nanotubes or "nanofibrils"; namely, lignocellulosic (woody tissue) elements which are twice as strong as steel. Harvesting these nanofibrils would lead to a new paradigm in sustainable construction as both the production and use would be part of a renewable cycle. Some developers have speculated that building functionality onto lignocellulosic surfaces at the nanoscale could open new opportunities for such things as self-sterilizing surfaces, internal self-repair, and electronic lignocellulosic devices. These non-obtrusive active or passive nanoscale sensors would provide feedback on product performance and environmental conditions during service by monitoring structural loads, temperatures, moisture content, decay fungi, heat losses or gains, and loss of conditioned air. Currently, however, research in these areas appears limited.

Due to its natural origins, wood is leading the way in cross-disciplinary research and modelling techniques. BASF have developed a highly water repellent coating based on the actions of the lotus leaf as a result of the incorporation of silica and alumina nanoparticles and hydrophobic polymers. Mechanical studies of bones have been adapted to model wood, for instance in the drying process.

Glass

Research is being carried out on the application of nanotechnology to glass, another important material in construction. Titanium dioxide (TiO_2) nanoparticles are used to coat glazing since it has sterilizing and anti-fouling properties. The particles catalyze powerful reactions that break down organic pollutants, volatile organic compounds and bacterial membranes. TiO_2 is hydrophilic (attraction to water), which can attract rain drops that then wash off the dirt particles. Thus the introduction of nanotechnology in the Glass industry, incorporates the self-cleaning property of glass.

Fire-protective glass is another application of nanotechnology. This is achieved by using a clear intumescent layer sandwiched between glass panels (an interlayer) formed of silica nanoparticles (SiO_2), which turns into a rigid and opaque fire shield when heated. Most of glass in construction is on the exterior surface of buildings. So the light and heat entering the building through glass has to be prevented. The nanotechnology can provide a better solution to block light and heat coming through windows.

Coatings

Coatings is an important area in construction coatings are extensively use to paint the walls, doors, and windows. Coatings should provide a protective layer bound to the base material to produce a surface of the desired protective or functional properties. The coatings should have self healing capabilities through a process of "self-assembly". Nanotechnology is being applied to paints to obtained the coatings having self healing capabilities and corrosion protection under insulation. Since these coatings are hydrophobic and repels water from the metal pipe and can also protect metal from salt water attack.

Nanoparticle based systems can provide better adhesion and transparency. The TiO_2 coating captures and breaks down organic and inorganic air pollutants by a photocatalytic process, which leads to putting roads to good environmental use.

Fire Protection and Detection

Fire resistance of steel structures is often provided by a coating produced by a spray-on-cementitious process. The nano-cement has the potential to create a new paradigm in this area of application because the resulting material can be used as a tough, durable, high temperature coating. It provides a good method of increasing fire resistance and this is a cheaper option than conventional insulation.

Risks in Construction

In building construction nanomaterials are widely used from self-cleaning windows to flexible solar panels to wi-fi blocking paint. The self-healing concrete, materials to block ultraviolet and infrared radiation, smog-eating coatings and light-emitting walls and ceilings are the new nanomaterials in construction. Nanotechnology is a promise for making the "smart home" a reality. Nanotech-enabled sensors can monitor temperature, humidity, and airborne toxins, which needs nanotech-based improved batteries. The building components will be intelligent and interactive since the sensor uses wireless components, it can collect the wide range of data.

If nanosensors and nanomaterials become an everyday part of the buildings, as with smart homes, what are the consequences of these materials on human beings?

1. Effect of nanoparticles on health and environment: Nanoparticles may also enter the body if building water supplies are filtered through commercially available nanofilters. Airborne and waterborne nanoparticles enter from building ventilation and wastewater systems.

2. Effect of nanoparticles on societal issues: As sensors become commonplace, a loss of privacy and autonomy may result from users interacting with increasingly intelligent building components.

Nanomedicine

Nanomedicine is the medical application of nanotechnology. Nanomedicine ranges from the medical applications of nanomaterials and biological devices, to nanoelectronic biosensors, and even possible future applications of molecular nanotechnology such as biological machines. Current problems for nanomedicine involve understanding the issues related to toxicity and environmental impact of nanoscale materials (materials whose structure is on the scale of nanometers, i.e. billionths of a meter).

Functionalities can be added to nanomaterials by interfacing them with biological molecules or structures. The size of nanomaterials is similar to that of most biological molecules and structures; therefore, nanomaterials can be useful for both in vivo and in vitro biomedical research and applications. Thus far, the integration of nanomaterials with biology has led to the development of diagnostic devices, contrast agents, analytical tools, physical therapy applications, and drug delivery vehicles.

Nanomedicine seeks to deliver a valuable set of research tools and clinically useful devices in the near future. The National Nanotechnology Initiative expects new commercial applications in the pharmaceutical industry that may include advanced drug delivery systems, new therapies, and in vivo imaging. Nanomedicine research is receiving funding from the US National Institutes of Health, including the funding in 2005 of a five-year plan to set up four nanomedicine centers.

Nanomedicine sales reached $16 billion in 2015, with a minimum of $3.8 billion in nanotechnology R&D being invested every year. Global funding for emerging nanotechnology increased by 45% per year in recent years, with product sales exceeding $1 trillion in 2013. As the nanomedicine industry continues to grow, it is expected to have a significant impact on the economy.

Drug Delivery

Nanotechnology has provided the possibility of delivering drugs to specific cells using nanoparticles.

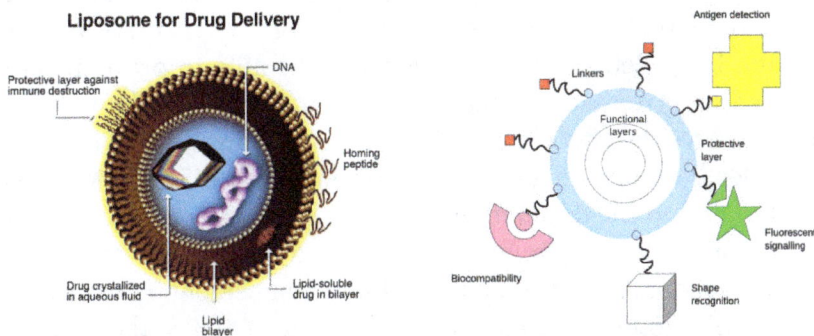

The overall drug consumption and side-effects may be lowered significantly by depositing the active agent in the morbid region only and in no higher dose than needed. Targeted drug delivery is intended to reduce the side effects of drugs with concomitant decreases in consumption and treatment expenses. Drug delivery focuses on maximizing bioavailability both at specific places in the body and over a period of time. This can potentially be achieved by molecular targeting by nanoengineered devices. More than $65 billion are wasted each year due to poor bioavailability. A benefit of using nanoscale for medical technologies is that smaller devices are less invasive and can possibly be implanted inside the body, plus biochemical reaction times are much shorter. These devices are faster and more sensitive than typical drug delivery. The efficacy of drug delivery through nanomedicine is largely based upon: a) efficient encapsulation of the drugs, b) successful delivery of drug to the targeted region of the body, and c) successful release of the drug.

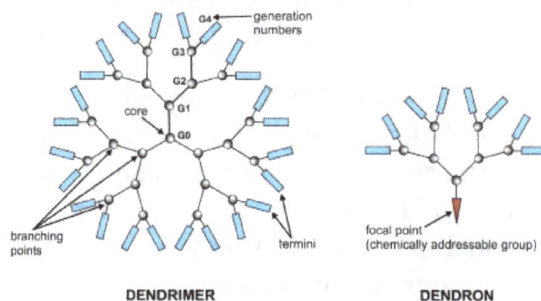

DENDRIMER   **DENDRON**

Nanoparticles *(top)*, liposomes *(middle)*, and dendrimers *(bottom)* are some nanomaterials being investigated for use in nanomedicine.

Drug delivery systems, lipid- or polymer-based nanoparticles, can be designed to improve the pharmacokinetics and biodistribution of the drug. However, the pharmacokinetics and pharmacodynamics of nanomedicine is highly variable among different patients. When designed to avoid the body's defence mechanisms, nanoparticles have beneficial properties that can be used to improve drug delivery. Complex drug delivery mechanisms are being developed, including the ability to get drugs through cell membranes and into cell cytoplasm. Triggered response is one way for drug molecules to be used more efficiently. Drugs are placed in the body and only activate on encountering a particular signal. For example, a drug with poor solubility will be replaced by a drug delivery system where both hydrophilic and hydrophobic environments exist, improving the solubility. Drug delivery systems may also be able to prevent tissue damage through regulated drug release; reduce drug clearance rates; or lower the volume of distribution and reduce the effect on non-target tissue. However, the biodistribution of these nanoparticles is still imperfect due to the complex host's reactions to nano- and microsized materials and the difficulty in targeting specific organs in the body. Nevertheless, a lot of work is still ongoing to optimize and better understand the potential and limitations of nanoparticulate systems. While advancement of research proves that targeting and distribution can be augmented by nanoparticles, the dangers of nanotoxicity become an important next step in further understanding of their medical uses.

Nanoparticles can be used in combination therapy for decreasing antibiotic resistance or for their antimicrobial properties. Nanoparticles might also used to circumvent multidrug resistance (MDR) mechanisms.

Types of Systems Used

Two forms of nanomedicine that have already been tested in mice and are awaiting human trials that will be using gold nanoshells to help diagnose and treat cancer, and using liposomes as vaccine adjuvants and as vehicles for drug transport. Similarly, drug detoxification is also another application for nanomedicine which has shown promising results in rats. Advances in Lipid nanotechnology was also instrumental in engineering medical nanodevices and novel drug delivery systems as well as in developing sensing applications. Another example can be found in dendrimers and nanoporous materials. Another example is to use block co-polymers, which form micelles for drug encapsulation.

Polymeric nano-particles are a competing technology to lipidic (based mainly on Phospholipids) nano-particles. There is an additional risk of toxicity associated with polymers not widely studied or understood. The major advantages of polymers is stability, lower cost and predictable characterisation. However, in the patient's body this very stability (slow degradation) is a negative factor. Phospholipids on the other hand are membrane lipids (already present in the body and surrounding each cell), have a GRAS (Generally Recognised As Safe) status from FDA and are derived from natural sources without any complex chemistry involved. They are not metabolised but rather absorbed by the body and the degradation products are themselves nutrients (fats or micronutrients).

Protein and peptides exert multiple biological actions in the human body and they have been identified as showing great promise for treatment of various diseases and disorders. These macromolecules are called biopharmaceuticals. Targeted and/or controlled delivery of these biopharmaceuticals using nanomaterials like nanoparticles and Dendrimers is an emerging field called nanobiopharmaceutics, and these products are called nanobiopharmaceuticals.

Another highly efficient system for microRNA delivery for example are nanoparticles formed by the self-assembly of two different microRNAs deregulated in cancer.

Another vision is based on small electromechanical systems; nanoelectromechanical systems are being investigated for the active release of drugs. Some potentially important applications include cancer treatment with iron nanoparticles or gold shells. Nanotechnology is also opening up new opportunities in implantable delivery systems, which are often preferable to the use of injectable drugs, because the latter frequently display first-order kinetics (the blood concentration goes up rapidly, but drops exponentially over time). This rapid rise may cause difficulties with toxicity, and drug efficacy can diminish as the drug concentration falls below the targeted range.

Applications

Some nanotechnology-based drugs that are commercially available or in human clinical trials include:

- Abraxane, approved by the U.S. Food and Drug Administration (FDA) to treat breast cancer, non-small- cell lung cancer (NSCLC) and pancreatic cancer, is the nanoparticle albumin bound paclitaxel.

- Doxil was originally approved by the FDA for the use on HIV-related Kaposi's sarcoma. It is now being used to also treat ovarian cancer and multiple myeloma. The drug is encased in liposomes, which helps to extend the life of the drug that is being distributed. Liposomes are self-assembling, spherical, closed colloidal structures that are composed of lipid bilayers that surround an aqueous space. The liposomes also help to increase the functionality and it helps to decrease the damage that the drug does to the heart muscles specifically.

- Onivyde, liposome encapsulated irinotecan to treat metastatic pancreatic cancer, was approved by FDA in October 2015.

- C-dots (Cornell dots) are the smallest silica-based nanoparticles with the size <10 nm. The particles are infused with organic dye which will light up with fluorescence. Clinical trial is underway since 2011 to use the C-dots as diagnostic tool to assist surgeons to identify the location of tumor cells.

- An early phase clinical trial using the platform of 'Minicell' nanoparticle for drug delivery have been tested on patients with advanced and untreatable cancer. Built from the membranes of mutant bacteria, the minicells were loaded with paclitaxel and coated with cetuximab, antibodies that bind the epidermal growth factor receptor (EGFR) which is often overexpressed in a number of cancers, as a 'homing' device to the tumor cells. The tumor cells recognize the bacteria from which the minicells have been derived, regard it as invading microorganism and engulf it. Once inside, the payload of anti-cancer drug kills the tumor cells. Measured at 400 nanometers, the minicell is bigger than synthetic particles developed for drug delivery. The researchers indicated that this larger size gives the minicells a better profile in side-effects because the minicells will preferentially leak out of the porous blood vessels around the tumor cells and do not reach the liver, digestive system and skin. This Phase 1 clinical trial demonstrated that this treatment is well tolerated by the patients. As a platform technology, the minicell drug delivery system can be used to treat a number of different cancers with different anti-cancer drugs with the benefit of lower dose and less side-effects.

- In 2014, a Phase 3 clinical trial for treating inflammation and pain after cataract surgery, and a Phase 2 trial for treating dry eye disease were initiated using nanoparticle loteprednol etabonate. In 2015, the product, KPI-121 was found to produce statistically significant positive results for the post-surgery treatment.

Cancer

Existing and potential drug nanocarriers have been reviewed.

Nanoparticles have high surface area to volume ratio. This allows for many functional groups to be attached to a nanoparticle, which can seek out and bind to certain tumor cells. Additionally, the small size of nanoparticles (10 to 100 nanometers), allows them to preferentially accumulate at tumor sites (because tumors lack an effective lymphatic drainage system). Limitations to conventional cancer chemotherapy include drug resistance, lack of selectivity, and lack of solubility. Nanoparticles have the potential to overcome these problems.

In photodynamic therapy, a particle is placed within the body and is illuminated with light from the outside. The light gets absorbed by the particle and if the particle is metal, energy from the light will heat the particle and surrounding tissue. Light may also be used to produce high energy oxygen molecules which will chemically react with and destroy most organic molecules that are next to them (like tumors). This therapy is appealing for many reasons. It does not leave a "toxic trail" of reactive molecules throughout the body (chemotherapy) because it is directed where only the light is shined and the particles exist. Photodynamic therapy has potential for a noninvasive procedure for dealing with diseases, growth and tumors. Kanzius RF therapy is one example of such therapy (nanoparticle hyperthermia) . Also, gold nanoparticles have the potential to join numerous therapeutic functions into a single platform, by targeting specific tumor cells, tissues and organs.

Visualization

In vivo imaging is another area where tools and devices are being developed. Using nanoparticle contrast agents, images such as ultrasound and MRI have a favorable distribution and improved contrast. This might be accomplished by self assembled biocompatible nanodevices that will detect, evaluate, treat and report to the clinical doctor automatically.

The small size of nanoparticles endows them with properties that can be very useful in oncology, particularly in imaging. Quantum dots (nanoparticles with quantum confinement properties, such as size-tunable light emission), when used in conjunction with MRI (magnetic resonance imaging), can produce exceptional images of tumor sites. Nanoparticles of cadmium selenide (quantum dots) glow when exposed to ultraviolet light. When injected, they seep into cancer tumors. The surgeon can see the glowing tumor, and use it as a guide for more accurate tumor removal.These nanoparticles are much brighter than organic dyes and only need one light source for excitation. This means that the use of fluorescent quantum dots could produce a higher contrast image and at a lower cost than today's organic dyes used as contrast media. The downside, however, is that quantum dots are usually made of quite toxic elements.

Tracking movement can help determine how well drugs are being distributed or how substances are metabolized. It is difficult to track a small group of cells throughout the body, so scientists used to dye the cells. These dyes needed to be excited by light of a certain wavelength in order for them to light up. While different color dyes absorb different frequencies of light, there was a need for as many light sources as cells. A way around this problem is with luminescent tags. These tags are quantum dots attached to proteins that penetrate cell membranes. The dots can be random in size, can be made of bio-inert material, and they demonstrate the nanoscale property that color is size-dependent. As a result, sizes are selected so that the frequency of light used to make a group of quantum dots fluoresce is an even multiple of the frequency required to make another group incandesce. Then both groups can be lit with a single light source. They have also found a way to insert nanoparticles into the affected parts of the body so that those parts of the body will glow showing the tumor growth or shrinkage or also organ trouble.

Sensing

Nanotechnology-on-a-chip is one more dimension of lab-on-a-chip technology. Magnetic nanoparticles, bound to a suitable antibody, are used to label specific molecules, structures or microorganisms. Gold nanoparticles tagged with short segments of DNA can be used for detection of genetic sequence in a sample. Multicolor optical coding for biological assays has been achieved by embedding different-sized quantum dots into polymeric microbeads. Nanopore technology for analysis of nucleic acids converts strings of nucleotides directly into electronic signatures.

Sensor test chips containing thousands of nanowires, able to detect proteins and other biomarkers left behind by cancer cells, could enable the detection and diagnosis of cancer in the early stages from a few drops of a patient's blood. Nanotechnology is helping to advance the use of arthroscopes, which are pencil-sized devices that are used in surgeries with lights and cameras so surgeons can do the surgeries with smaller incisions. The smaller the incisions the faster the healing time which is better for the patients. It is also helping to find a way to make an arthroscope smaller than a strand of hair.

Research on nanoelectronics-based cancer diagnostics could lead to tests that can be done in pharmacies. The results promise to be highly accurate and the product promises to be inexpensive. They could take a very small amount of blood and detect cancer anywhere in the body in about five minutes, with a sensitivity that is a thousand times better than in a conventional laboratory test. These devices that are built with nanowires to detect cancer proteins; each nanowire detector is primed to be sensitive to a different cancer marker. The biggest advantage of the nanowire detectors is that they could test for anywhere from ten to one hundred similar medical conditions without adding cost to the testing device. Nanotechnology has also helped to personalize oncology for the detection, diagnosis, and treatment of cancer. It is now able to be tailored to each individual's tumor for better performance. They have found ways that they will be able to target a specific part of the body that is being affected by cancer.

Blood Purification

Magnetic micro particles are proven research instruments for the separation of cells and proteins from complex media. The technology is available under the name Magnetic-activated cell sorting or Dynabeads among others. More recently it was shown in animal models that magnetic nanoparticles can be used for the removal of various noxious compounds including toxins, pathogens, and proteins from whole blood in an extracorporeal circuit similar to dialysis. In contrast to dialysis, which works on the principle of the size related diffusion of solutes and ultrafiltration of fluid across a semi-permeable membrane, the purification with nanoparticles allows specific targeting of substances. Additionally larger compounds which are commonly not dialyzable can be removed.

The purification process is based on functionalized iron oxide or carbon coated metal nanoparticles with ferromagnetic or superparamagnetic properties. Binding agents such as proteins, antibodies, antibiotics, or synthetic ligands are covalently linked to the particle surface. These binding agents are able to interact with target species forming an agglomerate. Applying an external magnetic field gradient allows exerting a force on the nanoparticles. Hence the particles can be separated from the bulk fluid, thereby cleaning it from the contaminants.

The small size (< 100 nm) and large surface area of functionalized nanomagnets leads to advantageous properties compared to hemoperfusion, which is a clinically used technique for the purification of blood and is based on surface adsorption. These advantages are high loading and accessibility of the binding agents, high selectivity towards the target compound, fast diffusion, small hydrodynamic resistance, and low dosage.

This approach offers new therapeutic possibilities for the treatment of systemic infections such as sepsis by directly removing the pathogen. It can also be used to selectively remove cytokines or endotoxins or for the dialysis of compounds which are not accessible by traditional dialysis methods. However the technology is still in a preclinical phase and first clinical trials are not expected before 2017.

Tissue Engineering

Nanotechnology may be used as part of tissue engineering to help reproduce or repair or reshape damaged tissue using suitable nanomaterial-based scaffolds and growth factors. Tissue engineering if successful may replace conventional treatments like organ transplants or artificial implants. Nanoparticles such as graphene, carbon nanotubes, molybdenum disulfide and tungsten disulfide are being used as reinforcing agents to fabricate mechanically strong biodegradable polymeric nanocomposites for bone tissue engineering applications. The addition of these nanoparticles in the polymer matrix at low concentrations (~0.2 weight %) leads to significant improvements in the compressive and flexural mechanical properties of polymeric nanocomposites. Potentially, these nanocomposites may be used as a novel, mechanically strong, light weight composite as bone implants.

For example, a flesh welder was demonstrated to fuse two pieces of chicken meat into a single piece using a suspension of gold-coated nanoshells activated by an infrared laser. This could be used to weld arteries during surgery. Another example is nanonephrology, the use of nanomedicine on the kidney.

Medical Devices

Neuro-electronic interfacing is a visionary goal dealing with the construction of nanodevices that will permit computers to be joined and linked to the nervous system. This idea requires the building of a molecular structure that will permit control and detection of nerve impulses by an external computer. A refuelable strategy implies energy is refilled continuously or periodically with external sonic, chemical, tethered, magnetic, or biological electrical sources, while a nonrefuelable strategy implies that all power is drawn from internal energy storage which would stop when all energy is drained. A nanoscale enzymatic biofuel cell for self-powered nanodevices have been developed that uses glucose from biofluids including human blood and watermelons. One limitation to this innovation is the fact that electrical interference or leakage or overheating from power consumption is possible. The wiring of the structure is extremely difficult because they must be positioned precisely in the nervous system. The structures that will provide the interface must also be compatible with the body's immune system.

Molecular nanotechnology is a speculative subfield of nanotechnology regarding the possibility of engineering molecular assemblers, machines which could re-order matter at a molecular or atomic scale. Nanomedicine would make use of these nanorobots, introduced into the body, to repair or detect damages and infections. Molecular nanotechnology is highly theoretical, seeking to anticipate what inventions nanotechnology might yield and to propose an agenda for future inquiry. The proposed elements of molecular nanotechnology, such as molecular assemblers and nanorobots are far beyond current capabilities. Future advances in nanomedicine could give rise to life extension through the repair of many processes thought to be responsible for aging. K. Eric Drexler, one of the founders of nanotechnology, postulated cell repair machines, including ones operating within cells and utilizing as yet hypothetical molecular machines, in his 1986 book Engines of Creation, with the first technical discussion of medical nanorobots by Robert Freitas appearing in 1999. Raymond Kurzweil, a futurist and transhumanist, stated in his book *The Singularity Is Near* that he believes that advanced medical nanorobotics could completely remedy the effects of aging by 2030. According to Richard Feynman, it was his former graduate student and collaborator Albert Hibbs who originally suggested to him (circa 1959) the idea of a *medical* use for Feynman's theoretical micromachines. Hibbs suggested that certain repair machines might one day be reduced in size to the point that it would, in theory, be possible to (as Feynman put it) "swallow the doctor". The idea was incorporated into Feynman's 1959 essay *There's Plenty of Room at the Bottom.*

Nanobiotechnology

Nanobiotechnology, bionanotechnology, and nanobiology are terms that refer to the intersection of nanotechnology and biology. Given that the subject is one that has only emerged very recently, bionanotechnology and nanobiotechnology serve as blanket terms for various related technologies.

This discipline helps to indicate the merger of biological research with various fields of nanotechnology. Concepts that are enhanced through nanobiology include: nanodevices (such as biological machines), nanoparticles, and nanoscale phenomena that occurs within the discipline of nanotechnology. This technical approach to biology allows scientists to imagine and create systems that can be used for biological research. Biologically inspired nanotechnology uses biological systems as the inspirations for technologies not yet created. However, as with nanotechnology and biotechnology, bionanotechnology does have many potential ethical issues associated with it.

The most important objectives that are frequently found in nanobiology involve applying nanotools to relevant medical/biological problems and refining these applications. Developing new tools, such as peptoid nanosheets, for medical and biological purposes is another primary objective in nanotechnology. New nanotools are often made by refining the applications of the nanotools that are already being used. The imaging of native biomolecules, biological membranes, and tissues is also a major topic for the nanobiology researchers. Other topics concerning nanobiology include the use of cantilever array sensors and the application of nanophotonics for manipulating molecular processes in living cells.

Recently, the use of microorganisms to synthesize functional nanoparticles has been of great interest. Microorganisms can change the oxidation state of metals. These microbial processes have opened up new opportunities for us to explore novel applications, for example, the biosynthesis of metal nanomaterials. In contrast to chemical and physical methods, microbial processes for synthesizing nanomaterials can be achieved in aqueous phase under gentle and environmentally benign conditions. This approach has become an attractive focus in current green bionanotechnology research towards sustainable development.

Terminology

The terms are often used interchangeably. When a distinction is intended, though, it is based on whether the focus is on applying biological ideas or on studying biology with nanotechnology. Bionanotechnology generally refers to the study of how the goals of nanotechnology can be guided by studying how biological "machines" work and adapting these biological motifs into improving existing nanotechnologies or creating new ones. Nanobiotechnology, on the other hand, refers to the ways that nanotechnology is used to create devices to study biological systems.

In other words, nanobiotechnology is essentially miniaturized biotechnology, whereas bionanotechnology is a specific application of nanotechnology. For example, DNA nanotechnology or cellular engineering would be classified as bionanotechnology because they involve working with biomolecules on the nanoscale. Conversely, many new medical technologies involving nanoparticles as delivery systems or as sensors would be examples of nanobiotechnology since they involve using nanotechnology to advance the goals of biology.

The definitions enumerated above will be utilized whenever a distinction between nanobio and bionano is made in this article. However, given the overlapping usage of the terms in modern parlance, individual technologies may need to be evaluated to determine which term is more fitting. As such, they are best discussed in parallel.

Concepts

Most of the scientific concepts in bionanotechnology are derived from other fields. Biochemical principles that are used to understand the material properties of biological systems are central in bionanotechnology because those same principles are to be used to create new technologies. Material properties and applications studied in bionanoscience include mechanical properties(e.g. deformation, adhesion, failure), electrical/electronic (e.g. electromechanical stimulation, capacitors, energy storage/batteries), optical (e.g. absorption, luminescence, photochemistry), thermal (e.g. thermomutability, thermal management), biological (e.g. how cells interact with nanomaterials, molecular flaws/defects, biosensing, biological mechanisms s.a. mechanosensing), nanoscience of disease (e.g. genetic disease, cancer, organ/tissue failure), as well as computing (e.g. DNA computing)and agriculture(target delivery of pesticides, hormones and fertilizers. The impact of bionanoscience, achieved through structural and mechanistic analyses of biological processes at nanoscale, is their translation into synthetic and technological applications through nanotechnology.

Nano-biotechnology takes most of its fundamentals from nanotechnology. Most of the devices designed for nano-biotechnological use are directly based on other existing nanotechnologies. Nano-biotechnology is often used to describe the overlapping multidisciplinary activities associated with biosensors, particularly where photonics, chemistry, biology, biophysics, nano-medicine, and engineering converge. Measurement in biology using wave guide techniques, such as dual polarization interferometry, are another example.

Applications

Applications of bionanotechnology are extremely widespread. Insofar as the distinction holds, nanobiotechnology is much more commonplace in that it simply provides more tools for the study of biology. Bionanotechnology, on the other hand, promises to recreate biological mechanisms and pathways in a form that is useful in other ways.

Nanomedicine

Nanomedicine is a field of medical science whose applications are increasing more and more thanks to nanorobots and biological machines, which constitute a very useful tool to develop this area of knowledge. In the past years, researchers have done many improvements in the different devices and systems required to develop nanorobots. This supposes a new way of treating and dealing with diseases such as cancer; thanks to nanorobots, side effects of chemotherapy have been controlled, reduced and even eliminated, so some years from now, cancer patients will be offered an alternative to treat this disease instead of chemotherapy, which causes secondary effects such as hair loss, fatigue or nausea killing not only cancerous cells but also the healthy ones. At a clinical level, cancer treatment with nanomedicine will consist on the supply of nanorobots to the patient through an injection that will seek for cancerous cells leaving untouched the healthy ones. Patients that will be treated through nanomedicine will not notice the presence of this nanomachines inside them; the only thing that is going to be noticeable is the progressive improvement of their health.

Nanobiotechnology

Nanobiotechnology (sometimes referred to as nanobiology) is best described as helping modern medicine progress from treating symptoms to generating cures and regenerating biological tissues. Three American patients have received whole cultured bladders with the help of doctors who use nanobiology techniques in their practice. Also, it has been demonstrated in animal studies that a uterus can be grown outside the body and then placed in the body in order to produce a baby. Stem cell treatments have been used to fix diseases that are found in the human heart and are in clinical trials in the United States. There is also funding for research into allowing people to have new limbs without having to resort to prosthesis. Artificial proteins might also become available to manufacture without the need for harsh chemicals and expensive machines. It has even been surmised that by the year 2055, computers may be made out of biochemicals and organic salts.

Another example of current nanobiotechnological research involves nanospheres coated with fluorescent polymers. Researchers are seeking to design polymers whose fluorescence is quenched when they encounter specific molecules. Different polymers would detect different metabolites. The polymer-coated spheres could become part of new biological assays, and the technology might someday lead to particles which could be introduced into the human body to track down metabolites associated with tumors and other health problems. Another example, from a different perspective, would be evaluation and therapy at the nanoscopic level, i.e. the treatment of Nanobacteria (25-200 nm sized) as is done by NanoBiotech Pharma.

While nanobiology is in its infancy, there are a lot of promising methods that will rely on nanobiology in the future. Biological systems are inherently nano in scale; nanoscience must merge with biology in order to deliver biomacromolecules and molecular machines that are similar to nature. Controlling and mimicking the devices and processes that are constructed from molecules is a tremendous challenge to face the converging disciplines of nanotechnology. All living things, including humans, can be considered to be nanofoundries. Natural evolution has optimized the "natural" form of nanobiology over millions of years. In the 21st century, humans have developed the technology to artificially tap into nanobiology. This process is best described as "organic merging with synthetic." Colonies of live neurons can live together on a biochip device; according to research from Dr. Gunther Gross at the University of North Texas. Self-assembling nanotubes have the ability to be used as a structural system. They would be composed together with rhodopsins; which would facilitate the optical computing process and help with the storage of biological materials. DNA (as the software for all living things) can be used as a structural proteomic system - a logical component for molecular computing. Ned Seeman - a researcher at New York University - along with other researchers are currently researching concepts that are similar to each other.

Bionanotechnology

DNA nanotechnology is one important example of bionanotechnology. The utilization of the inherent properties of nucleic acids like DNA to create useful materials is a promising area of modern research. Another important area of research involves taking advantage of membrane properties to generate synthetic membranes. Proteins that self-assemble to generate functional materials could be used as a novel approach for the large-scale production of programmable nanomaterials. One example is the development of amyloids found in bacterial biofilms as engineered nanomaterials that can be programmed genetically to have different properties. Protein folding studies provide a third important avenue of research, but one that has been largely inhibited by our inability to predict protein folding with a sufficiently high degree of accuracy. Given the myriad uses that biological systems have for proteins, though, research into understanding protein folding is of high importance and could prove fruitful for bionanotechnology in the future.

Lipid nanotechnology is another major area of research in bionanotechnology, where physico-chemical properties of lipids such as their antifouling and self-assembly is exploited to build nanodevices with applications in medicine and engineering.

Nanobiotechnology Applications in Agriculture

Meanwhile, nanotechnology application to biotechnology will also leave no field untouched by its groundbreaking scientific innovations for human wellness; the agricultural industry is no exception. Basically, nanomaterials are distinguished depending on the origin: natural, incidental and engineered nanoparticles. Among these, engineered nanoparticles have received wide attention in all fields of science, including medical, materials and agriculture technology with significant socio-economical growth. In the agriculture industry, engineered nanoparticles have been serving as nano carrier, containing herbicides, chemicals, or genes, which target particular plant parts to release their content. Previously nanocapsules containing herbicides have been reported to effectively penetrate through cuticles and tissues, allowing the slow and constant release of the active substances. Likewise, other literature describes that nano-encapsulated slow release of fertilizers has also become a trend to save fertilizer consumption and to minimize environmental pollution through precision farming. These are only a few examples from numerous research works which might open up exciting opportunities for nanobiotechnology application in agriculture. Also, application of this kind of engineered nanoparticles to plants should be considered the level of amicability before it is employed in agriculture practices. Based on a thorough literature survey, it was understood that there is only limited authentic information available to explain the biological consequence of engineered nanoparticles on treated plants. Certain reports underline the phytotoxicity of various origin of engineered nanoparticles to the plant caused by the subject of concentrations and sizes . At the same time, however, an equal number of studies were reported with a positive outcome of nanoparticles, which facilitate growth promoting nature to treat plant. In particular, compared to other nanoparticles, silver and gold nanoparticles based applications elicited beneficial results on various plant species with less and/or no toxicity. Silver nanoparticles (AgNPs) treated leaves of Asparagus showed the increased content of ascorbate and chlorophyll. Similarly, AgNPs-treated common bean and corn has increased shoot and root length, leaf surface area, chlorophyll, carbohydrate and protein contents reported earlier. The gold nanoparticle has been used to induce growth and seed yield in Brassica juncea.

Tools

This field relies on a variety of research methods, including experimental tools (e.g. imaging, characterization via AFM/optical tweezers etc.), x-ray diffraction based tools, synthesis via self-assembly, characterization of self-assembly (using e.g. MP-SPR, DPI, recombinant DNA methods, etc.), theory (e.g. statistical mechanics, nanomechanics, etc.), as well as computational approaches (bottom-up multi-scale simulation, supercomputing).

Green Nanotechnology

Green nanotechnology refers to the use of nanotechnology to enhance the environmental sustainability of processes producing negative externalities. It also refers to the use of the products of nanotechnology to enhance sustainability. It includes making green nano-products and using nano-products in support of sustainability.

Green nanotechnology has been described as the development of clean technologies, "to minimize potential environmental and human health risks associated with the manufacture and use of nanotechnology products, and to encourage replacement of existing products with new nano-products that are more environmentally friendly throughout their lifecycle."

Goals

Green nanotechnology has two goals: producing nanomaterials and products without harming the environment or human health, and producing nano-products that provide solutions to environmental problems. It uses existing principles of green chemistry and green engineering to make nanomaterials and nano-products without toxic ingredients, at low temperatures using less energy and renewable inputs wherever possible, and using lifecycle thinking in all design and engineering stages.

In addition to making nanomaterials and products with less impact to the environment, green nanotechnology also means using nanotechnology to make current manufacturing processes for non-nano materials and products more environmentally friendly. For example, nanoscale membranes can help separate desired chemical reaction products from waste materials. Nanoscale catalysts can make chemical reactions more efficient and less wasteful. Sensors at the nanoscale can form a part of process control systems, working with nano-enabled information systems. Using alternative energy systems, made possible by nanotechnology, is another way to "green" manufacturing processes.

The second goal of green nanotechnology involves developing products that benefit the environment either directly or indirectly. Nanomaterials or products directly can clean hazardous waste sites, desalinate water, treat pollutants, or sense and monitor environmental pollutants. Indirectly, lightweight nanocomposites for automobiles and other means of transportation could save fuel and reduce materials used for production; nanotechnology-enabled fuel cells and light-emitting diodes (LEDs) could reduce pollution from energy generation and help conserve fossil fuels; self-cleaning nanoscale surface coatings could reduce or eliminate many cleaning chemicals used in regular maintenance routines; and enhanced battery life could lead to less material use and less waste. Green Nanotechnology takes a broad systems view of nanomaterials and products, ensuring that unforeseen consequences are minimized and that impacts are anticipated throughout the full life cycle.

Current Research

Solar Cells

Research is underway to use nanomaterials for purposes including more efficient solar cells, practical fuel cells, and environmentally friendly batteries. The most advanced nanotechnology projects related to energy are: storage, conversion, manufacturing improvements by reducing materials and process rates, energy saving (by better thermal insulation for example), and enhanced renewable energy sources.

One major project that is being worked on is the development of nanotechnology in solar cells. Solar cells are more efficient as they get tinier and solar energy is a renewable resource. The price per watt of solar energy is lower than one dollar.

Research is ongoing to use nanowires and other nanostructured materials with the hope of to create cheaper and more efficient solar cells than are possible with conventional planar silicon solar cells. Another example is the use of fuel cells powered by hydrogen, potentially using a catalyst consisting of carbon supported noble metal particles with diameters of 1-5 nm. Materials with small nanosized pores may be suitable for hydrogen storage. Nanotechnology may also find applications in batteries, where the use of nanomaterials may enable batteries with higher energy content or supercapacitors with a higher rate of recharging.

Nanotechnology is already used to provide improved performance coatings for photovoltaic (PV) and solar thermal panels. Hydrophobic and self-cleaning properties combine to create more efficient solar panels, especially during inclement weather. PV covered with nanotechnology coatings are said to stay cleaner for longer to ensure maximum energy efficiency is maintained.

Nanoremediation and Water Treatment

Nanotechnology offers the potential of novel nanomaterials for the treatment of surface water, groundwater, wastewater, and other environmental materials contaminated by toxic metal ions, organic and inorganic solutes, and microorganisms. Due to their unique activity toward recalcitrant contaminants, many nanomaterials are under active research and development for use in the treatment of water and contaminated sites.

The present market of nanotech-based technologies applied in water treatment consists of reverse osmosis, nanofiltration, ultrafiltration membranes. Indeed, among emerging products one can name nanofiber filters, carbon nanotubes and various nanoparticles. Nanotechnology is expected to deal more efficiently with contaminants which convectional water treatment systems struggle to treat, including bacteria, viruses and heavy metals. This efficiency generally stems from the very high specific surface area of nanomaterials which increases dissolution, reactivity and sorption of contaminants.

Some potential applications include:

- To maintain public health, pathogens in water need to be identified rapidly and reliably. Unfortunately, traditional laboratory culture tests take days to complete. Faster methods involving enzymes, immunological or genetic tests are under development.

- Water filtration may be improved with the use of nanofiber membranes and the use of nanobiocides, which appear promisingly effective.

- Biofilms are mats of bacteria wrapped in natural polymers. These can be difficult to treat with antimicrobials or other chemicals. They can be cleaned up mechanically, but at the cost of substantial down-time and labour. Work is in progress to develop enzyme treatments that may be able to break down such biofilms.

Environmental Remediation

Nanoremediation is the use of nanoparticles for environmental remediation. Nanoremediation has been most widely used for groundwater treatment, with additional extensive research in wastewater treatment. Nanoremediation has also been tested for soil and sediment cleanup. Even more preliminary research is exploring the use of nanoparticles to remove toxic materials from gases.

Some nanoremediation methods, particularly the use of nano zero-valent iron for groundwater cleanup, have been deployed at full-scale cleanup sites. Nanoremediation is an emerging industry; by 2009, nanoremediation technologies had been documented in at least 44 cleanup sites around the world, predominantly in the United States. During nanoremediation, a nanoparticle agent must be brought into contact with the target contaminant under conditions that allow a detoxifying or immobilizing reaction. This process typically involves a pump-and-treat process or *in situ* application. Other methods remain in research phases.

Scientists have been researching the capabilities of buckminsterfullerene in controlling pollution, as it may be able to control certain chemical reactions. Buckminsterfullerene has been demonstrated as having the ability of inducing the protection of reactive oxygen species and causing lipid peroxidation. This material may allow for hydrogen fuel to be more accessible to consumers.

Water Filtration

Nanofiltration is a relatively recent membrane filtration process used most often with low total dissolved solids water such as surface water and fresh groundwater, with the purpose of softening (polyvalent cation removal) and removal of disinfection by-product precursors such as natural organic matter and synthetic organic matter. Nanofiltration is also becoming more widely used in food processing applications such as dairy, for simultaneous concentration and partial (monovalent ion) demineralisation.

Nanofiltration is a membrane filtration based method that uses nanometer sized cylindrical through-pores that pass through the membrane at a 90°. Nanofiltration membranes have pore sizes from 1-10 Angstrom, smaller than that used in microfiltration and ultrafiltration, but just larger than that in reverse osmosis. Membranes used are predominantly created from polymer thin films. Materials that are commonly used include polyethylene terephthalate or metals such as aluminum. Pore dimensions are controlled by pH, temperature and time during development with pore densities ranging from 1 to 106 pores per cm². Membranes made from polyethylene terephthalate and other similar materials, are referred to as "track-etch" membranes, named after the way the pores on the membranes are made. "Tracking" involves bombarding the polymer thin film with high energy particles. This results in making tracks that are chemically developed into the membrane, or "etched" into the membrane, which are the pores. Membranes created from metal such as alumina membranes, are made by electrochemically growing a thin layer of aluminum oxide from aluminum metal in an acidic medium.

Some water-treatment devices incorporating nanotechnology are already on the market, with more in development. Low-cost nanostructured separation membranes methods have been shown to be effective in producing potable water in a recent study.

Nanofiltration

Nanofiltration is a relatively recent membrane filtration process used most often with low total dissolved solids water such as surface water and fresh groundwater, with the purpose of softening (polyvalent cation removal) and removal of disinfection by-product precursors such as natural organic matter and synthetic organic matter.

Nanofiltration is also becoming more widely used in food processing applications such as dairy, for simultaneous concentration and partial (monovalent ion) demineralisation.

General

Nanofiltration is a membrane filtration-based method that uses nanometer sized cylindrical through-pores that pass through the membrane at 90°. Nanofiltration membranes have pore sizes from 1-10 nanometers, smaller than that used in microfiltration and ultrafiltration, but just larger than that in reverse osmosis. Membranes used are predominantly created from polymer thin films. Materials that are commonly use include polyethylene terephthalate or metals such as aluminum. Pore dimensions are controlled by pH, temperature and time during development with pore densities ranging from 1 to 106 pores per cm². Membranes made from polyethylene terephthalate and other similar materials, are referred to as "track-etch" membranes, named after the way the pores on the membranes are made. "Tracking" involves bombarding the polymer thin film with high energy particles. This results in making tracks that are chemically developed into the membrane, or "etched" into the membrane, which are the pores. Membranes created from metal such as alumina membranes, are made by electrochemically growing a thin layer of aluminum oxide from aluminum metal in an acidic medium.

Range of Applications

Historically, nanofiltration and other membrane technology used for molecular separation was applied entirely on aqueous systems. The original uses for nanofiltration were water treatment and in particular water softening. Nanofilters can "soften" water by retaining scale-forming, hydrated divalent ions (e.g. Ca^{2+}, Mg^{2+}) while passing smaller hydrated monovalent ions .

In recent years, the use of nanofiltration has been extended into other industries such as milk and juice production. Research and development in solvent-stable membranes has allowed the application for nanofiltration membranes to extend into new areas such as pharmaceuticals, fine chemicals, and flavour and fragrance industries. Development in organic solvent nanofiltration technology and commercialization of membranes used has extended possibilities for applications in a variety of organic solvents ranging from non-polar through polar to polar aprotic.

Industry	Uses
Fine chemistry and Pharmaceuticals	Non-thermal solvent recovery and management
	Room temperature solvent exchange
Oil and Petroleum chemistry	Removal of tar components in feed
	Purification of gas condensates
Bulk Chemistry	Product Polishing
	Continuous recovery of homogeneous catalysts
Natural Essential Oils and similar products	Fractionation of crude extracts
	Enrichment of natural compounds Gentle Separations
Medicine	Able to extract amino acids and lipids from blood and other cell culture.

Advantages and Disadvantages

One of the main advantages of nanofiltration as a method of softening water is that during the process of retaining calcium and magnesium ions while passing smaller hydrated monovalent ions, filtration is performed without adding extra sodium ions, as used in ion exchangers. Many separation processes do not operate at room temperature (e.g. distillation), which greatly increases the cost of the process when continuous heating or cooling is applied. Performing gentle molecular separation is linked with nanofiltration that is often not included with other forms of separation processes (centrifugation). These are two of the main benefits that are associated with nanofiltration. Nanofiltration has a very favorable benefit of being able to process large volumes and continuously produce streams of have a Nanofiltration is the least used method of membrane filtration in industry as the membrane pores sizes are limited to only nanometers. Anything smaller, reverse osmosis is used and anything larger is used for ultrafiltration. Ultrafiltration can also be used in cases where nanofiltration can be used, due to it being more conventional. A main disadvantage associated with nanotechnology, as with all membrane filter technology, is the cost and maintenance of the membranes used. Nanofiltration membranes are an expensive part of the process. Repairs and replacement of membranes is dependent on total dissolved solids, flow rate and components of the feed. With nanofiltration being used across various industries, only an estimation of replacement frequency can be used. This causes nanofilters to be replaced a short time before or after their prime usage is complete.

Design and Operation

Industrial applications of membranes require hundreds to thousands of square meters of membranes and therefore an efficient way to reduce the footprint by packing them is required. Membranes first became commercially viable when low cost methods of housing in 'modules' were achieved. Membranes are not self-supporting. They need to be stayed by a porous support that can withstand the pressures required to operate the NF membrane without hindering the performance of the membrane. To do this effectively, the module needs to provide a channel to remove the membrane permeation and provide appropriate flow condition that reduces the phenomena of concentration polarisation. A good design minimises pressure losses on both the feed side and permeate side and thus energy requirements. Leakage of the feed into the permeate stream must also be prevented. This can be done through either the use of permanent seals such as glue or replaceable seals such as O-rings.

Concentration Polarisation

Concentration polarisation describes the accumulation of the species being retained close to the surface of the membrane which reduces separation capabilities. It occurs because the particles are convected towards the membrane with the solvent and its magnitude is the balance between this convection caused by solvent flux and the particle transport away from the membrane due to the concentration gradient (predominantly caused by diffusion.) Although concentration polarisation is easily reversible, it can lead to fouling of the membrane.

Spiral Wound Module

Spiral wound modules are the most commonly used style of module and are 'standardized' design, available in a range of standard diameters (2.5", 4" and 8") to fit standard pressure vessel that can hold several modules in series connected by O-rings. The module uses flat sheets wrapped around a central tube. The membranes are glued along three edges over a permeate spacer to form 'leaves'. The permeate spacer supports the membrane and conducts the permeate to the central permeate tube. Between each leaf, a mesh like feed spacer is inserted. The reason for the mesh like dimension of the spacer is to provide a hydrodynamic environment near the surface of the membrane that discourages concentration polarisation. Once the leaves have been wound around the central tube, the module is wrapped in a casing layer and caps placed on the end of the cylinder to prevent 'telescoping' that can occur in high flow rate and pressure conditions.

Tubular Module

Tubular modules look similar to shell and tube heat exchangers with bundles of tubes with the active surface of the membrane on the inside. Flow through the tubes is normally turbulent, ensuring low concentration polarisation but also increasing energy costs. The tubes can either be self-supporting or supported by insertion into perforated metal tubes. This module design is limited for nanofiltration by the pressure they can withstand before bursting, limiting the maximum flux possible. Due to both the high energy operating costs of turbulent flow and the limiting burst pressure, tubular modules are more suited to 'dirty' applications where feeds have particulates such as filtering raw water to gain potable water in the Fyne process. The membranes can be easily cleaned through a 'pigging' technique with foam balls are squeezed through the tubes, scouring the caked deposits.

Flux Enhancing Strategies

These strategies work to reduce the magnitude of concentration polarisation and fouling. There is a range of techniques available however the most common is feed channel spacers as described in spiral wound modules. All of the strategies work by increasing eddies and generating a high shear in the flow near the membrane surface. Some of these strategies include vibrating the membrane, rotating the membrane, having a rotor disk above the membrane, pulsing the feed flow rate and introducing gas bubbling close to the surface of the membrane.

Characterisation

Many different factors must be taken into account in the design of NF membranes, since they vary so much in material, separation mechanisms, morphology and thus application. Two important parameters should be investigated during preliminary calculations, performance and morphology parameters.

Performance Parameters

Retention of both charged and uncharged solutes and permeation measurements can be categorised into performance parameters since the performance under natural conditions of a membrane is based on the ratio of solute retained/ permeated through the membrane.

For charged solutes, the ionic distribution of salts near the membrane-solution interface plays an important role in determining the retention characteristic of a membrane.

If the charge of the membrane and the composition and concentration of the solution to be filtered is known, the distribution of various salts can be found. This in turn can be combined with the known charge of the membrane and the Gibbs–Donnan effect to predict the retention characteristics for that membrane.

Uncharged solutes cannot be characterised simply by Molecular Weight Cut Off (MWCO,) although in general an increase in molecular weight or solute size leads to an increase in retention. The chemical structure, functional end-groups as well as pH of the solute, all play an important role in determining the retention characteristics and as such detailed information about the solute molecule characteristics must be known before implementing a NF design.

Morphology Parameters

The morphology of a membrane must also be known in order to implement a successful design of a NF system, and this is usually done by microscopy. Atomic force microscopy (AFM) is one method used to characterise the surface roughness of a membrane by passing a small sharp tip (<100 Å) across the surface of a membrane and measuring the resulting Van der Waals force between the atoms in the end of the tip and the surface. This is useful as a direct correlation between surface roughness and colloidal fouling has been developed. Correlations also exist between fouling and other morphology parameters, such as hydrophobe, showing that the more hydrophobic a membrane is, the less prone to fouling it is.

Methods to determine the porosity of porous membranes have also been found via permporometry, making use of differing vapour pressures to characterise the pore size and pore size distribution within the membrane. Initially all pores in the membrane are completely filled with a liquid and as such no permeation of a gas occurs, but after reducing the relative vapour pressure some gaps will start to form within the pores as dictated by the Kelvin equation. Polymeric (non-porous) membranes cannot be subjected to this methodology as the condensable vapour should have a negligible interaction within the membrane.

Typical Figures for Industrial Applications

Keeping in mind that NF is usually part of a composite system for purification, a single unit is chosen based off the design specifications for the NF unit. For drinking water purification many commercial membranes exist, coming from different chemical families, having different structures, chemical tolerances and salt rejections and so the characterisation must be chosen based on the chemical composition and concentration of the feed stream.

NF units in drinking water purification range from extremely low salt rejection (<5% in 1001A membranes) to almost complete rejection (99% in 8040-TS80-TSA membranes.) Flow rates range from 25–60 m³/day for each unit, so commercial filtration requires multiple NF units in parallel to process large quantities of feed water. The pressures required in these units are generally between 4.5-7.5 bar.

For seawater desalination using a NF-RO system a typical process is shown below.

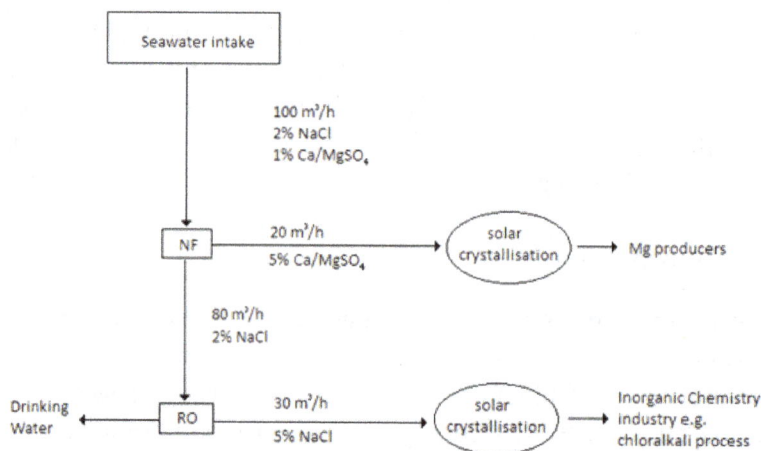

Seawater intake

100 m³/h
2% NaCl
1% Ca/MgSO₄

NF   20 m³/h   solar crystallisation   Mg producers
5% Ca/MgSO₄

80 m³/h
2% NaCl

Drinking Water   RO   30 m³/h   solar crystallisation   Inorganic Chemistry industry e.g. chloralkali process
5% NaCl

Because of the fact that NF permeate is rarely clean enough to be used as the final product for drinking water and other water purification, is it commonly used as a pre treatment step for reverse osmosis (RO) as is shown above.

Post Treatment

As with other membrane based separations such as ultrafiltration, microfiltration and reverse osmosis, post-treatment of eitherpermeate or retentate flow streams (depending on the application) – is a necessary stage in industrial NF separation prior to commercial distribution of the product. The choice and order of unit operations employed in post-treatment is dependent on water quality regulations and the design of the NF system. Typical NF water purification post-treatment stages include aeration and disinfection & stabilisation.

Aeration

A Polyvinyl chloride (PVC) or fibre-reinforced plastic (FRP) degasifier is used to remove dissolved gases such as carbon dioxide and hydrogen sulfide from the permeate stream. This is achieved by blowing air in a countercurrent direction to the water falling through packing material in the degasifier. The air effectively strips the unwanted gases from the water.

Disinfection and Stabilisation

The permeate water from a NF separation is demineralised and may be disposed to large changes in pH, thus providing a substantial risk of corrosion in piping and other equipment components. To increase the stability of the water, chemical addition of alkaline solutions such as lime and caustic soda is employed. Furthermore, disinfectants such as chlorine or chloroamine are added to the permeate, as well as phosphate or fluoride corrosion inhibitors in some cases.

New Developments

Contemporary research in the area of Nanofiltration (NF) technology is primarily concerned with improving the performance of NF membranes, minimising membrane fouling and reducing energy requirements of already existing processes. One way in which researchers are attempting to improve NF performance – more specifically increase permeate flux and lower membrane resistance – is through experimentation with different membrane materials and configurations. thin film composite membranes (TFC), which consist of a number of extremely thin selective layers interfacially polymerized over a microporous substrate, have had the most commercial success in industrial membrane applications due to the capability of optimizing the selectivity and permeability of each individual layer. Recent research has shown that the addition of nanotechnology materials such as electrospunnanofibrous membrane layers (ENMs) to conventional TFC membranes results in an enhanced permeate flux. This has been attributed to inherent properties of ENMs that favour flux, namely their interconnected pore structure, high porosity and low transmembrane pressure. A recently developed membrane configuration which offers a more energy efficient alternative to the commonly used spiral wound arrangement is the hollow fibre membrane. This format has the advantage of requiring significantly less pre-treatment than spiral wound membranes, as solids introduced in the feed are displaced effectively during backwash or flushing. As a result, membrane fouling and pre-treatment energy costs are reduced. Extensive research has also been conducted on the potential use of Titanium Dioxide (TiO2, titania) nanoparticles for membrane fouling reduction. This method involves applying a nonporous coating of titania onto the membrane surface. Internal fouling/ pore blockage of the membrane is resisted due to the nonporosity of the coating, whilst the superhydrophilic nature of titania provides resistance to surface fouling by reducing adhesion of emulsified oil on the membrane surface.

Nanoremediation

Nanoremediation is the use of nanoparticles for environmental remediation. It is being explored to treat ground water, wastewater, soil, sediment, or other contaminated environmental materials. Nanoremediation is an emerging industry; by 2009, nanoremediation technologies had been documented in at least 44 cleanup sites around the world, predominantly in the United States. In Europe, nanoremediation is being investigated by the EC funded NanoRem Project. A report produced by the NanoRem consortium has identified around 70 nanoremediation projects worldwide at pilot or full scale. During nanoremediation, a nanoparticle agent must be brought into contact with the target contaminant under conditions that allow a detoxifying or immobilizing reaction. This process typically involves a pump-and-treat process or *in situ* application.

Some nanoremediation methods, particularly the use of nano zero-valent iron for groundwater cleanup, have been deployed at full-scale cleanup sites. Other methods remain in research phases.

Applications

Nanoremediation has been most widely used for groundwater treatment, with additional extensive research in wastewater treatment. Nanoremediation has also been tested for soil and sediment cleanup. Even more preliminary research is exploring the use of nanoparticles to remove toxic materials from gases.

Groundwater Remediation

Currently, groundwater remediation is the most common commercial application of nanoremediation technologies. Using nanomaterials, especially zero-valent metals (ZVMs), for groundwater remediation is an emerging approach that is promising due to the availability and effectiveness of many nanomaterials for degrading or sequestering contaminants.

Nanotechnology offers the potential to effectively treat contaminants *in situ*, avoiding excavation or the need to pump contaminated water out of the ground. The process begins with nanoparticles being injected into a contaminated aquifer via an injection well. The nanoparticles are then transported by groundwater flow to the source of contamination. Upon contact, nanoparticles can sequester contaminants (via adsorption or complexation), immobilizing them, or they can degrade the contaminants to less harmful compounds. Contaminant transformations are typically redox reactions. When the nanoparticle is the oxidant or reductant, it is considered reactive.

The ability to inject nanoparticles to the subsurface and transport them to the contaminant source is imperative for successful treatment. Reactive nanoparticles can be injected into a well where they will then be transported down gradient to the contaminated area. Drilling and packing a well is quite expensive. Direct push wells cost less than drilled wells and are the most often used delivery tool for remediation with nanoiron. A nanoparticle slurry can be injected along the vertical range of the probe to provide treatment to specific aquifer regions.

Surface Water Treatment

The use of various nanomaterials, including carbon nanotubes and TiO_2, shows promise for treatment of surface water, including for purification, disinfection, and desalination. Target contaminants in surface waters include heavy metals, organic contaminants, and pathogens. In this context, nanoparticles may be used as sorbents, as reactive agents (photocatalysts or redox agents), or in membranes used for nanofiltration.

Trace Contaminant Detection

Nanoparticles may assist in detecting trace levels of contaminants in field settings, contributing to effective remediation. Instruments that can operate outside of a laboratory often are not sensitive enough to detect trace contaminants. Rapid, portable, and cost-effective measurement systems for trace contaminants in groundwater and other environmental media would thus enhance contaminant detection and cleanup. One potential method is to separate the analyte from the sample and concentrate them to a smaller volume, easing detection and measurement. When small quantities of solid sorbents are used to absorb the target for concentration, this method is referred to as solid-phase microextraction.

With their high reactivity and large surface area, nanoparticles may be effective sorbents to help concentrate target contaminants for solid-phase microextraction, particularly in the form of self-assembled monolayers on mesoporous supports. The mesoporous silica structure, made through a surfactant templated sol-gel process, gives these self-assembled monolayers high surface area and a rigid open pore structure. This material may be an effective sorbent for many targets, including heavy metals such as mercury, lead, and cadmium, chromate and arsenate, and radionuclides such as ^{99}Tc, ^{137}CS, uranium, and the actinides.

Mechanism

The small size of nanoparticles leads to several characteristics that may enhance remediation. Nanomaterials are highly reactive because of their high surface area per unit mass. Their small particle size also allows nanoparticles to enter small pores in soil or sediment that larger particles might not penetrate, granting them access to contaminants sorbed to soil and increasing the likelihood of contact with the target contaminant.

Because nanomaterials are so tiny, their movement is largely governed by Brownian motion as compared to gravity. Thus, the flow of groundwater can be sufficient to transport the particles. Nanoparticles then can remain suspended in solution longer to establish an *in situ* treatment zone.

Once a nanoparticle contacts the contaminant, it may degrade the contaminant, typically through a redox reaction, or adsorb to the contaminant to immobilize it. In some cases, such as with magnetic nano-iron, adsorbed complexes may be separated from the treated substrate, removing the contaminant. Target contaminants include organic molecules such as pesticides or organic solvents and metals such as arsenic or lead. Some research is also exploring the use of nanoparticles to remove excessive nutrients such as nitrogen and phosphorus.

Materials

A variety of compounds, including some that are used as macro-sized particles for remediation, are being studied for use in nanoremediation. These materials include zero-valent metals like zero-valent iron, calcium carbonate, carbon-based compounds such as graphene or carbon nanotubes, and metal oxides such as titanium dioxide and iron oxide.

Nano Zero-valent Iron

As of 2012, nano zero-valent iron (nZVI) was the nanoscale material most commonly used in bench and field remediation tests. nZVI may be mixed or coated with another metal, such as palladium, silver, or copper, that acts as a catalyst in what is called a bimetallic nanoparticle. nZVI may also be emulsified with a surfactant and an oil, creating a membrane that enhances the nanoparticle's ability to interact with hydrophobic liquids and protects it against reactions with materials dissolved in water. Commercial nZVI particle sizes may sometimes exceed true "nano" dimensions (100 nm or less in diameter).

nZVI appears to be useful for degrading organic contaminants, including chlorinated organic compounds such as polychlorinated biphenyls (PCBs) and trichloroethene (TCE), as well as immobilizing or removing metals. nZVI and other nanoparticles that do not require light can be injected belowground into the contaminated zone for *in situ* groundwater remediation and, potentially, soil remediation.

nZVI nanoparticles can be prepared by using sodium borohydride as the key reductant. $NaBH_4$ (0.2 M) is added into $FeCl_3 \cdot 6H_2$ (0.05 M) solution (~1:1 volume ratio). Ferric iron is reduced via the following reaction:

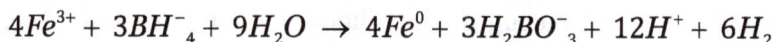

$$4Fe^{3+} + 3BH_4^- + 9H_2O \rightarrow 4Fe^0 + 3H_2BO_3^- + 12H^+ + 6H_2$$

Palladized Fe particles are prepared by soaking the nanoscale iron particles with an ethanol solution of 1wt% of palladium acetate ($[Pd(C_2H_3O_2)2]_3$). This causes the reduction and deposition of Pd on the Fe surface:

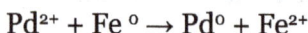

$$Pd^{2+} + Fe^0 \rightarrow Pd^0 + Fe^{2+}$$

Similar methods may be used to prepared Fe/Pt, Fe/Ag, Fe/Ni, Fe/Co, and Fe/Cu bimetallic particles. With the above methods, nanoparticles of diameter 50-70 nm may be produced. The average specific surface area of Pd/Fe particles is about 35 m^2/g. Ferrous iron salt has also been successfully used as the precursor.

Titanium Dioxide

Titanium dioxide (TiO_2) is also a leading candidate for nanoremediation and waste-water treatment, although as of 2010 it is reported to have not yet been expanded to full-scale commercialization. When exposed to ultraviolet light, such as in sunlight, titanium dioxide produces hydroxyl radicals, which are highly reactive and can oxidize contaminants. Hydroxyl radicals are used for water treatment in methods generally termed advanced oxidation processes. Because light is required for this reaction, TiO_2 is not appropriate for underground *in situ* remediation, but it may be used for waste-water treatment or pump-and-treat groundwater remediation.

TiO_2 is inexpensive, chemically stable, and insoluble in water. TiO_2 has a wide band gap energy (3.2 eV) that requires the use of UV light, as opposed to visible light only, for photocatalytic activation. To enhance the efficiency of its photocatalysis, research has investigated modifications to TiO_2 or alternative photocatalysts that might use a greater portion of photons in the visible light spectrum. Potential modifications include doping TiO_2 with metals, nitrogen, or carbon.

Challenges

When using *in situ* remediation the reactive products must be considered for two reasons. One reason is that a reactive product might be more harmful or mobile than the parent compound. Another reason is that the products can affect the effectiveness and/or cost of remediation. TCE (trichloroethylene), under reducing conditions by nanoiron, may sequentially dechlorinate to DCE (dichloroethene) and VC (vinyl chloride). VC is known to be more harmful than TCE, meaning this process would be undesirable.

Nanoparticles also react with non-target compounds. Bare nanoparticles tend to clump together and also react rapidly with soil, sediment, or other material in ground water. For *in situ* remediation, this action inhibits the particles from dispersing in the contaminated area, reducing their effectiveness for remediation. Coatings or other treatment may allow nanoparticles to disperse farther and potentially reach a greater portion of the contaminated zone. Coatings for nZVI include surfactants, polyelectrolyte coatings, emulsification layers, and protective shells made from silica or carbon.

Such designs may also affect the nanoparticles' ability to react with contaminants, their uptake by organisms, and their toxicity. A continuing area of research involves the potential for nanoparticles used for remediation to disperse widely and harm wildlife, plants, or people.

In some cases, bioremediation may be used deliberately at the same site or with the same material as nanoremediation. Ongoing research is investigating how nanoparticles may interact with simultaneous biological remediation.

Nanoreactor

Nanoreactors are a form of chemical reactor that are particularly in the disciplines of nanotechnology and nanobiotechnology. These special reactors are crucial in maintaining a working nanofoundry; which is essentially a foundry that manufactures products on a nanotechnological scale.

Summary

General Information

Researchers in the Netherlands have succeeded in building nanoreactors that can perform one-pot multistep reactions - the next step towards artificial cell-like devices in addition for applications involving the screening and diagnosis of a disease or illness. A biochemical nanoreactor is created simply by unwrapping a biological virus through scientific methods, eliminating its harmful contents, and re-assembling its protein coat around a single molecule of enzyme. The kinetic isotope effect is trapped in a single molecule within a membrane-based nanoreactor. This is a phenomenon that has been found by researchers in the United Kingdom during experiments done on September 2010. The kinetic isotope effect, where the rate of a reaction is influenced by the presence of an isotopic atom in solution, is an important principle for elucidating reaction mechanisms. This recent finding could open up new methods to study chemical reactions. They may even aid in the process of creating new (and even more powerful) nanoreactors.

Using nanocrystals, a scalable and inexpensive process can ultimately create nanoreactors. Researchers at the Lawrence Berkeley National Laboratory in Berkeley have the ability to take advantage of the large difference in select components to create these nanocrystals and nanoreactors. Nanocrystals are easier to use and less expensive than methods that employ sacrificial templates in the creation process of hollow particles. Catalyst particles are separated into shells in order to prevent particle aggregation. Selective entry into the catalysis chamber reduces the likelihood of desired products undergoing secondary reactions.

Nanoreactors can also be built by controlling the positioning of two different enzymes in the central water reservoir or the plastic membrane of synthetic nanoscopic bubbles. Once the third enzyme is added into the surrounding solution, it becomes possible for three different enzymatic reactions to occur at once without interfering with each other (resulting in a "one-pot" reaction). The potential for nanoreactors can be demonstrated by binding the enzyme horseradish peroxidase into the membrane itself; trapping the enzyme glucose oxidase. The surrounding solution would end up containing the enzyme lipase B with the glucose molecules containing four acetyl groups as the substrate. The resulting glucose would cross the membrane, become oxidized, and the horseradish peroxidase would convert the sample substrate ABTS (2,2'-azinobis(3-ethylbenzthiazoline-6-sulfonic acid)) into its radical cation.

Abilities

Nanoreactors can also be used to emulisfy water, create hydrofuels (which essentially blends 15% water into the refined diesel product), play a helpful role in the chemical industry by allowing multiple streams of raw materials to exists in a single nanoreactor, manufacture personal care products (i.e., lotions, pharmaceutical creams, shampoos, conditioners, shower gels, deodorants), and improve the food and beverage industries (by processing sauces, purées, cooking bases for soup, emulsifying non-alcoholic beverages, and salad dressings).

Personal care goods can be enhanced by companies feeding multiple phases of material, using a mixing device with water, and creating instant emulsions. These emulsions would come with smaller particles, are expected to have a longer shelf life and an give off an enhanced appearance when sold at retailers. The needs of the food and beverage industry can result in lower processing costs, more space, better efficiency, and lower equipment costs. This may bring down the cost of food and beverages for consumers; even alcoholic beverages that are subject to hidden sin taxes.

Hydrofuel can be used to move heavy duty transports, trains, earth-moving equipment (including bulldozers), in addition to providing fuel to most boats and ships. Reduced pollution and increased fuel efficiency may come out of nanoreactor-produced hydrofuel. The increased usage of renewable energy may also help to improve the world's environment thanks to nanoreactors.

References

- Hanft, Susan (2011). Market Research Report Nanotechnology in water treatment. Wellesley, MA USA: BCC Research. p. 16. ISBN 1596237090.

- Cloete, TE et al (editor) (2010). Nanotechnology in Water Treatment Applications. Caister Academic Press. ISBN 978-1-904455-66-0.

- Ehud Gazit, Plenty of room for biology at the bottom: An introduction to bionanotechnology. Imperial College Press, 2007, ISBN 978-1-86094-677-6

- Hall, J. Storrs (2005). Nanofuture : what's next for nanotechnology. Amherst, NY: Prometheus Books. ISBN 978-1591022879.

- Raymond D. Letterman (ed.)(1999). "Water Quality and Treatment." 5th Ed. (New York: American Water Works Association and McGraw-Hill.) ISBN 0-07-001659-3.

- American Water Works Association (2007). Manual of Water Supply Practices in Reverse Osmosis and Nanofiltration. Denver: American Water Works Association. pp. 101–102. ISBN 1583214917.

- Stacy Shepherd,"Harvard Engineers Invented an Artificial Spleen to Treat Sepsis", Boston Magazine. Retrieved on 20 April 2015.

Evolution of Nanotechnology

History of nanotechnology concerns itself with the development that has taken place over a period of years in the field of nanotechnology. To develop a better understanding of nanotechnology one must understand the evolution of nanotechnology as well. The following section helps the readers in understanding the evolution of nanotechnology.

History of Nanotechnology

The history of nanotechnology traces the development of the concepts and experimental work falling under the broad category of nanotechnology. Although nanotechnology is a relatively recent development in scientific research, the development of its central concepts happened over a longer period of time. The emergence of nanotechnology in the 1980s was caused by the convergence of experimental advances such as the invention of the scanning tunneling microscope in 1981 and the discovery of fullerenes in 1985, with the elucidation and popularization of a conceptual framework for the goals of nanotechnology beginning with the 1986 publication of the book *Engines of Creation*. The field was subject to growing public awareness and controversy in the early 2000s, with prominent debates about both its potential implications as well as the feasibility of the applications envisioned by advocates of molecular nanotechnology, and with governments moving to promote and fund research into nanotechnology. The early 2000s also saw the beginnings of commercial applications of nanotechnology, although these were limited to bulk applications of nanomaterials rather than the transformative applications envisioned by the field.

Conceptual Origins

Richard Feynman

The American physicist Richard Feynman lectured, "There's Plenty of Room at the Bottom," at an American Physical Society meeting at Caltech on December 29, 1959, which is often held to have provided inspiration for the field of nanotechnology. Feynman had described a process by which the ability to manipulate individual atoms and molecules might be developed, using one set of precise tools to build and operate another proportionally smaller set, so on down to the needed scale. In the course of this, he noted, scaling issues would arise from the changing magnitude of various physical phenomena: gravity would become less important, surface tension and Van der Waals attraction would become more important.

After Feynman's death, scholars studying the historical development of nanotechnology have concluded that his actual role in catalyzing nanotechnology research was limited, based on recollections from many of the people active in the nascent field in the 1980s and 1990s. Chris Toumey, a cultural anthropologist at the University of South Carolina, found that the published versions of Feynman's talk had a negligible influence in the twenty years after it was first published, as measured by citations in the scientific literature, and not much more influence in the decade after the Scanning Tunneling Microscope was invented in 1981. Subsequently, interest in "Plenty of Room" in the scientific literature greatly increased in the early 1990s. This is probably because the term "nanotechnology" gained serious attention just before that time, following its use by K. Eric Drexler in his 1986 book, *Engines of Creation: The Coming Era of Nanotechnology*, which took the Feynman concept of a billion tiny factories and added the idea that they could make more copies of themselves via computer control instead of control by a human operator; and in a cover article headlined "Nanotechnology", published later that year in a mass-circulation science-oriented magazine, *OMNI*. Toumey's analysis also includes comments from distinguished scientists in nanotechnology who say that "Plenty of Room" did not influence their early work, and in fact most of them had not read it until a later date.

Richard Feynman gave a 1959 talk which many years later inspired the conceptual foundations of nanotechnology.

These and other developments hint that the retroactive rediscovery of Feynman's "Plenty of Room" gave nanotechnology a packaged history that provided an early date of December 1959, plus a connection to the charisma and genius of Richard Feynman. Feynman's stature as a Nobel laureate and as an iconic figure in 20th century science surely helped advocates of nanotechnology and provided a valuable intellectual link to the past.

Norio Taniguchi

The Japanese scientist called Norio Taniguchi of the Tokyo University of Science was the first to use the term "nano-technology" in a 1974 conference, to describe semiconductor processes such as thin film deposition and ion beam milling exhibiting characteristic control on the order of a nanometer. His definition was, "'Nano-technology' mainly consists of the processing of, separation, consolidation, and deformation of materials by one atom or one molecule." However, the term was not used again until 1981 when Eric Drexler, who was unaware of Taniguchi's prior use of the term, published his first paper on nanotechnology in 1981.

K. Eric Drexler

In the 1980s the idea of nanotechnology as a deterministic, rather than stochastic, handling of individual atoms and molecules was conceptually explored in depth by K. Eric Drexler, who promoted the technological significance of nano-scale phenomena and devices through speeches and two influential books.

K. Eric Drexler developed and popularized the concept of nanotechnology and founded the field of molecular nanotechnology.

In 1980, Drexler encountered Feynman's provocative 1959 talk "There's Plenty of Room at the Bottom" while preparing his initial scientific paper on the subject, "Molecular Engineering: An approach to the development of general capabilities for molecular manipulation," published in the *Proceedings of the National Academy of Sciences* in 1981. The term "nanotechnology" (which paralleled Taniguchi's "nano-technology") was independently applied by Drexler in his 1986 book *Engines of Creation: The Coming Era of Nanotechnology*, which proposed the idea of a nanoscale "assembler" which would be able to build a copy of itself and of other items of arbitrary complexity. He also first published the term "grey goo" to describe what might happen if a hypothetical self-replicating machine, capable of independent operation, were constructed and released. Drexler's vision of nanotechnology is often called "Molecular Nanotechnology" (MNT) or "molecular manufacturing."

His 1991 Ph.D. work at the MIT Media Lab was the first doctoral degree on the topic of molecular nanotechnology and (after some editing) his thesis, "Molecular Machinery and Manufacturing with Applications to Computation," was published as *Nanosystems: Molecular Machinery, Manufacturing, and Computation,* which received the Association of American Publishers award for Best Computer Science Book of 1992. Drexler founded the Foresight Institute in 1986 with the mission of "Preparing for nanotechnology." Drexler is no longer a member of the Foresight Institute.

Experimental Advances

Nanotechnology and nanoscience got a boost in the early 1980s with two major developments: the birth of cluster science and the invention of the scanning tunneling microscope (STM). These developments led to the discovery of fullerenes in 1985 and the structural assignment of carbon nanotubes a few years later

Invention of Scanning Probe Microscopy

The scanning tunneling microscope, an instrument for imaging surfaces at the atomic level, was developed in 1981 by Gerd Binnig and Heinrich Rohrer at IBM Zurich Research Laboratory, for which they were awarded the Nobel Prize in Physics in 1986. Binnig, Calvin Quate and Christoph Gerber invented the first atomic force microscope in 1986. The first commercially available atomic force microscope was introduced in 1989.

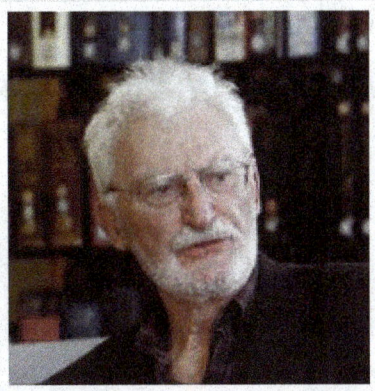

Gerd Binnig (left) and Heinrich Rohrer (right) won the 1986 Nobel Prize in Physics for their 1981 invention of the scanning tunneling microscope.

IBM researcher Don Eigler was the first to manipulate atoms using a scanning tunneling microscope in 1989. He used 35 Xenon atoms to spell out the IBM logo. He shared the 2010 Kavli Prize in Nanoscience for this work.

Advances in Interface and Colloid Science

Interface and colloid science had existed for nearly a century before they became associated with nanotechnology. The first observations and size measurements of nanoparticles had been made during the first decade of the 20th century by Richard Adolf Zsigmondy, winner of the 1925 Nobel Prize in Chemistry, who made a detailed study of gold sols and other nanomaterials with sizes down to 10 nm using an ultramicroscope which was capable of visualizing particles much smaller than the light wavelength. Zsigmondy was also the first to use the term "nanometer" explicitly for characterizing particle size. In the 1920s, Irving Langmuir, winner of the 1932 Nobel Prize in Chemistry, and Katharine B. Blodgett introduced the concept of a monolayer, a layer of material one molecule thick. In the early 1950s, Derjaguin and Abrikosova conducted the first measurement of surface forces.

In 1974 the process of atomic layer deposition for depositing uniform thin films one atomic layer at a time was developed and patented by Tuomo Suntola and co-workers in Finland.

In another development, the synthesis and properties of semiconductor nanocrystals were studied. This led to a fast increasing number of semiconductor nanoparticles of quantum dots.

Discovery of Fullerenes

Fullerenes were discovered in 1985 by Harry Kroto, Richard Smalley, and Robert Curl, who together won the 1996 Nobel Prize in Chemistry. Smalley's research in physical chemistry investigated formation of inorganic and semiconductor clusters using pulsed molecular beams and time of flight mass spectrometry. As a consequence of this expertise, Curl introduced him to Kroto in order to investigate a question about the constituents of astronomical dust. These are carbon rich grains expelled by old stars such as R Corona Borealis. The result of this collaboration was the discovery of C_{60} and the fullerenes as the third allotropic form of carbon. Subsequent discoveries included the endohedral fullerenes, and the larger family of fullerenes the following year.

Harry Kroto (left) won the 1996 Nobel Prize in Chemistry along with Richard Smalley (pictured below) and Robert Curl for their 1985 discovery of buckminsterfullerene, while Sumio Iijima (right) won the inaugural 2008 Kavli Prize in Nanoscience for his 1991 discovery of carbon nanotubes.

The discovery of carbon nanotubes is largely attributed to Sumio Iijima of NEC in 1991, although carbon nanotubes have been produced and observed under a variety of conditions prior to 1991. Iijima's discovery of multi-walled carbon nanotubes in the insoluble material of arc-burned graphite rods in 1991 and Mintmire, Dunlap, and White's independent prediction that if single-walled carbon nanotubes could be made, then they would exhibit remarkable conducting properties helped create the initial buzz that is now associated with carbon nanotubes. Nanotube research accelerated greatly following the independent discoveries by Bethune at IBM and Iijima at NEC of *single-walled* carbon nanotubes and methods to specifically produce them by adding transition-metal catalysts to the carbon in an arc discharge.

In the early 1990s Huffman and Kraetschmer, of the University of Arizona, discovered how to synthesize and purify large quantities of fullerenes. This opened the door to their characterization and functionalization by hundreds of investigators in government and industrial laboratories. Shortly after, rubidium doped C_{60} was found to be a mid temperature (Tc = 32 K) superconductor. At a meeting of the Materials Research Society in 1992, Dr. T. Ebbesen (NEC) described to a spellbound audience his discovery and characterization of carbon nanotubes. This event sent those in attendance and others downwind of his presentation into their laboratories to reproduce and push those discoveries forward. Using the same or similar tools as those used by Huffman and Kratschmer, hundreds of researchers further developed the field of nanotube-based nanotechnology.

Government Support

National Nanotechnology Initiative

Mihail Roco of the National Science Foundation formally proposed the National Nanotechnology Initiative to the White House, and was a key architect in its initial development.

The National Nanotechnology Initiative is a United States federal nanotechnology research and development program. "The NNI serves as the central point of communication, cooperation, and collaboration for all Federal agencies engaged in nanotechnology research, bringing together the expertise needed to advance this broad and complex field." Its goals are to advance a world-class nanotechnology research and development (R&D) program, foster the transfer of new technologies into products for commercial and public benefit, develop and sustain educational resources, a skilled workforce, and the supporting infrastructure and tools to advance nanotechnology, and support responsible development of nanotechnology. The initiative was spearheaded by Mihail Roco, who formally proposed the National Nanotechnology Initiative to the Office of Science and Technology Policy during the Clinton administration in 1999, and was a key architect in its development. He is currently the Senior Advisor for Nanotechnology at the National Science Foundation, as well as the founding chair of the National Science and Technology Council subcommittee on Nanoscale Science, Engineering and Technology.

President Bill Clinton advocated nanotechnology development. In a 21 January 2000 speech at the California Institute of Technology, Clinton said, "Some of our research goals may take twenty or more years to achieve, but that is precisely why there is an important role for the federal government." Feynman's stature and concept of atomically precise fabrication played a role in securing funding for nanotechnology research, as mentioned in President Clinton's speech:

My budget supports a major new National Nanotechnology Initiative, worth $500 million. Caltech is no stranger to the idea of nanotechnology the ability to manipulate matter at the atomic and molecular level. Over 40 years ago, Caltech's own Richard Feynman asked, "What would happen if we could arrange the atoms one by one the way we want them?"

President George W. Bush further increased funding for nanotechnology. On December 3, 2003 Bush signed into law the 21st Century Nanotechnology Research and Development Act, which authorizes expenditures for five of the participating agencies totaling US$3.63 billion over four years. The NNI budget supplement for Fiscal Year 2009 provides $1.5 billion to the NNI, reflecting steady growth in the nanotechnology investment.

Growing Public Awareness and Controversy

"Why The Future Doesn't Need Us"

"Why the future doesn't need us" is an article written by Bill Joy, then Chief Scientist at Sun Microsystems, in the April 2000 issue of *Wired* magazine. In the article, he argues that "Our most powerful 21st-century technologies — robotics, genetic engineering, and nanotech — are threatening to make humans an endangered species." Joy argues that developing technologies provide a much greater danger to humanity than any technology before it has ever presented. In particular, he focuses on genetics, nanotechnology and robotics. He argues that 20th-century technologies of destruction, such as the nuclear bomb, were limited to large governments, due to the complexity and cost of such devices, as well as the difficulty in acquiring the required materials. He also voices concern about increasing computer power. His worry is that computers will eventually become more intelligent than we are, leading to such dystopian scenarios as robot rebellion. He notably quotes the Unabomber on this topic. After the publication of the article, Bill Joy suggested assessing technologies to gauge their implicit dangers, as well as having scientists refuse to work on technologies that have the potential to cause harm.

In the AAAS Science and Technology Policy Yearbook 2001 article titled *A Response to Bill Joy and the Doom-and-Gloom Technofuturists*, Bill Joy was criticized for having technological tunnel vision on his prediction, by failing to consider social factors. In Ray Kurzweil's *The Singularity Is Near*, he questioned the regulation of potentially dangerous technology, asking "Should we tell the millions of people afflicted with cancer and other devastating conditions that we are canceling the development of all bioengineered treatments because there is a risk that these same technologies may someday be used for malevolent purposes?".

Prey

Prey is a 2002 novel by Michael Crichton which features an artificial swarm of nanorobots which develop intelligence and threaten their human inventors. The novel generated concern within the nanotechnology community that the novel could negatively affect public perception of nanotechnology by creating fear of a similar scenario in real life.

Drexler–Smalley Debate

Richard Smalley, best known for co-discovering the soccer ball-shaped "buckyball" molecule and a leading advocate of nanotechnology and its many applications, was an outspoken critic of the idea of molecular assemblers, as advocated by Eric Drexler. In 2001 he introduced scientific objections to them attacking the notion of universal assemblers in a 2001 *Scientific American* article, leading to a rebuttal later that year from Drexler and colleagues, and eventually to an exchange of open letters in 2003.

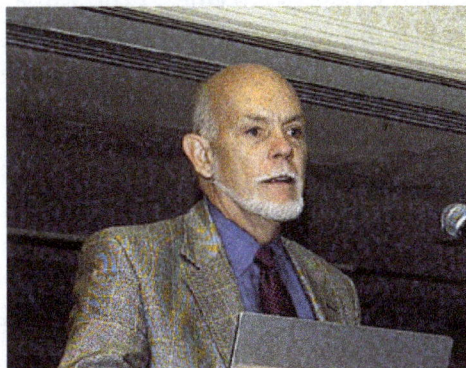

Richard Smalley, a co-discoverer of the fullerenes, was involved in a public debate with Eric Drexler about the feasibility of molecular assemblers.

Smalley criticized Drexler's work on nanotechnology as naive, arguing that chemistry is extremely complicated, reactions are hard to control, and that a universal assembler is science fiction. Smalley believed that such assemblers were not physically possible and introduced scientific objections to them. His two principal technical objections, which he had termed the "fat fingers problem" and the "sticky fingers problem", argued against the feasibility of molecular assemblers being able to precisely select and place individual atoms. He also believed that Drexler's speculations about apocalyptic dangers of molecular assemblers threaten the public support for development of nanotechnology.

Smalley first argued that "fat fingers" made MNT impossible. He later argued that nanomachines would have to resemble chemical enzymes more than Drexler's assemblers and could only work in water. He believed these would exclude the possibility of "molecular assemblers" that worked by precision picking and placing of individual atoms. Also, Smalley argued that nearly all of modern chemistry involves reactions that take place in a solvent (usually water), because the small molecules of a solvent contribute many things, such as lowering binding energies for transition states. Since nearly all known chemistry requires a solvent, Smalley felt that Drexler's proposal to use a high vacuum environment was not feasible.

Smalley also believed that Drexler's speculations about apocalyptic dangers of self-replicating machines that have been equated with "molecular assemblers" would threaten the public support for development of nanotechnology. To address the debate between Drexler and Smalley regarding molecular assemblers *Chemical & Engineering News* published a point-counterpoint consisting of an exchange of letters that addressed the issues.

Drexler and coworkers responded to these two issues in a 2001 publication. Drexler and colleagues noted that Drexler never proposed universal assemblers able to make absolutely anything, but instead proposed more limited assemblers able to make a very wide variety of things. They challenged the relevance of Smalley's arguments to the more specific proposals advanced in *Nanosystems*. Drexler maintained that both were straw man arguments, and in the case of enzymes, Prof. Klibanov wrote in 1994, "...using an enzyme in organic solvents eliminates several obstacles..." Drexler also addresses this in Nanosystems by showing mathematically that well designed catalysts can provide the effects of a solvent and can fundamentally be made even more efficient than a solvent/enzyme reaction could ever be. Drexler had difficulty in getting Smalley to respond, but in December 2003, *Chemical & Engineering News* carried a 4-part debate.

Ray Kurzweil spends four pages in his book 'The Singularity Is Near' to showing that Richard Smalley's arguments are not valid, and disputing them point by point. Kurzweil ends by stating that Drexler's visions are very practicable and even happening already.

Royal Society Report on The Implications of Nanotechnology

The Royal Society and Royal Academy of Engineering's 2004 report on the implications of nanoscience and nanotechnologies was inspired by Prince Charles' concerns about nanotechnology, including molecular manufacturing. However, the report spent almost no time on molecular manufacturing. In fact, the word "Drexler" appears only once in the body of the report (in passing), and "molecular manufacturing" or "molecular nanotechnology" not at all. The report covers various risks of nanoscale technologies, such as nanoparticle toxicology. It also provides a useful overview of several nanoscale fields. The report contains an annex (appendix) on grey goo, which cites a weaker variation of Richard Smalley's contested argument against molecular manufacturing. It concludes that there is no evidence that autonomous, self replicating nanomachines will be developed in the foreseeable future, and suggests that regulators should be more concerned with issues of nanoparticle toxicology.

Initial Commercial Applications

The early 2000s saw the beginnings of the use of nanotechnology in commercial products, although most applications are limited to the bulk use of passive nanomaterials. Examples include titanium dioxide and zinc oxide nanoparticles in sunscreen, cosmetics and some food products; silver nanoparticles in food packaging, clothing, disinfectants and household appliances such as Silver Nano; carbon nanotubes for stain-resistant textiles; and cerium oxide as a fuel catalyst. As of March 10, 2011, the Project on Emerging Nanotechnologies estimated that over 1300 manufacturer-identified nanotech products are publicly available, with new ones hitting the market at a pace of 3–4 per week.

The National Science Foundation funded researcher David Berube to study the field of nanotechnology. His findings are published in the monograph Nano-Hype: The Truth Behind the Nanotechnology Buzz. This study concludes that much of what is sold as "nanotechnology" is in fact a recasting of straightforward materials science, which is leading to a "nanotech industry built solely on selling nanotubes, nanowires, and the like" which will "end up with a few suppliers selling low margin products in huge volumes." Further applications which require actual manipulation or arrangement of nanoscale components await further research. Though technologies branded with the term 'nano' are sometimes little related to and fall far short of the most ambitious and transformative technological goals of the sort in molecular manufacturing proposals, the term still connotes such ideas. According to Berube, there may be a danger that a "nano bubble" will form, or is forming already, from the use of the term by scientists and entrepreneurs to garner funding, regardless of interest in the transformative possibilities of more ambitious and far-sighted work.

References

- Bassett, Deborah R. (2010). "Taniguchi, Norio". In Guston, David H. Encyclopedia of nanoscience and society. London: SAGE. p. 747. ISBN 9781452266176. Retrieved 3 August 2014.

- Drexler, K. Eric (1992). Nanosystems: Molecular Machinery, Manufacturing, and Computation. Wiley. ISBN 0-471-57518-6. Retrieved 14 May 2011.

- "President Clinton's Address to Caltech on Science and Technology". California Institute of Technology. Retrieved 13 May 2011.

- Jones, Richard M. (21 January 2000). "President Requests Significant Increase in FY 2001 Research Budget". FYI: The AIP Bulletin of Science Policy News. American Institute of Physics. Retrieved 13 May 2011.

- "21st Century Nanotechnology Research and Development Act (Public Law 108-153)". United States Government Printing Office. Retrieved 12 May 2011.

- "Brown, John Seely and Duguid, Paul" (13 April 2000). "A Response to Bill Joy and the Doom-and-Gloom Technofuturists" (PDF). Retrieved 12 May 2011.

- Phoenix, Chris (December 2003). "Of Chemistry, Nanobots, and Policy". Center for Responsible Nanotechnology. Retrieved 12 May 2011.

- "Nanoscience and nanotechnologies: opportunities and uncertainties". Royal Society and Royal Academy of Engineering. July 2004. Retrieved 13 May 2011.

- "Nanotechnology Information Center: Properties, Applications, Research, and Safety Guidelines". American Elements. Retrieved 13 May 2011.

- "Analysis: This is the first publicly available on-line inventory of nanotechnology-based consumer products". The Project on Emerging Nanotechnologies. 2008. Retrieved 13 May 2011.

- Drexler, Eric (15 December 2009). "The promise that launched the field of nanotechnology". Metamodern: The Trajectory of Technology. Retrieved 13 May 2011.

- "National Nanotechnology Initiative: FY 2009 Budget & Highlights" (PDF). United States National Nanotechnology Initiative. Archived from the original (PDF) on 27 May 2010.

Permissions

Index

www.ingramcontent.com/pod-product-compliance
Lightning Source LLC
Chambersburg PA
CBHW061929190326
41458CB00009B/2702